Diffraction for Materials Scientists

AMUNDSON *Mathematical Methods in Chemical Engineering: Matrices and Their Application*

BALZHISER, SAMUELS, AND ELIASSEN *Chemical Engineering Thermodynamics*

BRIAN *Staged Cascades in Chemical Processing*

BUTT *Reaction Kinetics and Reactor Design*

DENN *Process Fluid Mechanics*

FOGLER *The Elements of Chemical Kinetics and Reactor Calculations: A Self-Paced Approach*

HIMMELBLAU *Basic Principles and Calculations in Chemical Engineering, 4th edition*

HOLLAND *Fundamentals and Modeling of Separation Processes: Absorption, Distillation, Evaporation, and Extraction*

HOLLAND AND ANTHONY *Fundamentals of Chemical Reaction Engineering*

LEVICH *Physicochemical Hydrodynamics*

MODELL AND REID *Thermodynamics and its Applications*

MYERS AND SEIDER *Introduction to Chemical Engineering and Computer Calculations*

NEWMAN *Electrochemical Systems*

PRAUSNITZ *Molecular Thermodynamics of Fluid-Phase Equilibria*

PRAUSNITZ, ET AL. *Computer Calculations for Multicomponent Vapor-Liquid and Liquid-Liquid Equilibria*

RUDD ET AL. *Process Synthesis*

SCHULTZ *Diffraction for Materials Scientists*

SCHULTZ *Polymer Materials Science*

VILLADSEN AND MICHELSEN *Solution of Differential Equation Models by Polynomial Approximation*

WILLIAMS *Polymer Science and Engineering*

Diffraction for Materials Scientists

JEROLD M. SCHULTZ

Department of Chemical Engineering
University of Delaware

PRENTICE-HALL, INC.

Englewood Cliffs, New Jersey 07632

Library of Congress Cataloging in Publication Data

Schultz, Jerold
Diffraction for materials scientists.

(Prentice-Hall international series in the physical and chemical engineering sciences)
Includes bibliographies and index.
1. Materials—Optical properties. 2. Diffraction.
I. Title. II. Series.
TA418.62.S38 620.1'1295 81-8496
ISBN 0-13-211920-X AACR2

Editorial/production supervision by Lori Opre
Manufacturing buyer: Joyce Levatino

Printed in the United States of America

10 9 8 7 6 5 4 3 2 1

ISBN 0-13-211920-X

PRENTICE-HALL INTERNATIONAL, INC., *London*
PRENTICE-HALL OF AUSTRALIA PTY. LIMITED, *Sydney*
PRENTICE-HALL OF CANADA, LTD., *Toronto*
PRENTICE-HALL OF INDIA PRIVATE LIMITED, *New Delhi*
PRENTICE-HALL OF JAPAN, INC., *Tokyo*
PRENTICE-HALL OF SOUTHEAST ASIA PTE. LTD., *Singapore*
WHITEHALL BOOKS LIMITED, *Wellington, New Zealand*

To the students of MET 801, who keep me honest

Contents

Preface

The thrust of *Diffraction for Materials Scientists* is to convince the reader (student) of the universality and utility of the scattering method in solving structural problems in materials science. This textbook is aimed at teaching the fundamentals of scattering theory and the broad scope of applications in solving real problems. Consistent with this thrust, many of the details of the practice of diffraction experimentation have been omitted. It is intended that *Diffraction for Materials Scientists* be augmented by additional notes dealing with experimental practice; this is how the text has been used by the author.

The core of this text is the first chapter. There the underlying physical concepts and the mathematical tools of kinematic scattering theory are introduced. The remainder of the text could be thought of as the extension of those principles to general areas of application. (The exception to this scheme is Chapter 4, which deals with dynamical concepts.) An attempt has been made to make it clear that each application is a straightforward extension of the central theoretical framework.

The connection between theory and practice is made in two ways. Examples of the results of scattering investigations are used throughout the text, and numerous student problems are provided at the ends of chapters. Some of the problems are of a "drill" variety, while others require extension of the textual material.

Two departures from the approaches used by other authors demand comment. The first relates to the introduction of classical crystallography. The materials scientist is sometimes involved in establishing details of crystal structure, and consequently requires the crystallography background necessary to that purpose. It is the author's feeling that the introduction of classical crystallography early in a text can divert the student's attention and can distort the student's understanding of the position of symmetry relations in diffraction analysis. With this in mind, the concepts of classical crystallography have been deferred to the fifth chapter (of seven). There, the concepts are introduced very specifically as a vehicle for simplifying the work of crystal structure analysis, via characterization of redundancies of atomic position in the unit cell. By this point in the text, the student is comfortable with general diffraction theory and modes of its application and can handle this new set of concepts.

The second departure of the approach regards Chapter 4, on the dynamical theory. It was tempting to maintain textual continuity by omitting emphasis on dynamical concepts. However, the materials scientist is very often involved in questions of image formation in transmission electron microscopy and x-ray topography. It thus seemed important to deal with the mechanics of dynamical theory. The general aim of that material is toward the understanding of image formation. Concepts of primary and secondary extinction and of natural line breadth are made subservient to this aim.

The author is indebted to the many authors and publishers for permitting the republication of their illustrations; acknowledgments are presented in the figure captions. In addition, a large debt is owed to all of the students who had to deal with this book in manuscript form.

JEROLD SCHULTZ

Newark, Delaware

Diffraction for Materials Scientists

Diffraction Fundamentals 1

1.1 INTRODUCTION

The scattering of waves of submicron wavelength by matter provides a set of the most incredibly useful and versatile tools available to the materials scientist. A partial list of some of the types of information available from diffraction studies follows

- Determination of crystal structures
- Determination of liquid structures
- Identification of (crystalline) unknowns
- Orientation of single crystals
- Particle-size or grain-size analysis
- Measurement of stacking fault probabilities
- Internal (residual) strain measurement
- Determination of the matrix of elastic constants
- Determination of the Debye temperature
- Detection and measurement of ordering in alloys
- Measurement of degree of crystallinity in polymers
- Detailed composition determination in solid solutions
- Pole figure analysis
- Direct visual observation of lattice defects and local lattice strains
- Determination of dislocation Burgers' vectors

With the exception of the last two entries in the list, all of the above uses are based on a rather simple set of concepts which constitute the *kinematic theory of diffraction.* The last two entries depend on a refined treatment, the *dynamical theory.* This latter is more physically correct for all diffraction problems, but degenerates to kinematic results for all but relatively rather perfect crystals.

This text attempts to guide the reader through a unified approach to structure problems. Beginning with the first chapter, we set down the basic concepts and mathematics of the kinematic theory. For the most part, the remainder of the text becomes a compendium of specific applications of the basic theory. Thus, while the excitement lies ahead, it behooves the reader to pay particular attention to detail over the earliest portion of this text. The reader should have seen, by the time we have looked at a few applications, that the theory is an at least potentially rigorous one and one which possesses much power when properly understood. He or she should understand also by that time that great experimental imagination is needed to be able to produce truly meaningful results, even in so well-developed a science. That is, to utilize anything approaching the full power of the diffraction method, the investigator must have a thorough grasp of its implications and also of some of the barriers toward its utilization. The basic goal of this text is to foster such awareness.

1.2 GENERAL THEORY OF DIFFRACTION

The work of diffraction analysis proceeds in the following way. Consider in Fig. 1.1 a beam of radiation of wavelength λ incident on some "black box" body whose internal structure is unknown. Using a film or some type of radiation detector, the intensity of scattered radiation $I(\mathbf{s})$ is measured as a function of scattering angle 2θ. As we shall see, the vector $\mathbf{s}$, where $s = 2(\sin\theta)/\lambda$ is a more convenient variable than 2θ. The results of the scattering experiment may appear in different ways, as shown in Fig. 1.2, depending on the experimental conditions.

From the details of such patterns it is possible to reconstruct the structure of the scattering medium. The reason for this is as follows. Consider Fig. 1.3. A plane wave enters the medium and is then scattered spherically by each

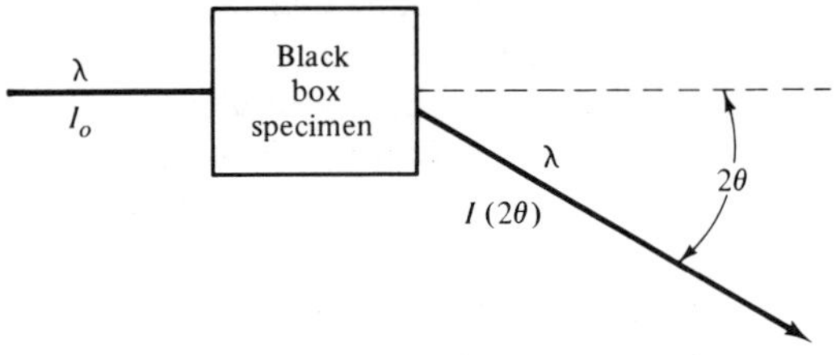

Figure 1.1. Representative experiment. An incident beam of wavelength λ and intensity I_o is scattered by the specimen. A ray elastically scattered in a direction 2θ from the direction of incidence, with intensity $I\,(2\theta)$, is shown.

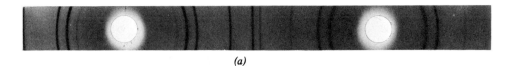

(a)

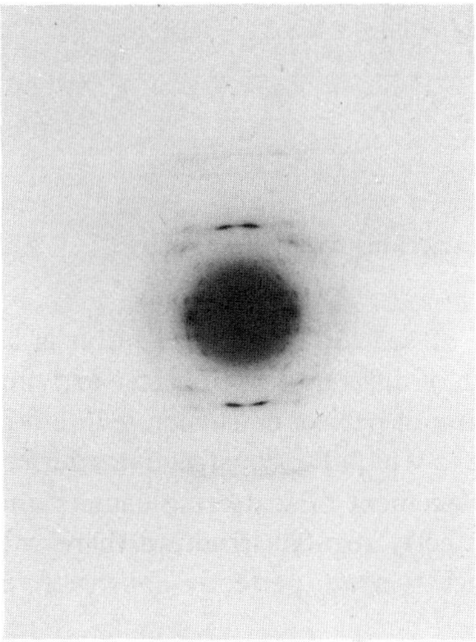

(b)

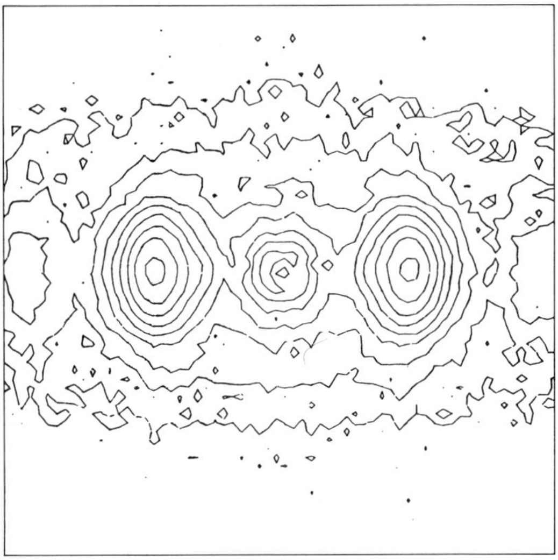

(c)

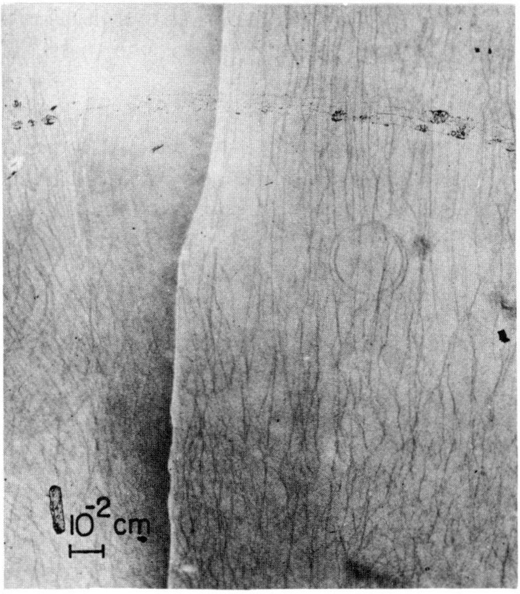

(d)

Figure 1.2. Examples of diffraction results: *(a)* Debye-Scherrer x-ray pattern from a copper powder (taken using CuK_α radiation); *(b)* electron diffraction pattern from a highly oriented poly(ethylene terephthalate) film (taken at 265°C, using 100-kV electrons); *(c)* small-angle x-ray scattering intensity contour plot from an "elastic hard film" of polypropylene (taken using the ORNL 10-meter SAXS instrument, using CuK_α radiation); and *(d)* enlargement of a $10\bar{1}3$ back-reflection diffraction spot from a zinc crystal (using FeK_α radiation). Dislocation lines and rings are clearly visible in *(d)*.

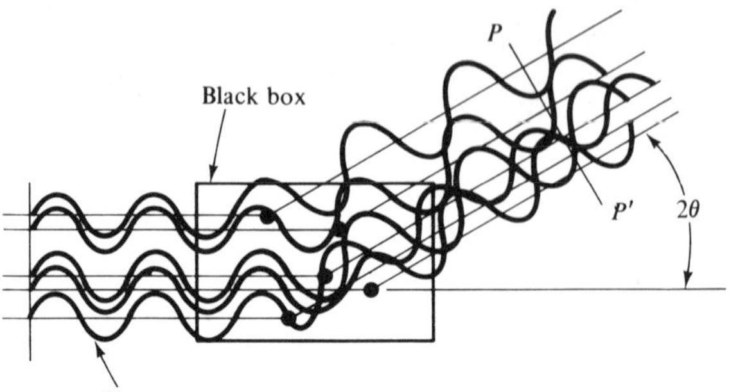

Figure 1.3. Interference effects developed by scattering from several centers.

scattering center. For a particular direction of scattering, the condition is as shown. Each scattered ray has traversed a path of different length before arriving at some common plane PP'. Thus the condition of perfect constructive interference has been destroyed. However, the degree to which the constructive interference has been destroyed depends upon the placement of scattering centers and on the scattering angle 2θ. In fact, for a perfectly regular structure there will be some angles $2\theta_{hkl}$ at which the interference is again perfectly constructive. These are the well-known Bragg angles

$$n\lambda = 2d_{hkl} \sin \theta_{hkl} \tag{1.1}$$

For imperfect crystals or amorphous materials, perfect constructive interference is an impossibility. Hence, the sharp diffraction spikes cannot be observed.

The general theory of scattering follows directly from this rudimentary introduction. Let us think of an incident wave of wavelength λ and unit propagation vector $\hat{\mathbf{s}}_o$. The wave scattered at angle 2θ will maintain its wavelength (coherent scattering) and will have its propagation direction characterized by the unit vector $\hat{\mathbf{s}}$. Let the vector distances of the several scattering centers be represented by a set $\mathbf{r}_i$ with origin anywhere within the "black box." The condition is now as indicated in Fig. 1.4.

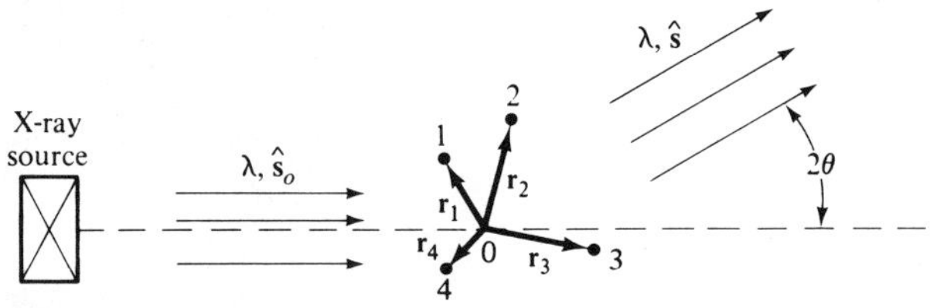

Figure 1.4. Schematic scattering experiment, to define vector quantities. $\hat{\mathbf{s}}_o$ and $\hat{\mathbf{s}}$ are unit vectors in the incident and scattered directions. The $\mathbf{s}_j$ are the vector distances from the origin to the jth scatterers.

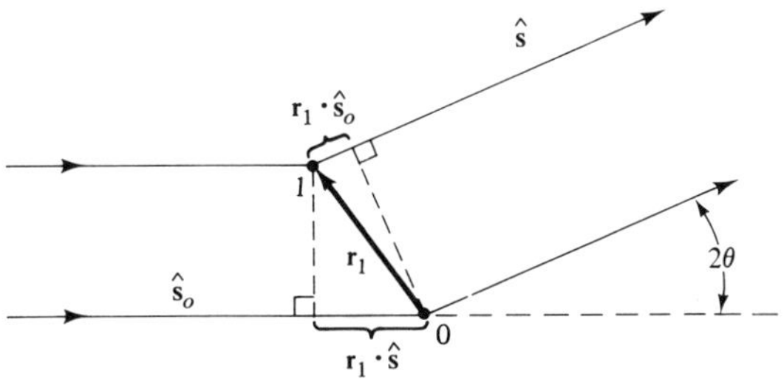

Figure 1.5. Path length difference geometry.

Consider the scattering by centers 1 and 2. Let the incident amplitude be

$$A = A_I \sin\left(\omega t - 2\pi \frac{x}{\lambda}\right) \tag{1.2}$$

Here A_I is the amplitude of scattering by one isolated scatterer. If the distance from the source, through O, to the detector (at scattering angle 2θ) is R_O, then the amplitudes of the two rays through centers 1 and 2 are, respectively

$$A_1 = A_I \sin\left\{\omega t - 2\pi \frac{R_O + \mathbf{r}_1 \cdot (\hat{\mathbf{s}} - \hat{\mathbf{s}}_O)}{\lambda}\right\} \tag{1.3}$$

$$A_2 = A_I \sin\left\{\omega t - 2\pi \frac{R_O + \mathbf{r}_2 \cdot (\hat{\mathbf{s}} - \hat{\mathbf{s}}_O)}{\lambda}\right\} \tag{1.4}$$

The construction leading to this result is shown in Fig. 1.5.

It is convenient at this point to introduce the scattering vector $\mathbf{s} = \dfrac{\hat{\mathbf{s}} - \hat{\mathbf{s}}_O}{\lambda}$. The construction for $\mathbf{s}$ is sketched in Fig. 1.6. Clearly

$$s = |\mathbf{s}| = \frac{2 \sin \theta}{\lambda} \tag{1.5}$$

(In the case of scattering by atomic planes within a crystal, $\mathbf{s}$ is normal to the diffracting plane.)

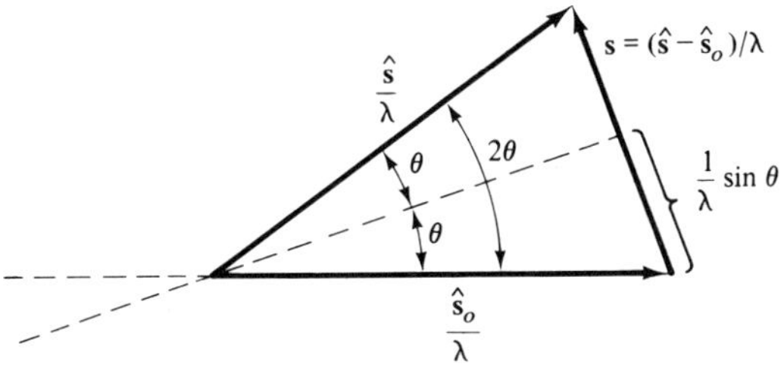

Figure 1.6. Geometrical representation of the scattering vector s.

The total amplitude is thus written

$$A = A_I \sum_i \sin\left[\left(\omega t - 2\pi\frac{R_O}{\lambda}\right) - 2\pi\mathbf{r}_i \cdot \mathbf{s}\right] \tag{1.6}$$

This simplifies a great deal. A result from the classical theory of electromagnetic waves tells us that the intensity of scattering is just the square of the amplitude (see Appendix A). Thus,

$$\begin{aligned} I(\mathbf{s}) = A_I^2 &\left\{ \sum_i \sin\left[\left(\omega t - 2\pi\frac{R_O}{\lambda}\right) - 2\pi\mathbf{r}_i \cdot \mathbf{s}\right]\right\} \\ &\left\{ \sum_j \sin\left[\left(\omega t - 2\pi\frac{R_O}{\lambda}\right) - 2\pi\mathbf{r}_j \cdot \mathbf{s}\right]\right\} \end{aligned} \tag{1.7}$$

This is of the form

$$I(\mathbf{s}) = A_I^2 \sum_i \sin\,(a + b_i) \sum_j \sin\,(a + b_j) \tag{1.8}$$

Using the identity

$$\sin\,(a + b) = \cos a \sin b + \sin a \cos b \tag{1.9}$$

the intensity formula expands to become

$$\begin{aligned} I(\mathbf{s}) = A_I^2 &\left(\sin^2 a \sum_i \sum_j \cos b_i \cos b_j + \cos^2 a \sum_i \sum_j \sin b_i \sin b_j\right) \\ &+ A_I^2 \sin a \cos a \left(\sum_i \sum_j \cos b_i \sin b_j + \sum_i \sum_j \sin b_i \cos b_j\right) \end{aligned} \tag{1.10}$$

Experimentally we average over many periods of the wave. That is, we take an average over time. Thus, we replace $\sin^2 a$ by its time average, $\cos^2 a$ by its time average, etc. These averages are

$$\begin{aligned} \overline{\sin^2 a} &= 1/2 \\ \overline{\cos^2 a} &= 1/2 \\ \overline{\sin a \cos a} &= 0 \end{aligned} \tag{1.11}$$

and we have for our experimental intensity

$$\begin{aligned} I(\mathbf{s}) &= \frac{A_I^2}{2}\left(\sum_i \sum_j \cos b_i \cos b_j + \sum_i \sum_j \sin b_i \sin b_j\right) \\ &= I_I\left(\sum_i \sum_j \cos b_i \cos b_j + \sum_i \sum_j \sin b_i \sin b_j\right) \end{aligned} \tag{1.12}$$

where $I_I = (A_I^2)/2$ is the mean squared, or measurable, intensity of the primary beam.

Some other interesting results derive from the above. First we shall see that the amplitude is more conveniently written

$$A' = A_I' \sum_i e^{-2\pi i \mathbf{r}_i \cdot \mathbf{s}} \tag{1.13}$$

where

$$A_I' = \sqrt{I_I}$$

This is shown as follows. Again let $2\pi\mathbf{r}_i \cdot \mathbf{s} = b_i$. We shall find that

$$I = A'^*A' \tag{1.14}$$

is an alternate form of (1.12). To do so, we insert (1.13) into (1.14)

$$I = I_I \sum_i e^{ib_i} \sum_j e^{-ib_j} = I_I \sum_i (\cos b_i + i \sin b_i) \sum (\cos b_j - i \sin b_j)$$

$$= I_I \sum_i \sum_j \cos b_i \cos b_j + I_I \sum_i \sum_j \sin b_i \sin b_j \tag{1.15}$$

$$+ iI_I \sum_i \sum_j \cos b_j \sin b_i - iI_I \sum_i \sum_j \cos b_i \sin b_j$$

The last two terms cancel, since for each $\cos b_j \sin b_i$ in the third term there exists the same term with a negative sign in the fourth position. Thus

$$I = I_I \sum_i \sum_j \cos b_i \cos b_j + I_I \sum_i \sum_j \sin b_i \sin b_j \tag{1.16}$$

which is exactly (1.12). Thus (1.6) and (1.13) are interchangeable. Equation (1.13) is generally the more tractable of the two forms and is the one customarily used.

Second, the summation form for the amplitude is more correctly replaced by an integral form. This is a result of the nature of the scattering center. For x-rays, the scatterers are electrons. Heisenberg's principle tells us that the position of the electron cannot be exactly specified. The probability $p(\mathbf{r}, t)$ of finding the electron at the position $\mathbf{r}$ at time t is, however, known and is expressed

$$p(\mathbf{r}, t) = |\psi(\mathbf{r}, t)|\ |\psi(\mathbf{r}, t)| \tag{1.17}$$

where $\psi(\mathbf{r}, t)$ is the state function for the electron. Let us then define a time average density of electrons as

$$\rho(\mathbf{r})dv_{\mathbf{r}} = \frac{\int_0^T p(\mathbf{r}, t)dt}{\int_0^T dt} \tag{1.18}$$

where $\rho(\mathbf{r})$ is the (time average) number of electrons per unit volume and $dv_{\mathbf{r}}$

is the increment of volume at a vector distance **r** from the origin. For this continuous state of matter, the amplitude expression must now read

$$A(\mathbf{s}) = \sqrt{I_1} \int \rho(\mathbf{r}) e^{-2\pi i \mathbf{r}\cdot\mathbf{s}} dv_{\mathbf{r}} \tag{1.19}$$

We see that $A(\mathbf{s})/\sqrt{I_1}$ is the Fourier transform of the electron density. The nature of the scattering entity changes as we consider the scattering of other types of waves (electrons, neutrons, visible light). However, the form of (1.19) is maintained for those cases also, with the definitions of I_1, and $\rho(\mathbf{r})$ altered to correspond with the nature of the scatterer. We shall return to both (1.10) and to the question of the scattering of electrons, neutrons, and visible light later.

Third, the intensity now becomes

$$\begin{aligned} I(\mathbf{s}) = A^*A &= I_I \int \rho(\mathbf{r}) e^{2\pi i \mathbf{r}\cdot\mathbf{s}} dv_{\mathbf{r}} \int \rho(\mathbf{r}') e^{-2\pi i \mathbf{r}'\cdot\mathbf{s}} dv_{\mathbf{r}'} \\ &= I_I \iint \rho(\mathbf{r})\rho(\mathbf{r}') e^{2\pi i (\mathbf{r}-\mathbf{r}')\cdot\mathbf{s}} dv_{\mathbf{r}} dv_{\mathbf{r}'} \end{aligned} \tag{1.20}$$

Now set $\mathbf{r}' = \mathbf{r} + \mathbf{u}$, as in Fig. 1.7. Thus

$$I(\mathbf{s}) = I_I \int P(\mathbf{u}) e^{-2\pi i \mathbf{u}\cdot\mathbf{s}} dv_{\mathbf{u}} \tag{1.21}$$

where

$$P(\mathbf{u}) = \int \rho(\mathbf{r})\rho(\mathbf{r}+\mathbf{u}) dv_{\mathbf{r}} \tag{1.22}$$

is the *autocorrelation function.* We shall examine the properties of $P(\mathbf{u})$ presently.

Equations (1.19) and (1.21) are the most important equations in the kinematic theory of diffraction. Equation (1.19) expresses the relationship between diffraction amplitude and the structure $\rho(\mathbf{r})$ within our "black box." We shall see later that in principle the autocorrelation function $P(\mathbf{u})$ is always available from an analysis of the intensity over scattering space **s**.

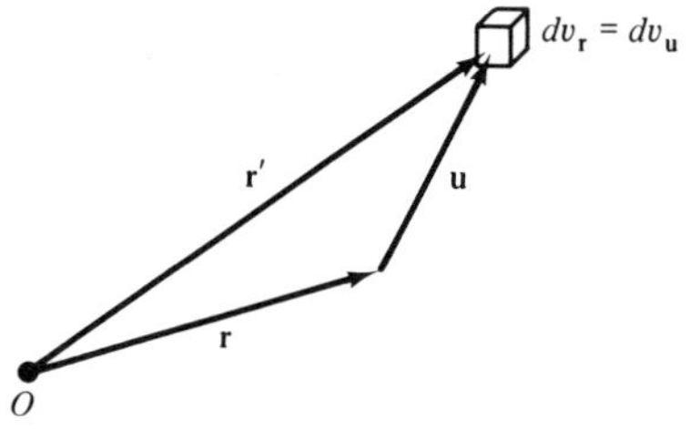

Figure 1.7. Interrelations among **r**, **r**′, and **u**.

1.3 IMPLICIT ASSUMPTIONS

If the reader at this point looks back carefully at what we have done, he or she will note that three major assumptions were made. These were:

1. The scattered wave maintains the wavelength of the incident beam; i.e., the scattering is elastic. We shall see that this assumption is partially justified; a portion of the wave scattered at angle 2θ is indeed elastic, but another portion is inelastic. Furthermore, the inelastic portion is incoherent. That is, its phase bears no relation to that of the incident photon. Thus there can be no interference effects among the once inelastically scattered rays. This problem, to which we shall return, causes us to replace I_l, the total intensity scattered by an electron by I_l', the coherent, or elastic portion.
2. The wave is not attenuated in the black box. That is, there is no absorption of the beam—only scattering. One need only compare diffracted intensities obtained in a reflective geometry with those obtained in a transmittive geometry to very quickly conclude that absorption need be taken into account. We shall see, however, that the problem, while a major one, is usually entirely geometrical and can be added as a correction to the intensity function derived for a nonabsorptive medium.
3. A photon, electron, or neutron can be scattered only once. For many real world situations this assumption fits nicely. The probability p_1 of a scattering event occurring for a given photon acting on a given electron is quite low. The probability of a given photon being scattered anywhere in the crystal is Np_1, where N is the number of electrons in the irradiated volume of the crystal. Thus the probability of two events for the same photon will be approximately $N^2p_1^2$. We shall see that in fact the N which we need consider is the domain over which a crystal is very perfect. Most natural systems are highly imperfect and multiple scattering is sufficiently small to be ignored. However, for very perfect crystals, N can be quite large and multiple scattering needs to be taken into account. It is not obvious from the above, but multiple scattering can also affect studies of scattering at small angles. Furthermore, the probability of a given *electron* being scattered is comparatively very high, relative to a photon. Thus multiple scattering is a common occurrence for the case of electron waves. In such cases, one needs an entirely new approach to diffraction phenomena. This approach, termed the *dynamical theory,* treats the interaction of waves multiply scattered within a material. Some very unusual, and sometimes useful, effects are predicted and observed. One of these is the ability to directly observe lattice defects and strain fields.

For the present we shall proceed as if our I_I were really I_I'—that is, as if we had already accounted for inelastic scattering. And we shall proceed also as if absorption and multiple scattering were not present. At a later point we shall see how absorption is treated and somewhat still further on we shall look at the structure of the dynamical theory.

1.4 MATHEMATICAL PROPERTIES USED IN DIFFRACTION THEORY

At this point it becomes very useful to digress from our systematic treatment of the physics of diffraction in order to develop the mathematical tools which are needed throughout. Time spent now in developing a few simple skills will enable us to apply (1.12), (1.13), and (1.21) to a large variety of useful situations. In particular, the mathematical tools we need to treat are the Fourier series, the Fourier transform, and the convolution theorem.

1.4.1 Fourier Series

In general, any function $f(x)$ defined between $x = -\lambda/2$ and $x = \lambda/2$ can be represented by a series of sine or cosine terms

$$f(x) = \sum_n a_n \sin \frac{2\pi nx}{\lambda} \tag{1.23}$$

The Fourier coefficients a_n are found in the following way. Multiply both sides of the equation by $\sin (2\pi mx)/\lambda$ and integrate over x from $-\lambda/2$ to $\lambda/2$

$$\int_{-\lambda/2}^{\lambda/2} f(x) \sin \frac{2\pi mx}{\lambda}\, dx = \sum_n a_n \int_{-\lambda/2}^{\lambda/2} \sin \frac{2\pi mx}{\lambda} \sin \frac{2\pi nx}{\lambda}\, dx \tag{1.24}$$

Now

$$\begin{aligned} \sin ax \sin bx &= \frac{e^{iax} - e^{-iax}}{2i} \cdot \frac{e^{ibx} - e^{-ibx}}{2i} \\ &= \frac{e^{i(a+b)x} + e^{-i(a+b)x}}{-4} - \frac{e^{i(a-b)x} + e^{-i(a-b)x}}{-4} \\ &= -\frac{1}{2}\cos(a+b)x + \frac{1}{2}\cos(a-b)x \end{aligned} \tag{1.25}$$

Thus

$$\int_{-\lambda/2}^{\lambda/2} f(x) \sin \frac{2\pi mx}{\lambda}\, dx = -\frac{1}{2}\sum_n a_n \left[\int_{-\lambda/2}^{\lambda/2} \cos \frac{2\pi (m+n)x}{\lambda}\, dx\right.$$

$$-\int_{-\lambda/2}^{\lambda/2} \cos\frac{2\pi(m-n)}{\lambda}dx\Bigg] = -\frac{1}{2}\sum_n a_n\left[\frac{\lambda}{2\pi(m+n)}\sin\frac{2\pi(m+n)x}{\lambda}\Bigg|_{-\lambda/2}^{\lambda/2}\right.$$
$$\left.-\frac{\lambda}{2\pi(m-n)}\sin\frac{2\pi(m-n)x}{\lambda}\Bigg|_{-\lambda/2}^{\lambda/2}\right] \tag{1.26}$$

The right side of this equation is identically zero, since sin (integer $\times\ \pi$) = 0. That is, it is zero unless $m = n$. Then the second term on the right is 0/0. We look now at the case for $m = n$

$$\int_{-\lambda/2}^{\lambda/2} \sin^2\frac{2\pi nx}{\lambda}dx = \frac{1}{2}\frac{\lambda}{2\pi n}x - \sin 2x\Bigg|_{x=-\pi n}^{x=\pi n} = \frac{\lambda}{2} \tag{1.27}$$

Thus

$$\int_{-\lambda/2}^{\lambda/2} f(x)\sin\frac{2\pi nx}{\lambda}dx = \frac{\lambda}{2}a_n \tag{1.28}$$

In the same way it can be demonstrated that the cosine Fourier series behaves identically as the sine series

$$g(x) = \sum_n b_n \cos\frac{2\pi nx}{\lambda}$$
$$b_n = \frac{2}{\lambda}\int_{-\lambda/2}^{\lambda/2} g(x)\cos\frac{2\pi nx}{\lambda}dx \tag{1.29}$$

Analogously, since $e^{i\theta} = \cos\theta + i\sin\theta$

$$h(x) = \sum_n c_n e^{i\frac{2\pi nx}{\lambda}}$$
$$c_n = \frac{2}{\lambda}\int_{-\lambda/2}^{\lambda/2} h(x)e^{i\frac{2\pi nx}{\lambda}}dx \tag{1.30}$$

1.4.2 The Fourier Integral

Suppose now that λ becomes very large, such that n/λ represents a set of very fine, nearly continuous, divisions. Let $q = n/\lambda_v$ in (1.30). Suppose also that c_n is now $c(q)$. Then, in the limit as $\Delta n \to 0$

$$h(x) = \int_{\infty}^{\infty} c(q)e^{i2\pi qx}dq \tag{1.31}$$

is a suitable representation of the arbitrary function $h(x)$. Now multiply both sides of this equation by $e^{-i2\pi q'x}$ and integrate over x from $-M$ to M

$$\int_{-M}^{+M} h(x)e^{-2\pi iq'x}dx = \int_{-\infty}^{\infty} c(q)\int_{-M}^{+M} e^{2\pi i(q-q')x}dx\,dq \tag{1.32}$$

Consider the inner integral on the right

$$\int_{-M}^{+M} e^{2\pi i(q-q')x}\,dx = \int_{-M}^{+M} \cos\,[2\pi(q-q')x]dx + i\int_{-M}^{+M} \sin\,[2\pi(q-q')x]dx \tag{1.33}$$

The second term on the right is clearly zero, since $\sin\theta$ is an odd function. Let us look carefully at the first term on the right. Using $Q = q - q'$, we have

$$\int_{-M}^{M} \cos 2\pi Qx dx = \frac{1}{2\pi Q}\sin 2\pi Qx \bigg|_{-M}^{M} = \frac{\sin 2\pi MQ}{\pi Q} \tag{1.34}$$

The maximum of this function occurs at $Q = 0$

$$\frac{\partial\left(\sin\dfrac{2\pi MQ}{\pi Q}\right)}{\partial Q} = \frac{1}{Q}2M\cos 2\pi MQ - \frac{1}{\pi Q^2}\sin 2\pi MQ = 0$$

$$2\pi MQ = \tan 2\pi MQ \tag{1.35}$$

$$Q = 0 \text{ rad}$$

The breadth of the function can be measured by its half-height. Its full height is

$$\lim_{Q=0}\frac{\sin 2\pi MQ}{\pi Q} = 2M\lim_{Q\to 0}\frac{\sin 2\pi MQ}{2\pi MQ} = 2M \tag{1.36}$$

At half-height, then

$$M = \frac{\sin 2\pi MQ}{\pi Q} \tag{1.37}$$

or

$$\frac{1}{2}(2\pi MQ_{1/2}) = \frac{1}{2}\psi = \sin\psi$$

$$\psi \approx 1.9 \text{ rad} = 2\pi MQ_{1/2} \tag{1.38}$$

$$Q_{1/2} \approx \frac{1}{\pi M}$$

Hence, as $M \to \infty$, $\sin 2\pi MQ/\pi Q$ becomes very narrow. It is effectively zero everywhere except at the origin. A plot of this function is shown in Fig. 1.8. Thus

$$\lim_{M\to\infty}\int_{-M}^{+M} e^{2\pi i(q-q')x}\,dx = 0 \tag{1.39}$$

for all q' except for $q' = q$. We now evaluate the integral at $q' = q$

$$\lim_{M\to\infty}\int_{-M}^{M} dx = \lim_{M\to\infty} 2M = \infty \tag{1.40}$$

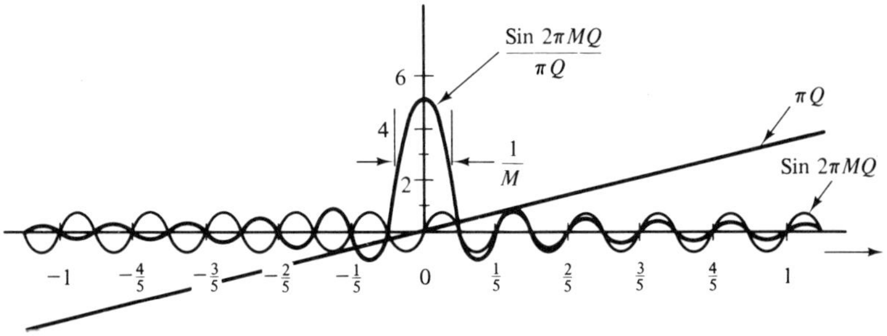

Figure 1.8. sin $(2\pi MQ)/(\pi Q)$, plotted here for $M = 5$. The central peak will become higher and narrower and the subsidiary maxima will die out faster as M increases.

Actually $\int_{-\infty}^{\infty} e^{2\pi i(q-q')x}dx$ has the properties of a Dirac delta function (δ) which are

$$\delta(q-q') = \begin{cases} 0 \text{ for all } q' \neq q \\ \infty \text{ for } q' = q \end{cases}$$
$$\int_{-\infty}^{\infty} \delta(q-q')dq' = 1 \tag{1.41}$$

We can see the latter, approximately, by replacing our peak by a rectangular function of the same height and area. Then the area A under the peak is, for all M

$$A \simeq (2M)\left(\frac{2}{\pi M}\right) = \frac{4}{\pi} \approx 1 \tag{1.42}$$

Thus we have

$$\int_{-\infty}^{\infty} h(x)e^{-2\pi iq'x}dx = \int_{-\infty}^{\infty} c(q)\delta(q-q')dq = c(q') \tag{1.43}$$

We now have the symmetrical situation

$$h(x) = \int_{-\infty}^{\infty} c(q)e^{2\pi iqx}dq$$
$$c(q) = \int_{-\infty}^{\infty} h(x)e^{-2\pi iqx}dx \tag{1.44}$$

In terms of our diffraction problem, we have

$$\frac{A(\mathbf{s})}{A_I} = \int \rho(\mathbf{r})e^{-2\pi i\mathbf{r}\cdot\mathbf{s}}dv_{\mathbf{r}} \tag{1.45}$$

$$\rho(\mathbf{r}) = \int \frac{A(\mathbf{s})}{A_I} e^{2\pi i\mathbf{r}\cdot\mathbf{s}}dv_{\mathbf{s}} \tag{1.46}$$

Thus in principal we could obtain the electron distribution in the black box directly from the measurement of the scattered amplitude $A(\mathbf{s})$ over all values of $\mathbf{s}$. However, $A(\mathbf{s})$ is never measured. We can only measure $I(\mathbf{s})$ and in $I(\mathbf{s})$ we lose much of the phase information inherent in $A(\mathbf{s})$. What we should always be able to get from a diffraction experiment is $P(\mathbf{u})$. Recall

$$\frac{I(\mathbf{s})}{I_I} = \int P(\mathbf{u}) e^{-2\pi i \mathbf{s}\cdot\mathbf{u}} dv_{\mathbf{u}} \tag{1.21}$$

Using the Fourier transform from above

$$P(\mathbf{u}) = \int \frac{I(\mathbf{s})}{I_I} e^{2\pi i \mathbf{s}\cdot\mathbf{u}} dv_{\mathbf{s}} \tag{1.47}$$

We shall come back later to this to see just how $P(\mathbf{u})$, the autocorrelation function, relates to the structure and how one evaluates the integral experimentally.

It is useful to note at this point that the scattering vector is often defined in a different way. In those cases, a scattering vector $\mathbf{k}$, where

$$\mathbf{k} = 2\pi\mathbf{s} \tag{1.48}$$

is used. The absolute magnitude of $\mathbf{k}$ is then

$$|\mathbf{k}| = \frac{4\pi \sin\theta}{\lambda} \tag{1.49}$$

Using this scattering vector, the amplitude and intensity functions appear simpler

$$\begin{aligned} \frac{A(\mathbf{k})}{A_I(k)} &= \int_{-\infty}^{\infty} \rho(\mathbf{r})_e{}^{-i\mathbf{k}\cdot\mathbf{r}} dv_{\mathbf{r}} \\ \frac{I(\mathbf{k})}{I_I(k)} &= \int_{-\infty}^{\infty} P(\mathbf{u}) e^{-i\mathbf{k}\cdot\mathbf{u}} dv_{\mathbf{u}} \end{aligned} \tag{1.50}$$

The inverse transforms, however, pick up a new constant multiplier

$$\begin{aligned} \rho(\mathbf{r}) &= \frac{1}{(2\pi)^3} \int \frac{A(\mathbf{k})}{A_I(k)} e^{i\mathbf{k}\cdot\mathbf{r}} dv_{\mathbf{k}} \\ P(\mathbf{u}) &= \frac{1}{(2\pi)^3} \int \frac{I(\mathbf{k})}{I_I(k)} e^{i\mathbf{k}\cdot\mathbf{u}} dv_{\mathbf{k}} \end{aligned} \tag{1.51}$$

The constant multipliers arise as follows. Using k in place of q, (1.44) becomes

$$c(k) = \int_{-\infty}^{\infty} h(x) e^{-ikx} dx$$

$$h(x) = \int_{-\infty}^{\infty} c(k) e^{ikx} d\left(\frac{k}{2\pi}\right) = \frac{1}{2\pi} \int_{-\infty}^{\infty} c(k) e^{ikx} dk$$

In (1.50) and (1.51) volume integrals are used. Thus, in (1.51) $dv_s = ds_1 ds_2 ds_3$ has been replaced by $dv_s = d(k_1/2\pi)d(k_2/2\pi)d(k_3/2\pi)$. We see that $dv_\mathbf{k}$ is then $(2\pi)^3$ larger than dv_s. It is this transformation from dv_s to $dv_\mathbf{k}$ that brings in the factor $(1/2\pi)^3$.

Another useful form of the Fourier integral transformation is

$$\begin{aligned} \tau(k) &= \int_{-\infty}^{\infty} t(r) \sin kr \, dr \\ t(r) &= \frac{2}{\pi}\int_{-\infty}^{\infty} \tau(k) \sin kr \, dk \end{aligned} \tag{1.52}$$

The proof of this sine transform is similar to the exponential proof we have just seen. We shall use this sine form in the analysis of scattering by liquids and amorphous (glassy) solids. Most of the scattering literature is written using **k.** We shall conform to **k**-usage in much of this text.

1.4.3 The Convolution Theorem

Consider the function

$$h(u) = \int f(x)g(u-x)dx \tag{1.53}$$

Such an integral is called a convolution integral. Note that

$$\int f(x)g(u-x)dx = -\int f(u-x)g(x)dx \tag{1.54}$$

This is seen in the following way. On the left let $x' = u - x$. Then

$$\int f(x)g(u-x)dx = -\int f(u-x')g(x')dx'$$

an equation identical to (1.53). Now our integral, $P(u)$

$$P(u) = \int \rho(x)\rho(u+x)dx \tag{1.55}$$

is similar to a convolution. The convolution operation itself, notwithstanding the similarity of the autocorrelation function, is often used in diffraction analysis.

The convolution theorem relates to Fourier integration of a convolution

$$H(k) = \int\int f(x)g(u-x)dx \, e^{-iku} du \equiv \mathscr{F}[f(x)*g(x)] \tag{1.56}$$

where $\mathscr{F}$ denotes Fourier integration and $*$ denotes the convolution operation. We now set $x' = u - x$

$$H(k)=\int \mathscr{F}\left[\int f(x)g(x')dx\right]e^{-ik(x+x')}dx'$$
$$=\left[\int f(x)e^{-ikx}dx\right]\left[\int g(x')e^{-ikx'}dx'\right] \qquad (1.57)$$
$$=\mathscr{F}[f(x)]\cdot\mathscr{F}[g(x)]$$

In shorthand

$$\mathscr{F}[f(x)*g(x)]\equiv\mathscr{F}[f(x)]\cdot\mathscr{F}[g(x)] \qquad (1.58)$$

Two examples of the use of convolutions are the following. The first example pertains to the probability of achieving a certain result in two tries. Suppose the probability of getting a certain result corresponding to the variable x is $p(x)$. Suppose this relates to the distance one can throw a discus. There will be some probability distribution which is the function $p(x)$. Supposing that every throw is independent of every other, what is the probability $Q(u)$ of getting a net distance u in two throws? This total is the sum (integral) of all combinations of two throws which could give a net distance u

$$Q(u)=\int p(x)p(u-x)dx \qquad (1.59)$$

A second example relates to the arrangement of a continuous function on a regular lattice. An example of such a function is shown as Fig. 1.9. Let the continuous function about $x=0$ be $g(x)$. The total function $f(x)$ is then

$$f(x)=\sum_n\int g(x)\delta[x-(x'+na)]dx \qquad (1.60)$$

where n is an integer.

Consider $n=3$. This element of the series is

$$\int g(x)\delta(x-x'-3a)dx \qquad (1.61)$$

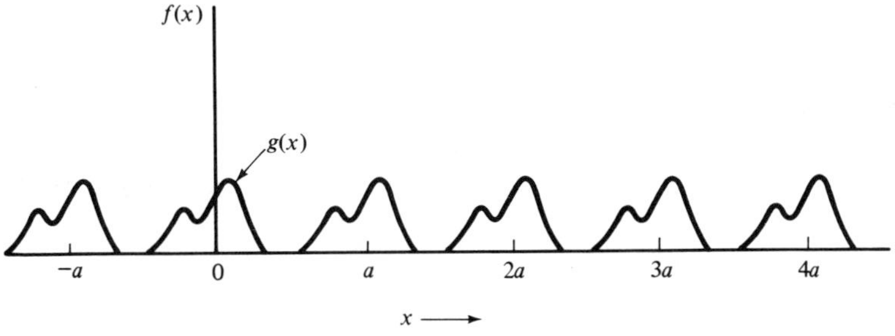

Figure 1.9. Arrangement of identical continuous functions on a one-dimensional periodic "lattice."

By the rules pertaining to the Dirac delta, the integral is

$$\int g(x)\delta(x - x' - 3a)dx = g(x' + 3a) \tag{1.62}$$

or the function $g(x)$ displaced three lattice spacings. The summation over n places $g(x)$ at every lattice site. Clearly this result can be extrapolated to three dimensions.

At this juncture it is useful to note that the convolution theorem can be represented in another form

$$\int \{\mathscr{F}[f(x)]\} \cdot \{\mathscr{F}[g(x)]\}dx = \{\mathscr{F}[f(x)]\}*\{\mathscr{F}[g(x)]\} \tag{1.63}$$

We see this in a straightforward manner. Let us write out the left side of (1.59)

$$\int \{\mathscr{F}[f(x)]\} \cdot \{\mathscr{F}[g(x)]\}dx$$

$$= \int \left\{\left[\int f(x)e^{ikx}dx\right]\left[\int g(x' - x)e^{ik(x'-x)}d(x' - x)\right]\right\}dx \tag{1.64}$$

Here it has been necessary to replace x by $x' - x$ in order that each x-term in $\mathscr{F}[f(x)]$ multiply all x-terms in $\mathscr{F}[g(x)]$. Thus we need to run the variable in $\mathscr{F}[g(x)]$ independently of that in $\mathscr{F}[f(x)]$. The means for doing that is to replace x formally by $x' - x$. Equation (1.64) is thus identically (1.63).

1.5 THE ATOMIC-SCATTERING FACTOR

1.5.1 X-rays

Within our black box the x-rays are scattered by the electrons. The x-ray wave, being merely a form of electromagnetic radiation, sets the electrons into motion, with a period given by the period of the incident wave. The electrons are accelerated in an oscillatory manner under this applied field. In general when a charged particle is accelerated it creates an electromagnetic field. In this case the field emanating from the electron (in all directions) has the periodicity of the exciting wave and in fact will continue the phase of that wave. The new periodic field emanating from the electron *is* the scattered wave.

For the case in which the electron has been caused by the incident wave to change to a different quantum state, the scattered wave will change wavelength slightly; some energy has been absorbed by the electron. This latter is the inelastically scattered wave. In principle, the incident photon could excite the charged nucleus in the same way, but the shear weight of the nucleus makes its small response undetectible.

In the present section we shall see that the *atom* acts in many ways as the effective scattering center, even though scattering is actually accomplished by the electrons. That is, the net effect of the continuous distribution of electrons about each atom is to scatter as if there were one center located at the center of mass of the atom. We begin by writing our amplitude

$$\frac{A(\mathbf{k})}{A_I(k)} = \int_{-\infty}^{\infty} \rho(\mathbf{r}) e^{-i\mathbf{k}\cdot\mathbf{r}} dv_{\mathbf{r}} \tag{1.50}$$

Now suppose that the centers of the several atoms are at $\mathbf{r}_1$, $\mathbf{r}_2$, $\mathbf{r}_3$ . . . and suppose that the density of the electron cloud about the nth atom is $\rho_n(\mathbf{r}-\mathbf{r}_n)$. Here we have merely located the center of coordinates at the center of the nth atom. The total electron density is then

$$\rho(\mathbf{r}) = \sum_n \rho_n(\mathbf{r}-\mathbf{r}_n) \tag{1.65}$$

Substituting into the amplitude expression, we now have

$$\frac{A(\mathbf{k})}{A_I(k)} = \int \sum_n \rho_n(\mathbf{r}-\mathbf{r}_n) e^{-i\mathbf{k}\cdot\mathbf{r}} dv_{\mathbf{r}} = \sum_n e^{-i\mathbf{k}\cdot\mathbf{r}_n} \left[\int \rho_n(\mathbf{r}-\mathbf{r}_n) e^{-i\mathbf{k}\cdot(\mathbf{r}-\mathbf{r}_n)} dv_{\mathbf{r}} \right] \tag{1.66}$$

The integral is immediately identified as the amplitude of scattering by the nth atom (independent of the other atoms). We shall consistently use f_n^x to denote this x-ray atomic scattering factor

$$\frac{A(\mathbf{k})}{A_I(k)} = \sum_n f_n^x(\mathbf{k}) e^{-i\mathbf{k}\cdot\mathbf{r}_n} \tag{1.67}$$

We note also that the value of f_n^x is nearly independent of the state of aggregation. That is, the electronic configuration of all but the valence electrons is fixed by the atomic type. Very often (atoms of higher atomic number) the difference in valence state exerts only a very small influence on the atomic scattering factor at any rate.

We note that (1.67) for the scattering amplitude is of the same form as (1.13), except that the $\mathbf{r}_n$ represent the positions of the atom centers rather than the electrons and that $f_n^x(\mathbf{k})$ is inserted to account for the specific scattering power of the nth atom. In this sense the atoms can be viewed as the effective scattering centers.

Let us now consider the form of the atomic scattering factor. It is written

$$f_n^x(k) = \int_{\substack{\text{one atom of} \\ \text{type } n}} \rho_n(r) e^{-i\mathbf{k}\cdot\mathbf{r}} \, dv_{\mathbf{r}} \tag{1.68}$$

Note that at $k = 0$ $(2\theta = 0)$

$$f_n^x(0) = \int \rho_n(\mathbf{r}) dv_{\mathbf{r}} = Z_n \tag{1.69}$$

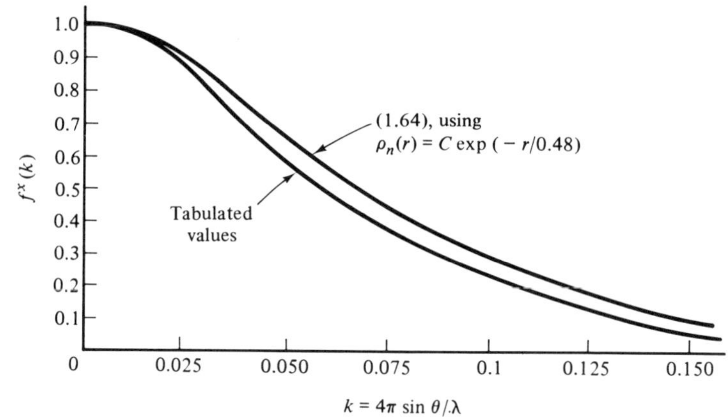

Figure 1.10. The atomic scattering factor of hydrogen. The upper curve is computed from (1.64). The lower curve is computed from a more precise wave function.

where Z_n is the atomic number of atoms of type n. At higher angles the magnitude of f_n^x drops off continuously, as shown in Fig. 1.10. This trend is explained in the following way. Let us assume that the electron density is spherically symmetric (as it is exactly in H and He). That is $\rho_n(\mathbf{r})$ becomes $\rho_n(r)$ and we have dropped all angular dependence in the mathematics. Let us now choose a spherical coordinate system, as in Fig. 1.11, where $\phi = 0$ occurs along $\mathbf{k}$ and the plane of θ is normal to $\mathbf{k}$. Thus $\mathbf{k} \cdot \mathbf{r} = \mathrm{kr} \cos\phi$ and $dv_r = r^2 \sin\phi \, d\theta \, d\phi \, dr$

$$f_n^x(k) = \int_0^\infty \int_0^\pi \int_0^{2\pi} \rho_n(r) e^{-ikr\cos\phi} r^2 \sin\phi \, d\theta \, d\phi \, dr \tag{1.70}$$

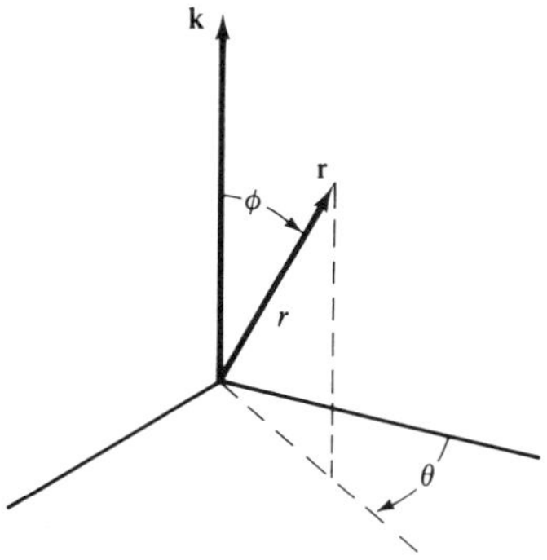

Figure 1.11. Spherical coordinate system.

Integrating over θ, we have

$$f_n^x(k) = 2\pi \int_0^\infty \int_0^\pi \rho_n(r) e^{-ikr\cos\phi}\, r^2 \sin\phi\, d\phi\, dr \tag{1.71}$$

Setting $u = \cos\phi$, and integrating over u, we have

$$\begin{aligned} f_n^x(k) &= -2\pi \int_0^\infty \int_{-1}^1 \rho_n(r) e^{-ikru}\, r^2 du\, dr \\ &= -2\pi \int_0^\infty r^2 \rho_n(r) \frac{-1}{ikr} (e^{ikr} - e^{-ikr})\, dr \end{aligned} \tag{1.72}$$

Rewriting, (1.72) becomes

$$f_n^x(k) = 4\pi \int_0^\infty r^2 \rho_n(r) \frac{\sin kr}{kr}\, dr \tag{1.73}$$

In (1.72) and (1.73) the electron density $\rho_n(r)$ can be approximated by $Ce^{-r/a}$ for an s orbital state. Here C and a are constants. Making this substitution (exact for H and He and a reasonable first approximation for other atoms) in (1.73), we have

$$\begin{aligned} f_n^x(k) &= \frac{2\pi C}{ik} \int_0^\infty r e^{-r/a} (e^{ikr} - e^{-ikr})\, dr \\ &= \frac{2\pi C}{ik} \int_0^\infty r e^{(-1/a+ik)r}\, dr - \int_0^\infty r e^{(-1/a-ik)r}\, dr \end{aligned} \tag{1.74}$$

Integrating by parts

$$\begin{aligned} f_n^x(k) = \frac{2\pi C}{ik} \Bigg\{ & -\frac{r}{b_1} e^{-b_1 r}\Big|_0^\infty - \frac{1}{b_1^2} e^{-b_1 r}\Big|_0^\infty \\ & + \frac{r}{b_2} e^{-b_2 r}\Big|_0^\infty + \frac{1}{b_2^2} e^{-b_2 r}\Big|_0^\infty \Bigg\} \end{aligned} \tag{1.75}$$

where

$$\begin{aligned} b_1 &= \frac{1}{a} - ik \\ b_2 &= \frac{1}{a} + ik \end{aligned} \tag{1.76}$$

Using l'Hospital's rule to evaluate the functions at their 0 and ∞ limits, we find that

$$f_n^x(k) = \frac{2\pi C}{ik} \cdot \left\{ 0 + \frac{1}{b_1^2} + 0 - \frac{1}{b_2^2} \right\}$$

$$= \frac{8\pi C}{a}\left(\frac{1}{a^2} + k^2\right)^{-2} \tag{1.77}$$

In Fig. 1.10 we have plotted $\left(\frac{1}{a^2} + k^2\right)^{-4}$ versus $= 4\pi \sin \theta/\lambda$, using for a the value 0.48, which is exact for the hydrogen atom. (In this computation C is found by setting $f_n^x(0) = 1$, since $Z_H = 1$.) We see that the atomic scattering factor we have computed in this simple manner is very close to that tabulated from a more precise computation.

In general, the atomic scattering factor for any material is computed in this way. That is, one begins with electron densities $\rho(\mathbf{r})$ derived quantum mechanically. In all cases $f^x(0) = Z$ and the curve of $f^x(k)$ against k drops off in a manner similar to that seen in Fig. 1.10.

It should be noted here that when the incident x-ray has a large probability of exciting a k or l shell atomic electron to a higher energy state, then the formalism becomes somewhat different. This problem exists only when λ is near certain critical values. In general, the x-ray atomic scattering factor is given by $f_0 + \Delta f' + i\Delta f''$. Here $\Delta f'$ and $\Delta f''$ are deviations from the classical atomic scattering factors f. They are called the *dispersion corrections.*

A tabulation of x-ray atomic scattering factors is found in Appendix Table B.2, while dispersion corrections may be found in Appendix Table B.3.

1.5.2 Electrons

Electrons are negatively charged and are sensitive to local electric fields. We can then write for one atom

$$f^e(\mathbf{k}) = B\int V(\mathbf{r})e^{i\mathbf{k}\cdot\mathbf{r}}\, dv_{\mathbf{r}'} \tag{1.78}$$

where $V(r)$ is the field felt by an electron at position $\mathbf{r}$ and B is a constant. Consider the effect of the charge in an increment of volume at $\mathbf{r}'$ on the electron at $\mathbf{r}$. See Fig. 1.12. The potential due to that volume element is $\rho_c(\mathbf{r}')dv_{\mathbf{r}'}/|\mathbf{r}-\mathbf{r}'|$, where $\rho_c(r')$ is the local charge density. The potential due to all volume elements, including those of the nucleus, is

$$V(r) = \int_{\text{atom}} \rho_c(\mathbf{r}')\frac{dv_{\mathbf{r}'}}{|\mathbf{r}-\mathbf{r}'|} \tag{1.79}$$

Inserting (1.79) into (1.78) we have

$$\begin{aligned} f^e(k) &= B\iint_{\text{atom}} \rho_c(\mathbf{r}')\frac{e^{i\mathbf{k}\cdot\mathbf{r}}}{|\mathbf{r}-\mathbf{r}'|}\, dv_{\mathbf{r}}\, dv_{\mathbf{r}'} \\ &= B\int\left[\int \frac{e^{i\mathbf{k}\cdot(\mathbf{r}-\mathbf{r}')}}{\mathbf{r}-\mathbf{r}'}\, dv_{\mathbf{r}}\right]\rho_c(\mathbf{r}')e^{i\mathbf{k}\cdot\mathbf{r}'}\, dv_{\mathbf{r}'} \end{aligned} \tag{1.80}$$

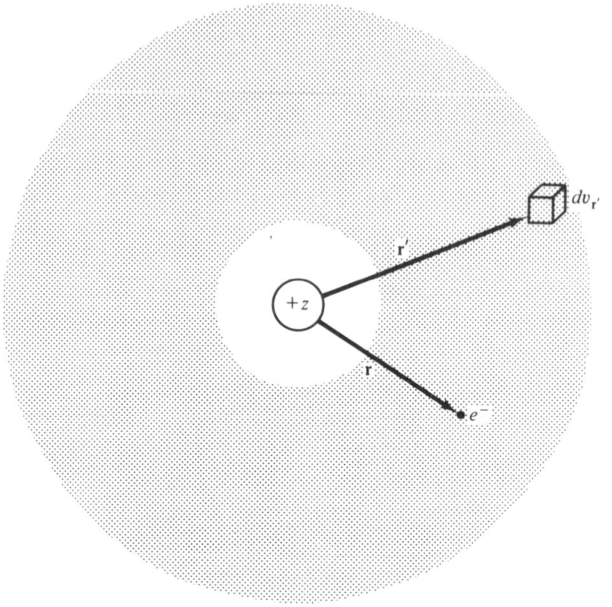

Figure 1.12. Geometry describing a charge contained in the volume elements $dv_{r'}$. located **r** from the atomic center.

The inner integral is $1/(\pi k^2)$. So

$$f^e(k) = \frac{B}{\pi k^2} \int \rho_c(\mathbf{r}')e^{i\mathbf{k}\cdot\mathbf{r}'}\, dv_{\mathbf{r}'} \tag{1.81}$$

This expression can profitably be further simplified. The charge density is

$$\rho_c(\mathbf{r}') = eZ\delta(\mathbf{r}') - e\rho_e(\mathbf{r}') \tag{1.82}$$

where the first term on the right represents the nucleus at $\mathbf{r}' = 0$, and the second term represents the atomic electrons, of local density $\rho_e(\mathbf{r}')$. Equation (1.80) thus becomes

$$\begin{aligned} f^e(k) &= \frac{Be}{\pi k^2}\left[Z\int \delta(\mathbf{r}')e^{i\mathbf{k}\cdot\mathbf{r}'}\, dv_{\mathbf{r}'} - \int \rho_e(\mathbf{r}')e^{i\mathbf{k}\cdot\mathbf{r}'}\, dv_{\mathbf{r}'}\right] \\ &= \frac{Be}{\pi k^2}\left[Z - \int \rho_e(\mathbf{r}')e^{i\mathbf{k}\cdot\mathbf{r}'}\, dv_{\mathbf{r}'}\right] \end{aligned} \tag{1.83}$$

We now note that the remaining integral is just the atomic scattering factor for x-rays, $f_i^x(k)$. Thus

$$f_i^e(k) = \frac{\text{Be}}{\pi}\frac{1}{k^2}[Z - f_i^x(k)] \tag{1.84}$$

The $1/k^2$ term dominates the k-dependence, and the magnitude of $f(k)$ decreases montonically with k. Figure 1.13 shows the atomic scattering amplitudes for carbon and silicon. In form, these are still similar to the atomic scattering factors for x-rays. And the overall magnitude of f^e increases with atomic number, again just as in the x-ray case.

A compilation of the electron scattering amplitudes of the elements is given in Appendix Table B.4.

The major differences between x-rays and electrons lie in these areas: wavelength, scattering probability, and absorption. Electron beams are created by the acceleration of electrons in a potential gradient. DeBroglie's law gives a relationship between wavelength and kinetic energy

$$\lambda = \frac{h}{p} = \frac{h}{\sqrt{2m_e E}} \tag{1.85}$$

where h is Planck's constant and m_e is the kinetic mass of the electron. The kinetic energy E is in turn related to the applied electric potential difference ϵ through $E = e\epsilon$. After taking account of relativistic effects, the wavelength expression becomes

$$\lambda = \frac{h}{\sqrt{2m_o e\epsilon\left(1 + \dfrac{e\epsilon}{2m_o c^2}\right)}} = \frac{12.26\text{Å}}{\sqrt{E}\sqrt{1 + 0.9788 \times 10^{-6}E}} \tag{1.86}$$

where m_o is the rest mass of the electron. The electron wavelength can thus be varied continuously. Table 1.1 gives wavelength against accelerating voltage over an accessible range. The relativistic wavelength λ and wave number λ^{-1}, are shown as functions of the accelerating potential E.

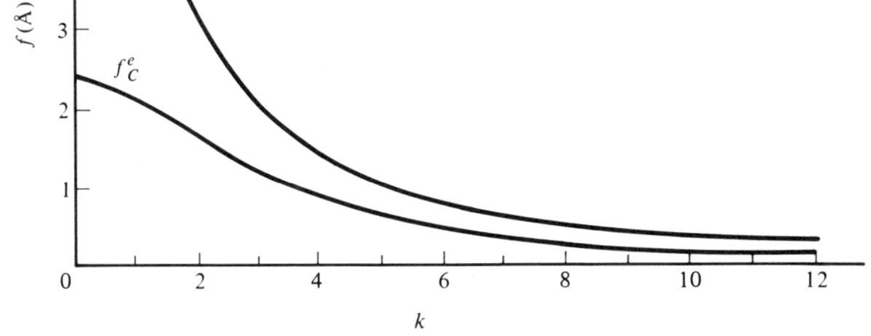

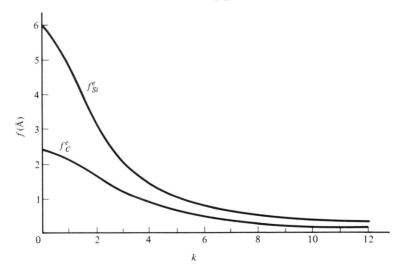

Figure 1.13. The electron scattering amplitudes for carbon and silicon.

TABLE 1.1
ELECTRON WAVELENGTH[a]

E (V)	λ $(\AA)$
1	12.26
10	3.878
100	1.226
500	0.5483
1,000	0.3876
2,000	0.2740
3,000	0.2236
4,000	0.1935
5,000	0.1730
6,000	0.1579
7,000	0.1461
8,000	0.1366
9,000	0.1287
10,000	0.1220
$2 \cdot 10^4$	0.0859
$3 \cdot 10^4$	0.0698
$4 \cdot 10^4$	0.0602
$5 \cdot 10^4$	0.0536
$6 \cdot 10^4$	0.0487
$7 \cdot 10^4$	0.0448
$8 \cdot 10^4$	0.0418
$9 \cdot 10^4$	0.0392
$1 \cdot 10^5$	0.0370
$2 \cdot 10^5$	0.0251
$3 \cdot 10^5$	0.0197
$4 \cdot 10^5$	0.0164
$5 \cdot 10^5$	0.0142
$6 \cdot 10^5$	0.0126
$7 \cdot 10^5$	0.0113
$8 \cdot 10^5$	0.0103
$9 \cdot 10^5$	0.0094
$1 \cdot 10^6$	0.0087
$2 \cdot 10^6$	0.0050
$4 \cdot 10^6$	0.0028
$6 \cdot 10^6$	0.0019
$8 \cdot 10^6$	0.0015
$1 \cdot 10^7$	0.0012

[a] After P. B. Hirsch, A. Howie, R. B. Nicholson, D. W. Pashley, and M. J. Whelan, *Electron Microscopy of Thin Crystals* (Plenum, N.Y., 1967).

Recall that x-rays were scattered indirectly. They caused atomic electrons to accelerate and these, in turn, produced a new, scattered wave. The case for electrons is more direct and appears as an increased probability of scattering (or, scattering cross-section). The probability of an x-ray photon being scattered by a given atom is orders of magnitude lower than the corresponding probability for an electron. This is why multiple scattering effects must generally be taken into account in analyzing typical ($\geq$ 50 kV) electron diffraction results.

Finally, electrons are absorbed much more readily than are x-rays. At very low voltages, electrons can be adsorbed in the first monolayer of a crystal surface. This effect gives the power of low energy electron diffraction (LEED) in studying the structure of surfaces. At voltages typical for the electron microscope, absorption distances, for metals, of some few hundred angstroms are typical.

1.5.3 Neutrons

A neutron is an uncharged elementary particle possessing atomic mass 1.01 and a magnetic moment with spin number ½. A neutron interacts directly with, and is scattered by, atomic nuclei. The neutron is also scattered through interaction of its magnetic moment with the magnetic moment of the electrons of the target. The neutron scattering factor f^n is then the sum of two terms: f^n_N, representing the nuclear interaction and f^n_E representing the magnetic interaction with electrons. The latter term is important for ferro- or antiferromagnetic materials, but is a relatively small contribution for other systems.

The atomic scattering factor for neutrons is independent of scattering angle, unlike the case for x-rays or electrons. For both x-rays and electrons, the scatterers within one atom are dispersed over a space with dimensions near those of the wavelength. As we have seen, it is the correlation of the rays scattered by these dispersed scatterers that produces an amplitude modulation and the variation of atomic scattering factor with scattering angle. This is not the case for the neutron–nucleus interaction. The diameter of the nucleus is approximately 10^{-13} cm; some five orders of magnitude smaller than the wavelengths of the thermal neutrons which are used in scattering studies.

The magnitude of the scattering amplitude due to the neutron–nucleus interaction does not depend on the atomic number in a simple way. The amplitude of the scattering depends on the spin number of the nucleus and that, in turn, depends on the number of neutrons it contains. Thus this amplitude depends on the details of the nuclear structure and not in a simple way on the atomic number. Further, the amplitude will be different for different isotopes. Figure 1.14 compares neutron and x-ray atomic scattering factors. One sees that the strong atomic number correlation present for x-rays does not hold for neutrons. Thus neutrons can give information about hydrides, for instance, whereas for x-rays the scattering from the hydrogen atom would be too weak to be detected.

The scattering reaction itself depends on the creation and release of a complex system comprising the nucleus and the neutron. The lifetime of this complex

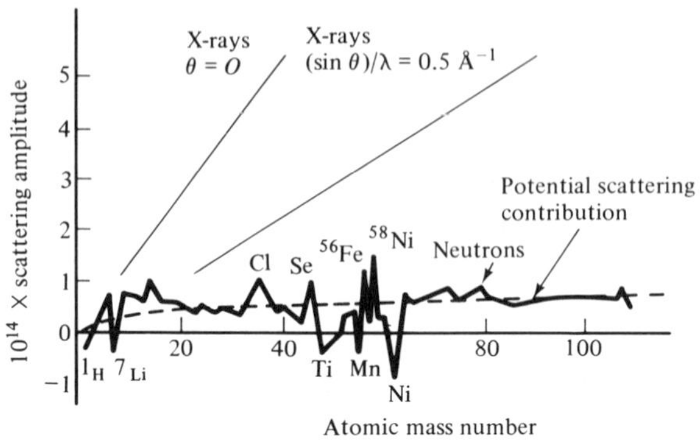

Figure 1.14. Scattering amplitudes for neutrons and x-rays as a function of atomic mass number. [G. E. Bacon, *Research,* **7,** 257 (1954)]

determines a phase shift for the scattered wave. The phase shift can be greater than 180° in some cases; consequently, the structure factor can possess *negative* values. Appendix Table B.4 is a compilation of neutron atomic scattering factors. It should be noted that in some cases, such as Mn and Fe, the neutron scattering factors are far apart (here due to electron magnetic interaction), whereas the atomic scattering factors for these atoms differ only slightly. In such cases, neutron diffraction can be more useful than x-rays in determining site occupancy by the two different atomic types.

Finally, it should be mentioned that the incoherent portion of the scattering of neutrons will be larger than that of either electrons or electromagnetic radiation. The mass of the neutron is similar to that of the scatterer. Thus the neutron can impart some of its momentum to the nucleus. The energy, and consequently the wavelength, of the neutron is changed in such inelastic scattering. This is a very useful property and can be used to measure lattice dynamics of the target material. In this text, however, we shall consider only elastic scattered neutrons, as only these cleanly carry structural information.

1.6 CONSERVATION OF ENERGY

What we shall show here is that the intensity integrated over all values of **k** is constant for a given mass, independent of its state of aggregation. This is a useful principle for the scaling of diffraction experiments, as we shall see later.

Recall that, for any system,

$$\frac{I(\mathbf{k})}{I_e(k)} = \int P(\mathbf{u}) e^{-i\mathbf{k}\cdot\mathbf{u}}\, dv_{\mathbf{u}} \tag{1.87}$$

Here the microstructure of the system is contained in the $P(\mathbf{u})$. Now integrate over all k-space

$$\int \frac{I(\mathbf{k})}{I_e(k)}\, dv_{\mathbf{k}} = \iint P(\mathbf{u}) e^{-i\mathbf{k}\cdot\mathbf{u}}\, dv_{\mathbf{u}} dv_{\mathbf{k}}$$
$$= \int \left[\int e^{-i\mathbf{k}\cdot\mathbf{u}}\, dv_{\mathbf{k}} \right] P(\mathbf{u}) dv_{\mathbf{u}} \tag{1.88}$$

The discussion on pp. 11–13 showed that the inner integral of (1.88) is a delta function. Hence

$$\int \frac{I(\mathbf{k})}{I_e(k)}\, dv_{\mathbf{k}} \int \delta(\mathbf{u}) P(\mathbf{u}) dv_{\mathbf{u}} = P(0) = \rho_e^2\, V \tag{1.89}$$

where V is the irradiated volume of the material. Since the right side of (1.89) is a constant, the value of the integrated intensity must be independent of structure.

PROBLEMS

1.1. Show that $\mathscr{F}[\delta(\mathbf{x}) + 1] = \delta(\mathbf{s}) + 1$.

1.2. Show that the autocorrelation function possesses all the transform properties of the convolution.

1.3. Find the Fourier tranforms of

(a) $\delta(x - a)$

(b) $\sum_{n=-\infty}^{n=\infty} \delta(x - na)$

(c) $p(x) = \begin{cases} 1 \text{ for } |x| < a/2 \\ 0 \text{ for } |x| > a/2 \end{cases}$

1.4. The Cauchy function is written

$$f(x) = \frac{b}{b^2 + x^2}$$

This function appears as shown in Fig. P1.4. The Fourier transform of the Cauchy function is

$$\mathscr{F}[f(x)] = \frac{\pi\, e^{-b|k|}}{b}$$

(a) Sketch what happens to the Cauchy function and to its transform as b increases. Comment on the properties of this transform.

(b) Find

$$\int_{-\infty}^{\infty} \int_{-\infty}^{\infty} \left[\sum_{n=0}^{\infty} \delta(k - na*) \right] \left[\frac{\pi}{b}\, e^{-b|k-k'|} \right] e^{2\pi i k' x} dk\, dk'$$

(Use the convolution theorem.) Sketch your result.

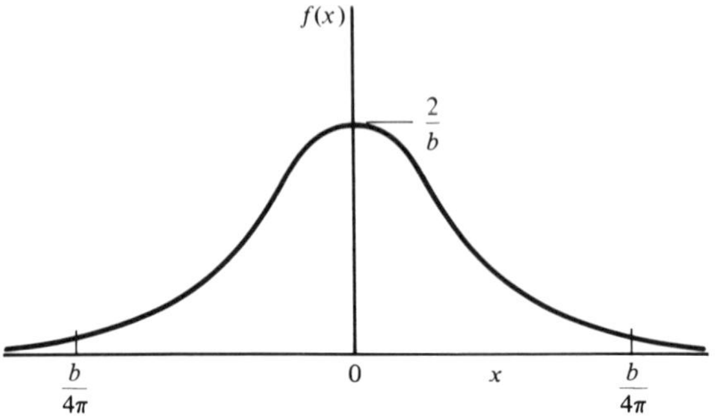

Figure P1.4.

1.5. (a) A completely structureless plate of thickness h and average electron density ρ_o is irradiated with x-rays, as shown in Fig. P1.5a. The lateral dimensions, Y and Z are very large compared with h. Derive a formula for the intensity function $I(2\theta)/I_e(2\theta)$. Ignore effects of absorption. Sketch $I(2\theta)/I_e(2\theta)$ versus $2 \sin \theta/\lambda$.

(b) Now consider an infinite regular stack of such plates ($N >> 1$), separated by vacuum ($\rho = 0$). Derive a formula for the scattering from this stack. Again ignore absorption. Sketch $I(2\theta)/I_e(2\theta)$ versus $2 \sin \theta/\lambda$. (Hint: you could express the density of each plate plus intervening vacuum by $\rho_o P_1$, where

$$P_1(x) = \begin{cases} 1 \text{ for } 0 \leqslant x < h \\ 0 \text{ for } h \leqslant x < 1 \end{cases}$$

Also, the convolution theorem will prove useful.)

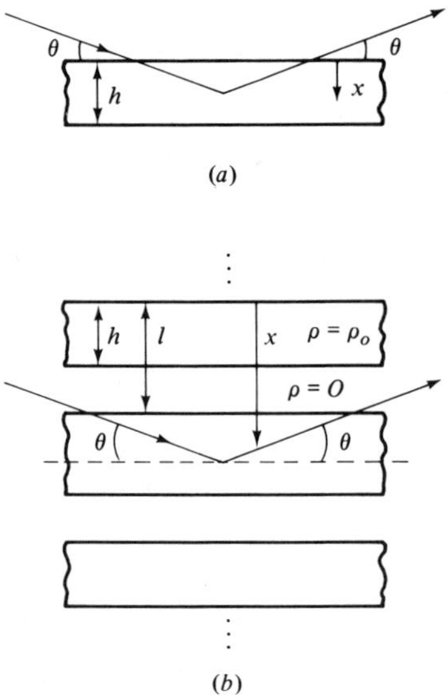

Figure P1.5.

BIBLIOGRAPHY

A. Basic Diffraction Theory

1. L. V. AZAROFF, R. KAPLOW, N. KATO, R. J. WEISS, A. J. C. WILSON, and R. A. YOUNG, *X-ray Diffraction,* Ch. 1, McGraw-Hill Book Company, N.Y. (1974).
2. J. B. COHEN, *Diffraction Methods in Materials Science,* Ch. 2, Macmillan, Inc., N.Y. (1966).
3. J. M. COWLEY, *Diffraction Physics,* Chs. 1–3 and 5, North-Holland, Amsterdam (1975).
4. A. GUINIER, *X-ray Diffraction in Crystals, Imperfect Crystals, and Amorphous Bodies,* Ch. 2, W. H. Freeman & Company Publishers, San Francisco (1963).
5. R. HOSEMANN, and S. N. BAGCHI, *Direct Analysis of Diffraction by Matter,* North-Holland, Amsterdam (1962).
6. H. S. LIPSON and C. A. TAYLOR, *Fourier Transforms and X-ray Diffraction,* Bell, London (1958).

B. Atomic Scattering Factors

1. A. GUINIER, *X-ray Crystallographic Technology,* Chs. 6 and 8, Hilger and Watts, London (1952).
2. L. V. AZAROFF, R. KAPLOW, N. KATO, R. J. WEISS, A. J. C. WILSON, and R. A. YOUNG, *X-ray Diffraction,* Ch. 1, McGraw-Hill Book Company, N.Y. (1974).
3. G. E. BACON, *Applications of Neutron Diffraction in Chemistry,* Ch. 1, Pergamon Press, Inc., N.Y. (1963).
4. G. E. BACON, *Neutron Diffraction,* Ch. 1, Clarendon, Oxford (1975).
5. J. B. COHEN, *Diffraction Methods in Materials Science,* Ch. 5, Macmillan, Inc., N.Y. (1966).
6. A. H. COMPTON and S. K. ALLISON, *X-rays in Theory and Experiment, 2nd Ed.,* Chs. 3 and 4, Van Nostrand Reinhold Company, N.Y. (1935).
7. J. M. COWLEY, *Diffraction Physics,* Ch. 4, North-Holland, Amsterdam (1975).
8. P. B. HIRSCH, A. HOWIE, R. B. NICHOLSON, D. W. PASHLEY, and M. J. WHELAN, *Electron Microscopy of Thin Crystals,* Ch. 4 and Appendix 3, Plenum Publishing Corporation, N.Y. (1967).
9. R. W. JAMES, *The Optical Principles of the Diffraction of X-rays,* Chs. 3 and 4, G. Bell & Sons, Ltd., London (1954).
10. LAUE, M. T. F. VON, *Röntgenstrahlinterferenzen, 3. Ausg.,* Ch. 2, Akademische Verlagsgesellschaft, Frankfurt (1960).
11. CAROLINE H. MACGILLVARY, G. D. RIECK, and KATHLEEN LONSDALE, eds., *International Tables for X-ray Crystallography, 2nd Ed., Vol. III,* The Kynoch Press, Birmingham, England (1965).
12. B. E. WARREN, *X-ray Diffraction,* Ch. 1, Addison-Wesley Publishing Company, Inc., Reading, MA (1969).

Scattering by Gases and Liquids

2

2.1. INTRODUCTION

In this chapter we shall obtain the basic relationships relating structure to scattering curves for gases and liquids. We shall see that there is a very direct relation between the structure of a molecule and the gas-phase scattering by the molecule. With regard to liquids, we shall see that diffraction data provides the investigator with a means of obtaining the radial distribution function—the primary structural characterization for liquids.

2.2. GASES

Let us begin by looking at the properties of a dilute monatomic gas. The results we obtain here will be directly extrapolable, with one modification, to the scattering by dilute molecular gases of any complexity.

First we show that the scattering by such a gas of N atoms is given exactly by

$$\frac{I(k)}{I_e(k)} = Nf_n^2(k) \tag{2.1}$$

To see this, we begin with (1.67), the amplitude of x-ray scattering by one atom

$$\frac{A(\mathbf{k})}{A_e(k)} = \sum_n f_n^x(k)e^{-i\mathbf{k}\cdot\mathbf{r}_n} \tag{1.67}$$

We now use (1.67) to write the intensity as

$$\frac{I(\mathbf{k})}{I_e(k)} = \left[\frac{A^*(\mathbf{k})}{A_e(k)}\right]\left[\frac{A(\mathbf{k})}{A_e(k)}\right] = \sum_m \sum_n f_m(k) f_n(k) e^{i\mathbf{k}\cdot(\mathbf{r}_n - \mathbf{r}_m)}$$

Since we have assumed all atoms to be identical

$$\frac{I(\mathbf{k})}{I_e(k)} = f_n^2(k) \sum_{m=1}^{N} \sum_{n=1}^{N} e^{i\mathbf{k}\cdot(\mathbf{r}_n - \mathbf{r}_m)} \tag{2.2}$$

We shall shortly see that it is convenient to separate out the terms for which $m = n$

$$\frac{I(\mathbf{k})}{I_e(k)} = f_n^2(k) \left[N + \sum_{m \neq n}^{N} \sum^{N} e^{i\mathbf{k}\cdot(\mathbf{r}_n - \mathbf{r}_m)} \right] \tag{2.3}$$

Since the molecules of a gas are randomly distributed and are in constant motion, and since we physically average our intensity measurement over a long time relative to the atomic velocity in a gas, we may replace the summation in (2.3) by an integral, using $\rho_a(\mathbf{r})$ to denote the (isotropic) atomic density from point to point. Thus, the second term on the right in (2.3) becomes

$$G_2(k) = \sum_{m \neq n} \sum e^{i\mathbf{k}\cdot(\mathbf{r}_n - \mathbf{r}_m)} = \iint \rho_a(\mathbf{r})\rho_a(\mathbf{r} + \mathbf{u}) e^{i\mathbf{k}\cdot\mathbf{u}}\, dv_{\mathbf{r}} dv_{\mathbf{u}} \tag{2.4}$$

For a dilute gas, $\rho_a(\mathbf{r})$ can be replaced by a constant ρ_a^o, because the distribution deviates from complete randomness only very near the center of an atom of the gas and we have taken the system to be dilute. Hence, using Cartesian coordinates x_1, x_2, and x_3

$$\begin{aligned} G_2(k) &= V(\rho_a^o)^2 \int_{-X}^{X} \int_{-Y}^{Y} \int_{-Z}^{Z} e^{i(k_1x_1 + k_2x_2 + k_3x_3)}\, dx_1 dx_2 dx_3 \\ &= V(\rho_a^o)^2 \int_{-M}^{M} e^{ik_1x_1} dx_1 \int_{-M}^{M} e^{ik_2x_2} dx_2 \int_{-M}^{M} e^{ik_3x_3} dx_3 \end{aligned} \tag{2.5}$$

where V is the irradiated volume.

Expand the first of the three integrals

$$\begin{aligned} \int_{-M}^{M} e^{ik_1x_1} dx_1 &= \int_{-M}^{M} \cos k_1x_1 dx_1 + i \int_{-M}^{M} \sin k_1x_1 dx_1 \\ &= \frac{2 \sin Mk_1}{k_1} \end{aligned} \tag{2.6}$$

Using the argument associated with Fig. 1.8, we see that $G_2(k)$ has shown itself to approach delta function behavior for a gas volume of finite size. Therefore $G_2(k)$ is a very sharp spike located at the origin. If our gas volume were 10^{-2} cm on a side—a very small sample—the breadth of the $G_2(k)$ contribution would be of the order of one second of arc! Consequently G_2 can be ignored as immeasurable. Thus for the measurable portion of the intensity

$$\frac{I(\mathbf{k})}{I_e(k)} = Nf_n^2(k) \tag{2.1}$$

The extension to molecular gases is easy. In a dilute molecular gas the physical assumptions we have used are unchanged. The only modification necessary is that we need to replace our atomic scattering factor $f(k)$ by a molecular scattering factor $g(k)$. Thus the intensity is written

$$\frac{I(k)}{I_e(k)} = Ng^2(k) \tag{2.7}$$

where $g^2(k)$ takes care of the interference relations among the N' atoms of the molecule

$$g^2(k) = < \sum_{m=1}^{N'} \sum_{n=1}^{N'} f_m f_n e^{i\mathbf{k}\cdot(\mathbf{r}_n - \mathbf{r}_m)} > \tag{2.8}$$

where the $<>$ symbol indicates averaging over all orientations of the molecule. Again separating out all $m = n$ terms and setting $\mathbf{r}_{mn} = \mathbf{r}_n - \mathbf{r}_m$, we have

$$g^2(k) = \sum_{n=1}^{N'} f_n^2(k) + \sum_{m \neq n} \sum f_m f_n <\cos \mathbf{k} \cdot \mathbf{r}_{mn}> \tag{2.9}$$

Here, $<e^{i\mathbf{k}\cdot\mathbf{r}} mn>$ is replaced by $<\cos \mathbf{k} \cdot \mathbf{r}_{mn}>$, because the average of sin $\mathbf{k} \cdot \mathbf{r}_{mn}$ is 0, the sine being an odd function.

We need now evaluate $<\cos \mathbf{k} \cdot \mathbf{r}_{mn}>$. To do this we observe the geometry of Fig. 2.1. The averaging of (2.9) denotes the average over all possible values of the angle α between $\mathbf{k}$ and $\mathbf{r}_{mn}$. Thus, using solid angle elements $\sin \alpha \, d\alpha \, d\beta$

$$<\cos \mathbf{k} \cdot \mathbf{r}_{mn}> = \frac{\int_{\alpha=0}^{\pi} \int_{\beta=0}^{2\pi} \cos (kr_{mn} \cos \alpha) \sin \alpha \, d\alpha \, d\beta}{\int_{\alpha=0}^{\pi} \int_{\phi=0}^{2\pi} \sin \alpha \, d\alpha \, d\beta}$$

$$= \frac{-\dfrac{1}{kr_{mn}} \int_{kr_{mn}}^{-kr_{mn}} \cos u \, du}{-2} = \frac{\sin kr_{mn}}{kr_{mn}} \tag{2.10}$$

And the molecular scattering factor, finally, is

$$g^2(k) = \sum_{n=1}^{N'} f_n^2 + \sum_{m \neq n}^{N'} \sum^{N'} f_m f_n \frac{\sin kr_{mn}}{kr_{mn}} \tag{2.11}$$

We are now in position to look at some examples.

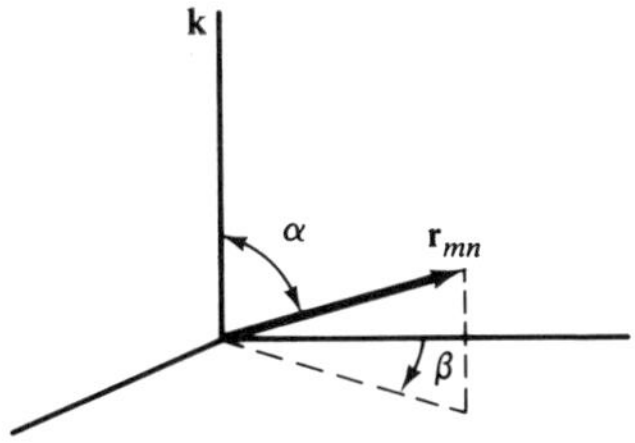

Figure 2.1. Geometrical relationships among the scattering vector **k**, the interatomic vector $\mathbf{r}_{mn}$, and the spherical coordinate angles α and β.

Example 1. *Diatomic Gas* (N_2)

Let $a = 1.06$ Å be the separation between the two nitrogen atoms of the molecule. In this case $f_m = f_n$ and can be removed from the summation

$$\frac{I(k)}{NI_e(k)} = 2f_N^2 + 2f_N^2 \frac{\sin ka}{ka} = 2f_N^2\left(1 + \frac{\sin ka}{ka}\right) \tag{2.12}$$

Figure 2.2 shows $f_N^2(k)$, $(1 + \sin ka/ka)$ and $I(k)/I_e(k)$. In this way we see separately the effects of the atomic scattering (via f^2) and the intramolecular interference [via $1 + (\sin ka/ka)$].

Example 2. AB_2 *Gas*

A typical molecular structure is shown as Fig. 2.3. In this case we must retain the f_n inside the summation. If the $A - B$ separation is a and the $B - B$ separation is b, then the molecular scattering factor is written

$$g(k) = 2f_B^2 + f_a^2 + 4f_A f_B \frac{\sin ka}{ka} + 2f_B^2 \frac{\sin kb}{kb} \tag{2.13}$$

Figure 2.4 shows the elements on the right-hand side of (2.12) and $g(k)$ for the CO_2 molecule.

2.3 LIQUIDS

2.3.1 Monatomic Pure Liquids

For monatomic liquids we begin with (2.2).

$$\frac{I(\mathbf{k})}{I_e(k)} = f_n^2(k)\left[\sum_m^N \sum_n^N e^{i\mathbf{k}\cdot(\mathbf{r}_{mn} - \mathbf{r}_m)}\right] \tag{2.2}$$

Again, as in the case of a gas, we consider a very mobile atomic system and replace the double sum by an integral, using the local atomic density function $\rho_a(\mathbf{r})$. Thus (2.2) becomes

$$\frac{I(\mathbf{k})}{I_e(k)f^2(k)} = \int\int \rho_a(\mathbf{r})\rho_a(\mathbf{r} + \mathbf{u})e^{i\mathbf{k}\cdot\mathbf{u}}\, dv_{\mathbf{r}}\, dv_{\mathbf{u}} \tag{2.14}$$

Unlike the gas case, however, we cannot take $\rho_a(\mathbf{r})$ as a constant. This is seen in the following way. A liquid is a condensed disordered system; the atoms

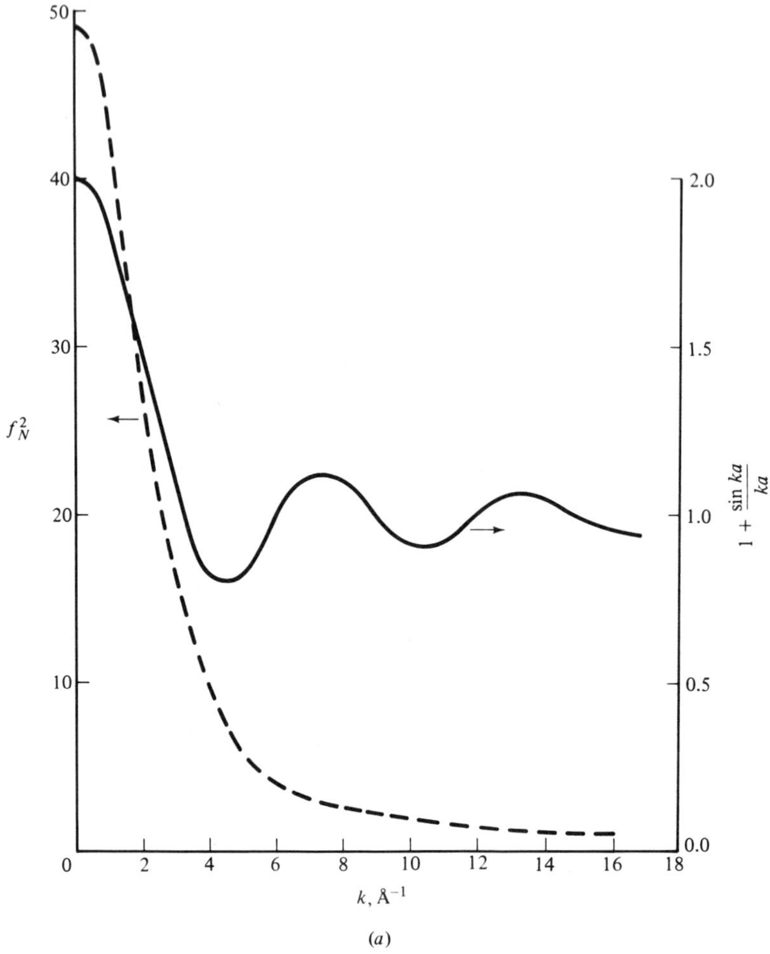

Figure 2.2. Components of the scattered intensity for the gaseous nitrogen molecule: *(a)* $f_N^2(k)$ and $(1 + (\sin ka)/ka)$; *(b)* $I(\mathbf{k})/I_e(k)$. The inset in *(b)* is the higher angle region of $I((\mathbf{k})/I_e(k)$ at larger scale.

are in contact (as in Fig. 2.5), but do not form a regular, periodic structure, as would a crystalline solid. Note in Fig. 2.5 that, as we look outward from the center of atom A, there is a sphere of diameter $4R$ which must be empty of other atom centers. Further, there must be near $|\mathbf{r}| = 2R$ a thin spherical shell with a considerable number of near neighbor atom centers. There would also be another shell of high occupancy for second nearest neighbors. And so forth. Thus the probability $\rho_a(r)$ of finding an atom in a given shell of radius r fluctuates greatly when r is small (within a small multiple of R). A plot of $\rho_a(r)$ versus r would appear schematically as in Fig. 2.6. Thus for a liquid we may not take $\rho_a(r)$ as a constant.

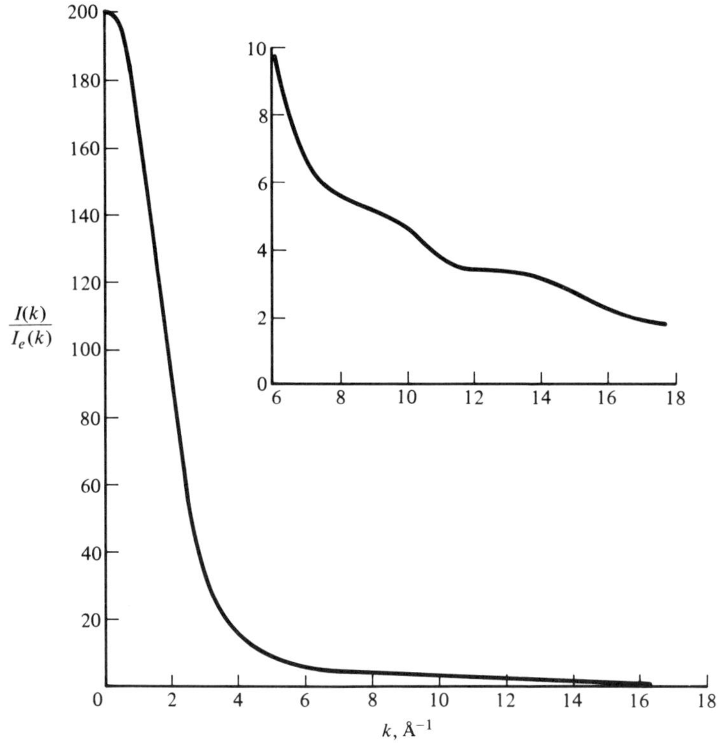

(*b*)

Figure 2.2. (Continued).

To analyze liquid diffraction data, it is convenient to work with the autocorrelation function for atomic density

$$P_a(\mathbf{u}) = \int \rho_a(\mathbf{r})\rho_a(\mathbf{u}-\mathbf{r})\, dv_{\mathbf{r}} \tag{2.15}$$

It is useful to note that $P_a(\mathbf{u})$ is an integration over all positions of the vector $\mathbf{r}$. But $P_a(\mathbf{u})$ has value only when $\mathbf{r}$ is at an atom center—and when $\mathbf{u} - \mathbf{r}$ is also at an atom center. Thus there are at any moment only N (the number of atoms in the specimen) sites at which $\rho_a(\mathbf{r})$ is finite. If we now take $\rho_a'(\mathbf{u})$ as the probability of finding another atom at $\mathbf{r} + \mathbf{u}$, given an atom at $\mathbf{r}$, we can write

$$\frac{I(\mathbf{k})}{I_e(k)f^2(k)} = N\int \delta(\mathbf{u})e^{i\mathbf{k}\cdot\mathbf{u}}\, dv_{\mathbf{u}} + N\int \rho_a'(\mathbf{u})e^{i\mathbf{k}\cdot\mathbf{u}}\, dv_{\mathbf{u}} = N + \mathscr{F}[\rho_a'(\mathbf{u})] \tag{2.16}$$

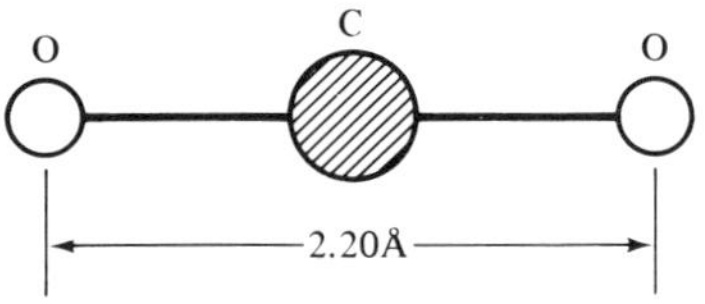

Figure 2.3. The linear CO_2 molecule.

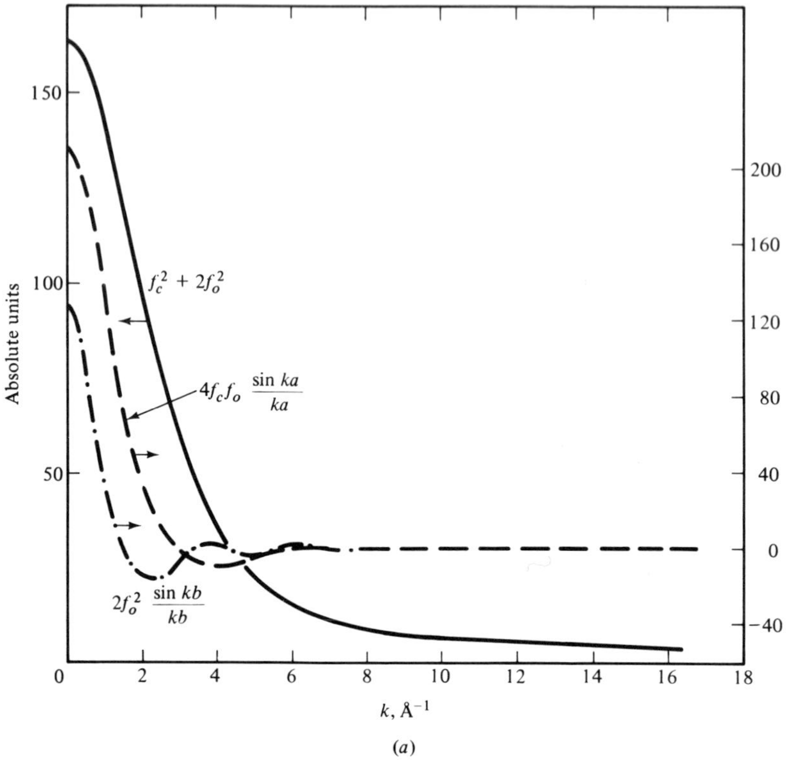

Figure 2.4. Components of the scattered intensity for gaseous CO_2: *(a)* $f_C^2 + 2f_O^2$, $4f_Cf_O$ (sin ka)/ka, and $2f_O$ (sin kb)/kb; *(b)* the molecular scattering factor $g(k)$. The inset shows the higher angle portion of $g(k)$ at larger scale.

The first term on the right of (1.16) relates to the case where $\mathbf{u} = 0$; i.e., $m = n$ in (2.2). As we shall see, it is convenient to add and subtract a term $N\mathscr{F}[\rho_o]$, where ρ_o is the average density of the liquid. Thus (2.16) becomes

$$\frac{I(\mathbf{k})}{NI_e(k)f^2(k)} = 1 + \mathscr{F}[\rho_a'(\mathbf{u}) - \rho_o] + \rho_o\mathscr{F}[1] \tag{2.17}$$

We have seen already that $\mathscr{F}[1]$ is the delta function, $\delta(\mathbf{k})$. Thus the third term in (2.17) is physically colinear with the incident beam and cannot be observed. We see now that the purpose of adding and subtracting $N\mathscr{F}[\rho_o]$ was to remove from consideration the unobservable portion of the $I(\mathbf{k})$. Thus the observable intensity $I'(\mathbf{k})$ is related to $\rho_a'(\mathbf{u})$ via

$$\frac{I'(\mathbf{k})}{NI_e(k)f^2(k)} - 1 = \mathscr{F}[\rho_a'(\mathbf{u}) - \rho_o] \tag{2.18}$$

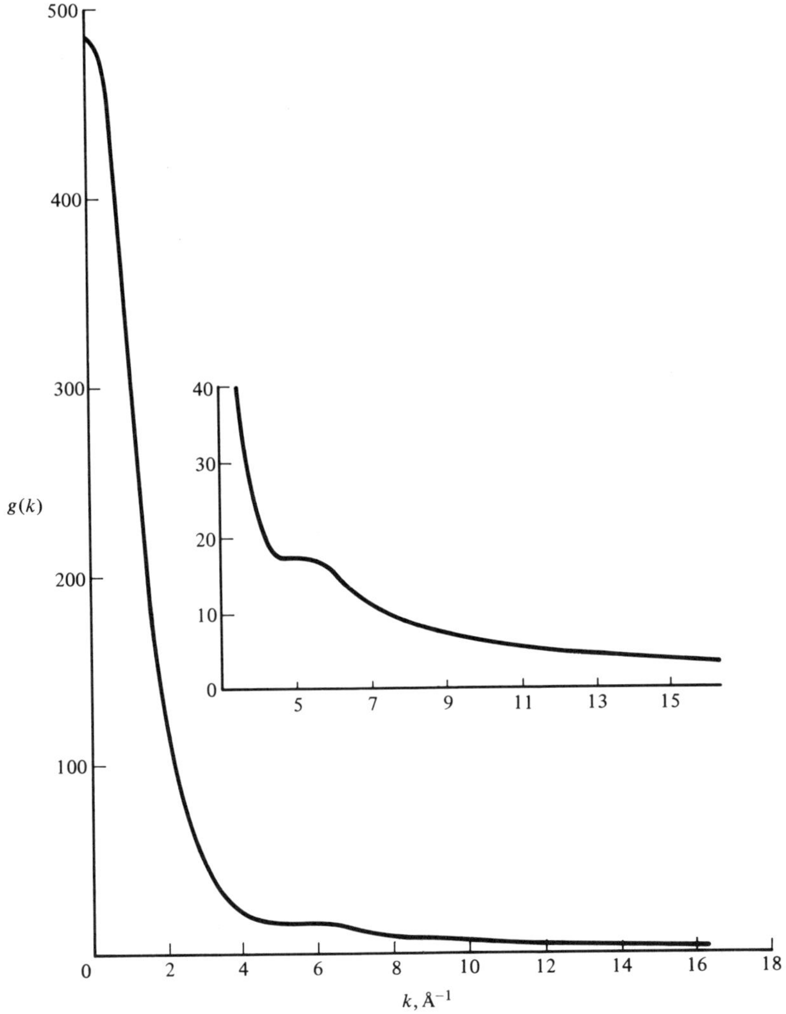

(b)

Figure 2.4. (Continued).

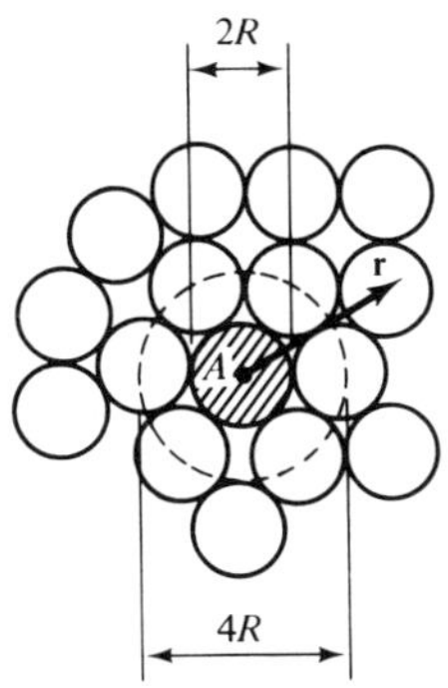

Figure 2.5. Model of a monatomic liquid. The atomic radius is R and the vector distance from a given atom is $\mathbf{r}$.

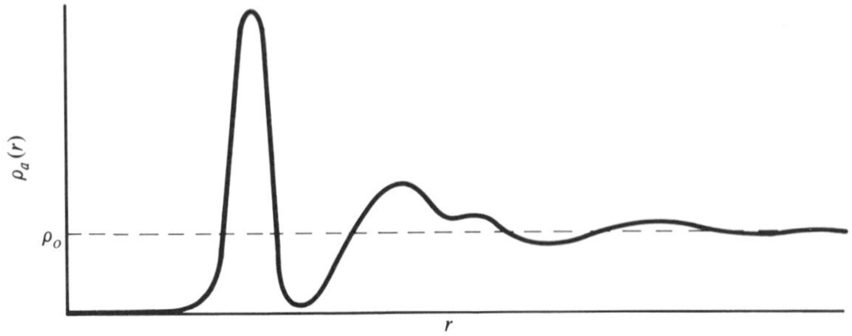

Figure 2.6. The probability $\rho_a(r)dr$ of finding another atom in a shell of radius r and width dr about a given atom.

Let us set

$$\mathscr{I}(\mathbf{k}) = \frac{I'(\mathbf{k})}{NI_e(k)f^2(k)} - 1 \tag{2.19}$$

We now use the isotropy of the liquid to our advantage. Let us rewrite (2.18), showing explicitly the isotropy

$$\mathscr{I}(k) = \int_{r=0}^{\infty} \int_{\theta=0}^{\pi} \int_{\phi=0}^{2\pi} [\rho_a'(r) - \rho_o] e^{ikr\cos\theta} r^2 \sin\theta \, d\phi \, d\theta \, dr \tag{2.20}$$

Proceeding exactly as in (1.70) to (1.73) to evaluate the integrals over ϕ and θ, we have

$$\mathscr{I}(k) = 4\pi \int_0^{\infty} r^2 [\rho_a'(r) - \rho_o] \frac{\sin kr}{kr} dr \tag{2.21}$$

Transposing the k on the right of (2.21) to the left and Fourier inverting, we find

$$\rho_a'(r) = \rho_o + \frac{1}{2\pi^2 r} \int_0^{\infty} \mathscr{I}(k) \, k \sin(kr) \, dk \tag{2.22}$$

Thus, at least in principle, one can find $\rho_a'(r)$ by measurement of $I(k)$. What is done is to evaluate the integral in (2.22) for each value of r. Then the $\rho_a'(r)$ values found can be plotted against the r values used, to obtain curves of the type shown in Fig. 2.6.

We shall now see that it is sometimes more useful to plot the quantity $4\pi r^2 \rho_a'(r)$ (rather than just $\rho_a'(r)$) against r. We observe that $4\pi r^2 \rho_a'(r) dr$ is the number of atomic centers in the spherical shell of incremental thickness dr located a distance r from our fiducial atom. We refer now to a plot of $4\pi r^2 \rho_a(r)$ versus r, shown for N_2 liquid in Fig. 2.7. In this figure the first hump represents the first neighbors of a given N atom. Thus the shaded area of thickness dr is the number of atom centers in the incremental shell at distance r from an average N atom, and the total area under the first hump is the

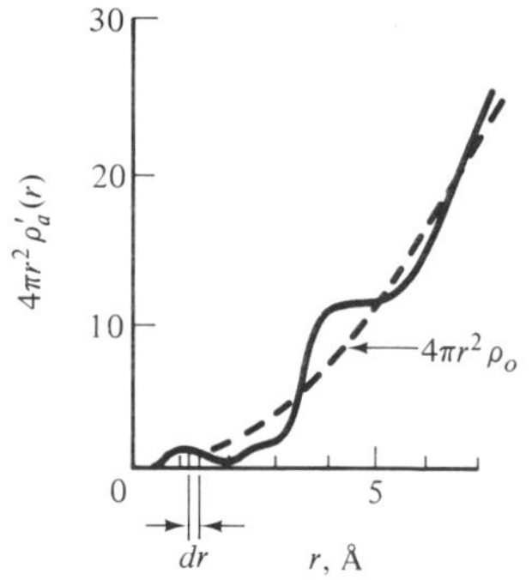

Figure 2.7. $4\pi r^2 \rho_a'$ *(r)* versus *r* for liquid chlorine at 2.5°C and 7.7 atm. [N. S. Gingrich, *Rev. Mod. Phys.*, **15,** 90 (1943)]

average total number of nearest neighbors about an N atom. The second and third neighbors may be counted in a similar way, although the overlapping of these humps renders the process more difficult. In the present case there is one nearest neighbor, as we expect from the molecular structure. The results of such analyses are less intuitive or trivial as we consider second neighbors in N_2 or first neighbors in monomolecular liquids.

In (2.22), the quantity to be evaluated from experiment is $\mathscr{I}(k) = I(k)/[NI_e(k)f^2(k)] - 1$. We shall see later (Chap. 4) that $I_e(k)$ is related to the angle of scattering 2θ and the incident intensity I_o through

$$I_e = r_e^2 \, P(2\theta) I_o / R_o^2 \tag{2.23}$$

where r_e is a universal constant, $P(2\theta)$ is a *polarization factor* (equal to $(1 + \cos^2 2\theta)/2$ for an unpolarized incident beam), I_o is the intensity (energy per unit area per second) incident on the specimen and R_o is the specimen-to-detector distance. In principal, all quantities can be measured. In practice, however, the exact value of N, the number of atoms in the irradiated volume is not easy to determine, since the incident beam is not sufficiently uniform. Further, I_o is very difficult to measure. This is so because I_o is usually several orders of magnitude larger than $I(k)$ and intensity measuring devices are not sensitive over so broad a range. In practice one uses an entirely different method to determine $I(k)$.

The experimental determination of $I(k)$ proceeds by means of a conservation principal. We shall now see how, in principle, this may be possible. We have already learned that scattered intensity integrated over all k-space (resp) is constant for a collection of N atoms of given type, independent of their state of aggregation (see Sect. 1.6). Note that for a monatomic gas the intensity per atom is given by $I_e(k)f^2(k)$. For our real system, the intensity per atom is $I(k)/N$. Therefore

$$\frac{1}{N}\int I(k)dv_{\mathbf{k}} = \int I_e(k)f^2(k)dv_{\mathbf{k}} = \frac{r_e^2}{R_o^2} I_o \int P(2\theta)f^2(k)dv_{\mathbf{k}} \tag{2.24}$$

For an isotropic system, $dv_{\mathbf{k}} = 4\pi k^2 dk$. Hence,

$$\int I(k)k^2dk = N\frac{r_e^2}{R_o^2}I_o\int P(k)f^2(k)k^2dk \tag{2.25}$$

Thus we see that the integral of $k^2[R_o^2I(k)/Nr_e^2I_o$ over all k must be identical with the integral of $k^2P(\theta)f^2(k)$ over all k. What can be done is to first plot $k^2P(\theta)f^2(k)$ against k. Then the function $I(k)$ is multiplied by a scale factor such that the area of $k^2I(k)$ over all k becomes equal to the area under $k^2P(\theta)f^2(k)$. The scale factor thus found must then be $[N(r_e^2/R_o^2)I_o]^{-1}$. Hence, in principle the appropriate scaling can be done without measurement of N or I_o. In practice a variant of this principle is usually used. It is observed that at large k a properly scaled $I(k)$ curve oscillates about $f^2(k)$, approaching $f^2(k)$ more and more closely as k increases. Thus what is done is to fit the tail of $I(k)$ to the tail of $f^2(k)$. A potential problem with real data is that the incoherently scattered portion of the scattered x-rays is ordinarily also measured. The incoherent radiation scattered per atoms, I_{inc}, is a computed quantity and is tabulated on the same scale as $f(k)$ in the International Tables for Crystallography. Since phase relationships do not obtain in the incoherent radiation, its variation over reciprocal space will be the same for all forms of aggregation of a given material. Experimentally, I_{inc} was included in the $I(k)$ measurement. Hence, I_{inc} must also be added to $f^2(k)$ before scaling is done. The sequence of events is seen with reference to Fig. 2.8. Part *(a)* shows the measured $I(k)$ plotted versus k. Part *(b)* shows the tabulated quantities $f^2(k)$ and $I_{inc}(k)$ as well as their sum. These quantities are in electron units. Part *(c)* shows the measured $I(k)$ multiplied by a constant factor such that its tail oscillates regularly about $f^2(k) + I_{inc}(k)$. The measured $I(k)$ data are now read in absolute electron units.

Using the above procedure, $I(k)$ is determined for a range of k. A problem still exists in performing the integration of (2.21) to obtain $\rho(r)$. The problem is that the integral is over all k, from 0 to ∞. But our measurement is limited by our experiment; i.e., in an experiment run at constant wavelength λ, the value $k = 4\pi \sin\theta/\lambda$ is limited to a maximum value of $4\pi/\lambda$, at $2\theta = 180°$. Consequently a truncation error is always present. The effect of the truncation error is seen in the analysis of Dejak et al of scattering data from liquid water (Fig. 2.9). They have computed $\rho(r)$ several times, truncating their data at different values each time. We note the effect of truncation to be in the form of spurious peaks. Note especially the peak at 4.1 Å. For many years this peak was believed to be a real structural manifestation.

Two methods of handling truncation error are indicated by Warren[1] and by Prober, Schultz, and Sandler.[2] In the former case the experimental $I(k)$ is multiplied in (2.22) by a factor $e^{-\alpha^2k^2}$, with a usually near 0.2 or 0.3 Å. Warren shows in a very nice proof that this procedure must eliminate spurious ripples, at small expense to the sharpness of the data. The procedure of Prober et al is to fit the tail of $I(k)$ by a damped sine wave, where the degree of damping is fit so that $\rho(r)$ is zero over the first portion of the $\rho(r)$ versus r curve.

[1] B. E. Warren, *X-ray Diffraction*, Addison-Wesley, Reading, Mass., 1969, chap. 10.

[2] J. M. Prober and J. M. Schultz, *J. Appl. Cryst.*, **8**, 405 (1975).

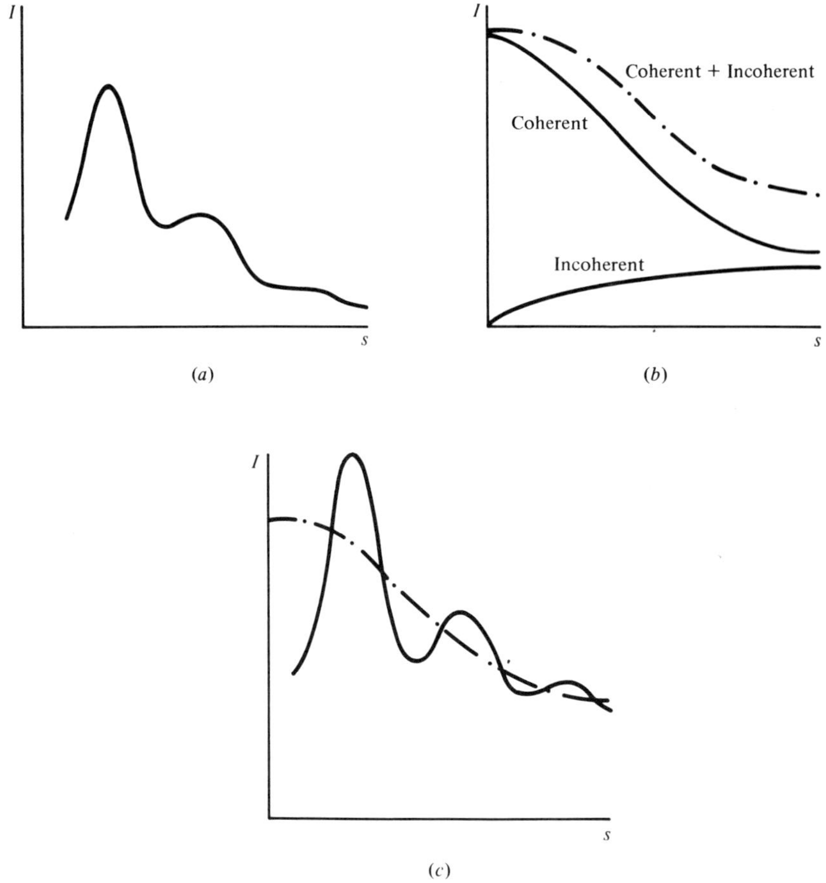

Figure 2.8. Calculation of $I(k)$ from experimental data: *(a)* experimental curve; *(b)* coherent scattering power per molecule, Compton, or incoherent, scattering power per molecule, and total (coherent plus incoherent) scattering power per molecule; *(c)* rescaling of the experimental data so as to oscillate regularly about the tail of the curve for scattering power per molecule. The experimental data are now in electron units.

2.3.2 Monatomic Liquids of Mixed Atomic Type

We now consider an alloy liquid. The analysis here parallels that of Sect. 2.3.1. We begin writing again the general intensity expression

$$\frac{I(\mathbf{k})}{I_e(k)} = \sum_{-N/2}^{N/2} f_n^2 + \sum_{\substack{-N/2 \\ m \neq n}}^{N/2} \sum_{-N/2}^{N/2} f_m f_n e^{i\mathbf{k}\cdot(\mathbf{r}_n - \mathbf{r}_m)} \tag{2.26}$$

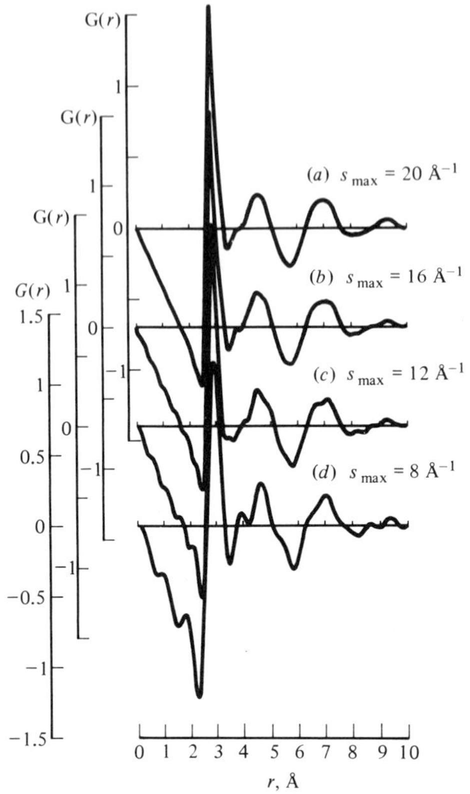

Figure 2.9. The effect of truncation on the radial distribution function for liquid water. Here $G(r) = 4\pi r[\rho_a'(r) - \rho_o]$ is plotted against r. This way of plotting enhances the magnitude of the function at larger r. [C. Dejak, G. Licheri, and G. Piccaluga, *Gazz. Chim. Ital.*, **101,** 159 (1971)]

Over the symmetric limits on the sum the sine portion of the complex exponential is clearly zero. Thus we rewrite (2.26)

$$\frac{I(\mathbf{k})}{I_e(k)} = N\overline{f^2} + \sum_{m \neq n}\sum f_m f_n \cos \mathbf{k} \cdot (\mathbf{r}_m - \mathbf{r}_m) \tag{2.27}$$

Averaging over all possible configurations, we have

$$\frac{I(k)}{I_e(k)} = N\overline{f^2} + \overline{\sum_{m \neq m}\sum f_m f_m \cos \mathbf{k} \cdot (\mathbf{r}_n - \mathbf{r}_m)} \tag{2.28}$$

where the overbar denotes an average over the system. The double sum can be simplified under certain restrictive assumptions. These are that the distribution of the two atomic types is random (no clustering or ordering) and that the atoms are about the same size ($\mathbf{f}_m \cdot \mathbf{f}_n$ not a function of $\mathbf{r}_n - \mathbf{r}_m$). In this case we note that (a) the f's are independent of the orientational average and (b) the average of a product is the product of the averages. We then have

$$\begin{aligned}\frac{I(k)}{I_e(k)} &= N\overline{f^2} + \sum_{m \neq n}\sum \bar{f}_m \bar{f}_n \langle\cos \mathbf{k} \cdot (\mathbf{r}_n - \mathbf{r}_m)\rangle \\ &= N\overline{f^2} + \bar{f}^2 \sum_{m \neq n}\sum \langle\cos \mathbf{k} \cdot (\mathbf{r}_n - \mathbf{r}_m)\rangle\end{aligned} \tag{2.29}$$

where $<>$ again denotes an average over all orientations of $(\mathbf{r}_n - \mathbf{r}_m)$ relative to $\mathbf{k}$. Using our previous result (1.73), we have now

$$\frac{I(k)}{I_e(k)} = N\overline{f^2} + \bar{f}^2 \sum_{m \neq n} \sum \frac{\sin k|\mathbf{r}_n - \mathbf{r}_m|}{k|\mathbf{r}_n - \mathbf{r}_m|} \tag{2.30}$$

As before, we replace the discrete atomic position $\mathbf{r}_m$ and $\mathbf{r}_n$ by a probabilistic continuum; the intensity function becomes

$$\frac{I(k)}{I_e(k)} = N\overline{f^2} + N\bar{f}^2\, 2\pi \int r^2 \,[\rho_a'(r) - \rho_o]\, \frac{\sin kr}{kr}\, dr \tag{2.31}$$

Here we have employed the steps (2.16) to (2.21). Inverting, the radial distribution function is

$$\rho_a'(r) = \rho_o + \frac{1}{2\pi^2 r} \int_0^\infty \left\{ \frac{[I(k)/NI_e(k)] - \overline{f^2}}{\bar{f}^2} \right\} k \sin kr \, dk \tag{2.32}$$

If either our assumption of randomness or of approximately equal size is untrue, then we must work directly from (2.28).

2.3.3 Molecular Liquids

There has been considerable interest over the years in liquids composed of fixed molecules (e.g., H_2O and CCl_4). In order to properly analyze the scattering by such liquids, one needs to do a great deal more work than was necessary in the monatomic cases. The reasons for this are that adjacent molecules may possess some relative orientation, thus obviating the steps leading from (2.29) to (2.32).

The first approximation to the scattering by such molecular liquids is to neglect the orientational dependences. In this case we merely replace the atomic scattering factors in (2.18) or (2.32) by molecular scattering factors (as in Sect. 2.2). A better approximation and a rigorous approach are both discussed by Warren.[3]

2.4 APPROXIMATE RELATION BETWEEN HUMP POSITION AND NEAREST NEIGHBOR DISTANCE

Recall that, in general, for any disordered system

$$\frac{I(k)}{I_e(k)} = \sum_n f_n^2(k) + \sum_{m \neq n} \sum f_m f_n \frac{\sin kr_{mn}}{kr_{mn}} \tag{2.33}$$

Recall also our work on gaseous systems. There the peak positions were controlled by the r_{mn} distances. In particular, the first peak was seen to occur

[3] B. E. Warren, *X-ray Diffraction*, Addison-Wesley, Reading, Mass., 1969, Chap. 10.

near the position $k = 2\pi/r_{mn}$. We shall now obtain a somewhat better relationship.

Suppose that the several peaks, due to different values of r_{mn}, are far enough apart that their tails do not overlap much. Then the peak for the first r_{mn} spacing is given by the first (nonzero-angle) maximum of the (sin kr)/(kr) function. Thus

$$\frac{d\left(\dfrac{\sin kr}{kr}\right)}{dk} = 0 = \frac{1}{kr} \cdot r \cos kr - \frac{\sin kr}{k^2 r} \tag{2.34}$$

or

$$kr = \tan (kr) \tag{2.35}$$

The two functions of (2.35) are plotted in Fig. 2.10. In this figure we see that the first root of (2.15) occurs at 0°. Root 2 represents a minimum. Root 3 is the first nonzero maximum.

The position of root 3 is $k_m r_m = 7.73 = 1.23 \times 2\pi$

$$k_m = \frac{1.23 \times 2\pi}{r_m} = 4\pi \frac{\sin \theta_m}{\lambda} \tag{2.36}$$

Thus, as a good approximation

$$\sin \theta_m = 1.23 \frac{\lambda}{2r_m} \tag{2.37}$$

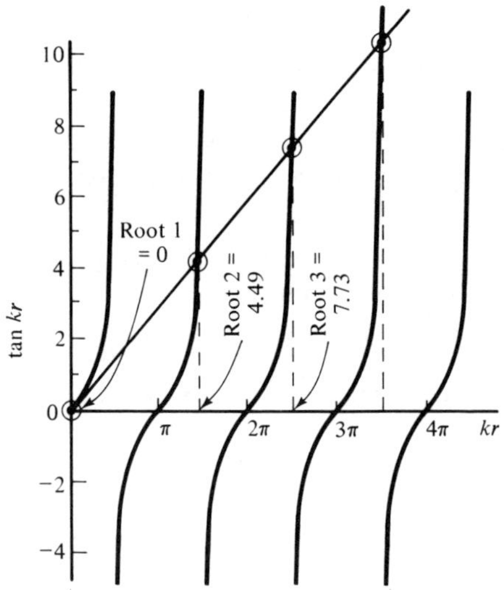

Figure 2.10. Tan kr versus kr.

Clearly when there are several interatomic distances to be considered, their sine portions will overlap and cause the peak to move, and 1.23 is not to be considered a magic value.

PROBLEMS

2.1. Draw, to scale, the intensity versus $\sin\theta/\lambda$ curve for sodium vapor, using CuK_α x-rays ($\lambda = 1.54$ Å) and for 8.0 Å neutrons. Note that I_e also has an angular dependence, $1 + \cos^2 2\theta$, termed the polarization factor.

2.2 Compute and sketch the x-ray scattering pattern for gaseous Cl_2. The atom–atom distance here is 2.0 Å. Let the radiation used be MoK_α ($\lambda = 0.71$ Å).

2.3. The structure of the water molecule is as shown in Fig. P2.3.

(a) Write an expression for the normalized scattered intensity $I(k)/I_e(k)$ for gaseous water.

(b) *Crudely* sketch this intensity versus k. It is not necessary to compute points in the curve.

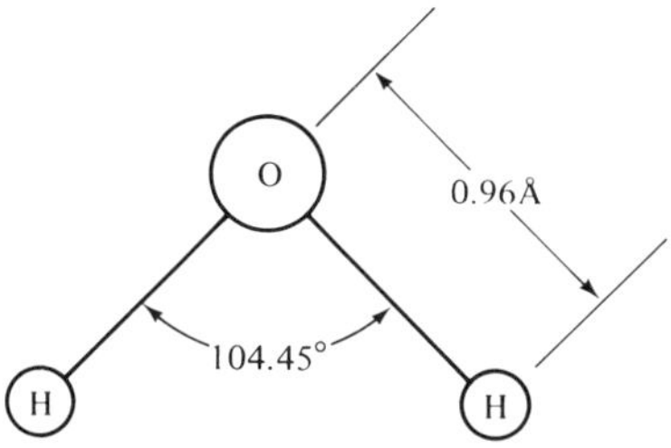

Figure P2.3.

2.4. Draw, in as much detail as you can predict, the x-ray scattering curve for carbon tetrachloride (CCl_4). In this molecule the C-Cl bond is 1.76 Å long and the distance between Cl neighbors is 2.87 Å. Let the x-ray wavelength be 1.94 Å (FeK_α).

2.5. Develop an expression for scattering by a monatomic liquid with an idealized density such that

$$\rho(r) = \delta(r) + P_1(r)$$

where

$$P_1 = \begin{cases} \dfrac{12}{4\pi a^2}\,\delta(r-a) \text{ for } 0 < r < 2a \\ \rho_o \text{ for } r \geq 2a \end{cases}$$

In this problem, let $a = 2$ Å and $\rho_o = 1.0 \times 10^{23}$ atoms/cm³. (This $\rho(r)$ gives 12 nearest neighbors to the central atom and then smears out the rest of the atoms into a constant density ρ_o). Roughly how will the scattering appear? (Sketch!)

2.6. One of the scattering curves shown in Fig. P2.6 is that for liquid mercury. The first neighbors in liquid Hg are 3.0 Å apart. Which curve represents this material?

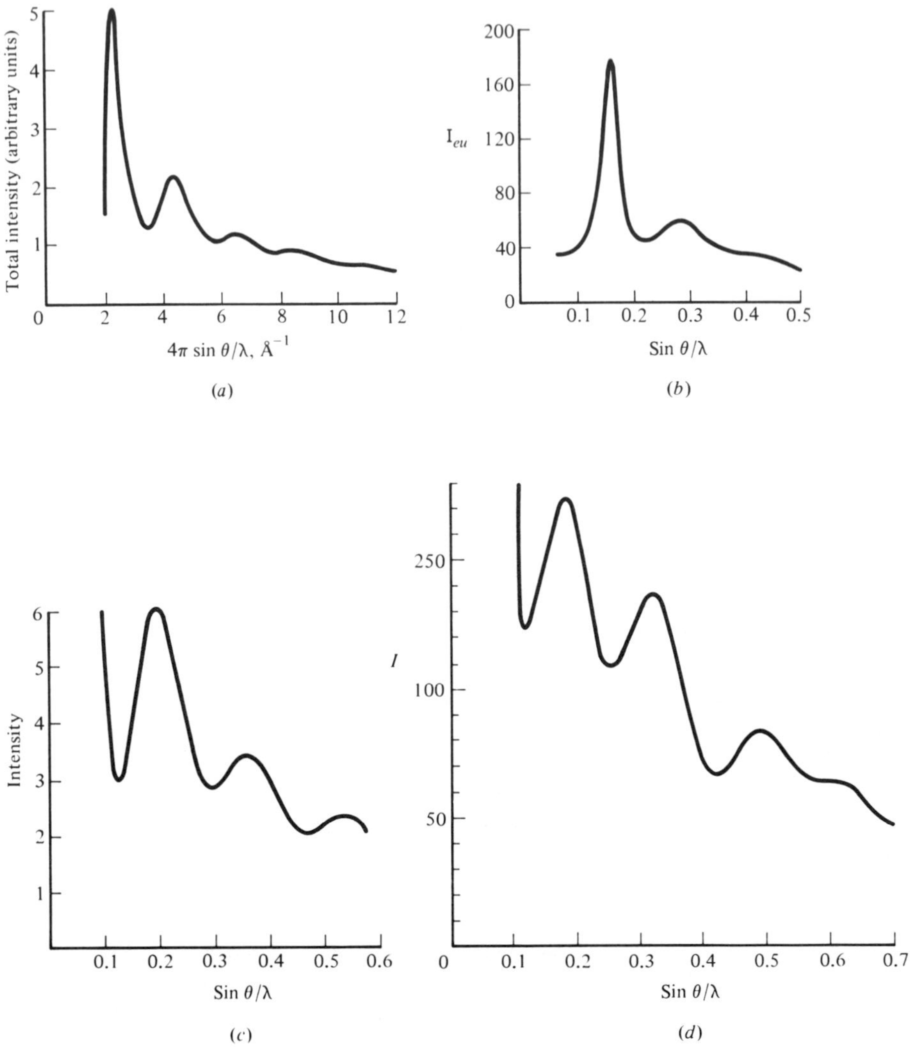

Figure P2.6.

BIBLIOGRAPHY

1. L. V. AZAROFF, *Elements of X-ray Crystallography,* Ch. 8, McGraw-Hill Book Company, N.Y. (1968).

2. L. V. AZAROFF, R. KAPLOW, N. KATO, R. J. WEISS, A. J. C. WILSON, and R. A. YOUNG, *X-ray Diffraction,* Ch. 2, McGraw-Hill Book Company, N.Y. (1974).

3. G. E. BACON, *Applications of Neutron Diffraction in Chemistry,* Ch. 8, Pergamon Press, Inc., N.Y. (1963).

4. A. H. COMPTON, and S. K. ALLISON, *X-ray in Theory and Experiment, 2nd Ed.,* Ch. 3, Van Nostrand Reinhold Company, N.Y. (1935).

5. P. A. EGELSTAFF, *Thermal Neutron Scattering,* Ch. 2, Academic Press, Inc., N.Y. (1965).

6. P. A. EGELSTAFF, *An Introduction to the Liquid State,* Ch. 6, Academic Press, Inc., N.Y. (1967).

7. R. HOSEMANN, and S. N. BAGCHI, *Direct Analysis of Diffraction by Matter,* Ch. 8, North-Holland, Amsterdam (1962).

8. R. W. JAMES, *The Optical Principles of the Diffraction of X-rays,* Ch. 2, G. Bell & Sons, Ltd., London (1954).

9. H. P. KLUG and L. E. ALEXANDER, *X-ray Diffraction Procedures for Polycrystalline and Amorphous Materials, 2nd Ed.,* Ch. 11, Wiley-Interscience, N.Y. (1959).

10. LAUE, M. T. F. VON, *Röntgenstrahlinterferenzen, 3. Ausg.,* Ch. 2, Akademische Verlagsgesellschaft, Frankfurt (1960).

11. Z. G. PINSKER, *Dynamical Scattering of X-rays in Crystals,* Ch. 11, Springer, Berlin (1978).

12. B. E. WARREN, *X-ray Diffraction,* Ch. 10, Addison-Wesley Publishing Co., Inc., Reading, MA (1969).

3 Reciprocal Space, the Reciprocal Lattice, and the Nature of Diffraction Experiments

3.1 FOUNDATIONS

This chapter introduces some very valuable tools for conceptualizing diffraction problems. Once we have established a working knowledge of reciprocal space (resp) and the reciprocal lattice (rel), we shall see that the results of experiments can often be intuited from an actual or mental figure in reciprocal dimensions.

In all the previous material, the variable which has been used to represent the scattering conditions is the scattering vector **s**. We have in fact performed formal integrations in the space of **s**—*reciprocal space.* Let us now examine the manner in which reciprocal space becomes the mapping domain for experiments. Consider an experiment performed with monochromatic radiation, as in Fig. 3.1. Here we take the origin of both real and reciprocal space as *O,* a point at the center of the specimen. We recall that with any intensity measurement is associated a particular value of **s**. In Fig. 3.1 we see that to each value of **s** there corresponds a magnitude $|\mathbf{s}| = (2 \sin \theta/)\lambda$, a scattering angle 2θ, and an azimuth angle ψ. In an experiment we can independently vary $\hat{\mathbf{s}}_o$, 2θ, and ψ. And we can assign to each point in reciprocal space (**s**-space) the predicted or measured intensity, given a set of experimental conditions. Thus in principle we could prepare a contour map of intensity in the space of **s**—$|\mathbf{s}|, 2\theta$, ψ in spherical coordinates. Such a map would be a representation of the results of experimentation on the specimen. It is in exactly this way that we shall continue to view reciprocal space *(resp).*

The reciprocal lattice (or *rel*) is a description of the results of a scattering experiment on a single crystal. We evolve the reciprocal lattice in the following

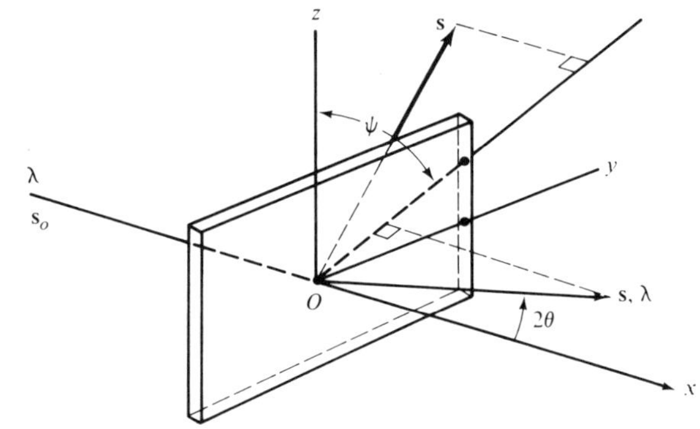

Figure 3.1. Experimental geometry: x is parallel to the incident beam; y and z are, respectively, horizontal and vertical directions through the specimen and normal to the incident beam. 2θ is the angle between incident and scattered unit vectors $\mathbf{s}_o$ and $\mathbf{s}$. ψ is the azimuthal angle between the vertical axis *(z)* and the *yz*-plane component of the scattering vector $\mathbf{s}$.

way. The perfect crystal is made up of a stacking of unit cells, as in Fig. 3.2. Let the density of electrons within each unit cell be, with origin $\mathbf{r} = 0$ in the cell, $\rho_c(\mathbf{r})$. The spatial positioning of the origins $\mathbf{r}_{pqr}$ of the unit cells is given by $p\mathbf{a} + q\mathbf{b} + r\mathbf{c}$. Mathematically, the loci of all lattice points is

$$z(\mathbf{r}) = \sum_{p=-\infty}^{\infty} \sum_{q=-\infty}^{\infty} \sum_{r=-\infty}^{\infty} \delta(\mathbf{r} - \mathbf{r}_{pqr}) \tag{3.1}$$

For a finite crystal we multiply $z(\mathbf{r})$ by a function $\sigma(\mathbf{r})$, a function which has value one within the crystal and zero everywhere else. Thus the electron density throughout an infinite, perfect crystal is given by

$$\rho(\mathbf{r}) = \rho_c(\mathbf{r}) * z(\mathbf{r}) \tag{3.2}$$

(Recall the property of convoluting a smooth function with a delta function). For the finite, perfect crystal this then becomes

$$\rho(\mathbf{r}) = \rho_c(\mathbf{r}) * [z(\mathbf{r}) \cdot \sigma(\mathbf{r})] \tag{3.3}$$

Inserting this value of $\rho(\mathbf{r})$ into the amplitude expression, we have

$$\frac{A(\mathbf{s})}{A_e(s)} = \mathscr{F}[\rho(\mathbf{r})] \cdot \mathscr{F}[z(\mathbf{r}) \cdot \sigma(\mathbf{r})] \tag{3.4}$$

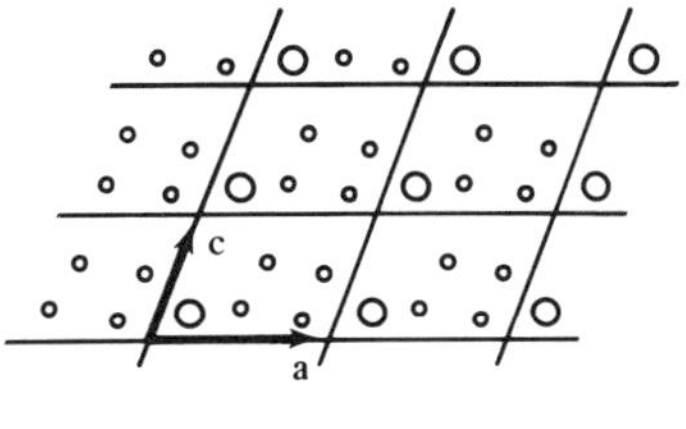

Figure 3.2. Two-dimensional section of a perfect crystal. Note the lattice vectors **a** and **c.**

Here, the effect of $\sigma(\mathbf{r})$ is merely to restrict the possible values of $\mathbf{r}$. We see that in this expression terms containing (a) the contents of the unit cell and (b) the distribution of unit cells have been separated. These two terms are called, respectively, the *structure factor* $F(\mathbf{s})$ and the *lattice factor* $Z(\mathbf{s})$. That is

$$\frac{A(\mathbf{s})}{A_e(s)} = F(\mathbf{s}) \cdot Z(\mathbf{s}) \tag{3.5}$$

We shall see that the effect of the lattice factor is to define the s-position which can have nonzero intensity while the structure factor determines the amplitude (or intensity) associated with each of those positions. In the following, we look first at the lattice factor.

What we need here is the transform of $z(\mathbf{r})$ under these restrictions. This is, now

$$
\begin{aligned}
Z(\mathbf{s}) &= \int_{\text{crystal}} z(\mathbf{r}) e^{-2\pi i \mathbf{s}\cdot\mathbf{r}} dv_{\mathbf{r}} \\
&= \int_{-\infty}^{\infty} \sum_{-N_1/2}^{N_1/2} \sum_{-N_2/2}^{N_2/2} \sum_{-N_3/2}^{N_3/2} [\delta(\mathbf{r} - \mathbf{r}_{pqr})] e^{-2\pi i \mathbf{s}\cdot\mathbf{r}} dv_{\mathbf{r}} \\
&= \underset{\text{crystal}}{\sum\sum\sum} e^{-2\pi i \mathbf{s}\cdot\mathbf{r}_{pqr}} \\
&= \sum_p \sum_q \sum_r e^{-2\pi i \mathbf{s}\cdot(p\mathbf{a} + q\mathbf{b} + r\mathbf{c})} \qquad (3.6) \\
&= \sum_{p=-N_1/2}^{N_1/2} e^{-2\pi i p \mathbf{s}\cdot\mathbf{a}} \sum_{q=-N_2/2}^{N_2/2} e^{-2\pi i q \mathbf{s}\cdot\mathbf{b}} \sum_{r=-N_3/2}^{N_3/2} e^{-2\pi i r \mathbf{s}\cdot\mathbf{c}}
\end{aligned}
$$

Using the result of problem 1.3*b*, we have

$$Z(\mathbf{s}) = \frac{\sin[(N_1+1)\pi\mathbf{s}\cdot\mathbf{a}]}{\sin \pi\mathbf{s}\cdot\mathbf{a}} \cdot \frac{\sin[(N_2+1)\pi\mathbf{s}\cdot\mathbf{b}]}{\sin \pi\mathbf{s}\cdot\mathbf{b}} \cdot \frac{\sin[(N_3+1)\pi\mathbf{s}\cdot\mathbf{c}]}{\sin \pi\mathbf{s}\cdot\mathbf{c}} \tag{3.7}$$

As N_1, N_2, N_3 become very large, the three terms approach delta function behavior

$$Z(\mathbf{s}) \cong N \cdot \sum_h \sum_k \sum_l \delta(\mathbf{s}\cdot\mathbf{a} - h)\, \delta(\mathbf{s}\cdot\mathbf{b} - k)\, \delta(\mathbf{s}\cdot\mathbf{c} - l) \tag{3.8}$$

where h, k, and l are integers and N is the total number of atoms in the crystal ($N = N_1N_2N_3$). In (3.8) we have defined a set of necessary conditions for nonzero intensity. This set of conditions (the *Laue conditions*) is expressed

$$\left.\begin{aligned} \mathbf{s}\cdot\mathbf{a} &= h \\ \mathbf{s}\cdot\mathbf{b} &= k \\ \mathbf{s}\cdot\mathbf{c} &= l \end{aligned}\right\} \; h,\ k,\ l \text{ integers} \tag{3.9}$$

We will now show that these describe a periodic point lattice in resp.

Consider three vectors $\mathbf{a}^*$, $\mathbf{b}^*$, and $\mathbf{c}^*$, of units $(\text{length})^{-1}$, such that

$$\begin{matrix} \mathbf{a}^* \cdot \mathbf{a} = 1 & \mathbf{a}^* \cdot \mathbf{b} = 0 & \mathbf{a}^* \cdot \mathbf{c} = 0 \\ \mathbf{b}^* \cdot \mathbf{a} = 0 & \mathbf{b}^* \cdot \mathbf{b} = 1 & \mathbf{b}^* \cdot \mathbf{c} = 0 \\ \mathbf{c}^* \cdot \mathbf{a} = 0 & \mathbf{c}^* \cdot \mathbf{a} = 0 & \mathbf{c}^* \cdot \mathbf{c} = 1 \end{matrix} \tag{3.10}$$

Then the following are also true

$$\begin{aligned} (h\mathbf{a}^* + k\mathbf{b}^* + l\mathbf{c}^*) \cdot \mathbf{a} &= h \\ (h\mathbf{a}^* + k\mathbf{b}^* + l\mathbf{c}^*) \cdot \mathbf{b} &= k \\ (h\mathbf{a}^* + k\mathbf{b}^* + l\mathbf{c}^*) \cdot \mathbf{c} &= l \end{aligned} \tag{3.11}$$

By the identity of (3.9) and (3.11), we may equate $\mathbf{s}$ with $h\mathbf{a}^* + k\mathbf{b}^* + l\mathbf{c}^*$. Thus we see that the values of $\mathbf{s}$ which correspond for finite diffracted intensity fall onto a lattice of points in reciprocal space as shown in Fig. 3.3.

The real lattice and the reciprocal lattice (rel) are mathematically reciprocal to each other

$$F\left[\sum \delta(\mathbf{r} - \mathbf{r}_{pqr})\right] = \sum \delta(\mathbf{s} - \mathbf{r}^*_{hkl}) \tag{3.12a}$$

$$F^{-1}\left[\sum \delta(\mathbf{s} - \mathbf{r}^*_{hkl})\right] = \sum \delta(\mathbf{r} - \mathbf{r}_{pqr}) \tag{3.12b}$$

The proof of (3.12a) was just shown. The inverse proof—(3.12b)—proceeds analogously.

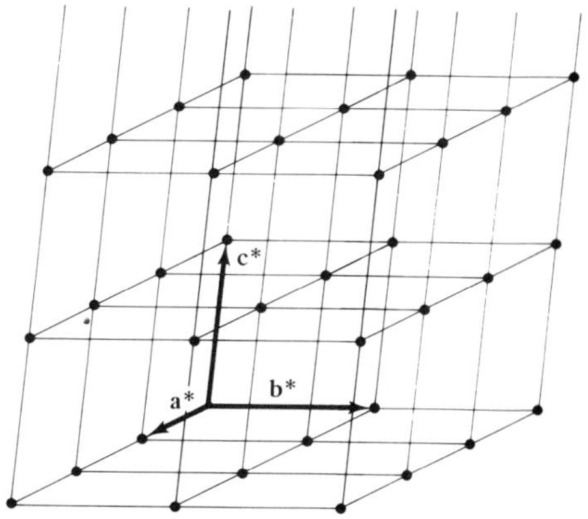

Figure 3.3. A portion of the reciprocal lattice (rel). Note the rel vectors $\mathbf{a}^*$, $\mathbf{b}^*$, and $\mathbf{c}^*$.

3.2 GEOMETRIC RELATIONS IN THE RECIPROCAL LATTICE

Below are given, in brief form, several useful geometric relations in the reciprocal lattice (rel). These are presented now for use in later sections.

1. Since $\mathbf{a}^*$ is perpendicular to both $\mathbf{b}$ and $\mathbf{c}$, $\mathbf{a}^*$ is perpendicular to the bc plane.
2. The magnitude of $\mathbf{a}^*$ is equal to $(a \cos \chi)^{-1}$, where χ is the angle between $\mathbf{b} \times \mathbf{c}$ and $\mathbf{a}$. This is seen in Fig. 3.4.
   Similarly, $|\mathbf{b}^*| = (b \cos \psi)^{-1}$ and $|\mathbf{c}^*| = (c \cos \omega)^{-1}$, where ψ and ω are, respectively, the angles between $\mathbf{c} \times \mathbf{a}$ and $\mathbf{b}$ and between $\mathbf{a} \times \mathbf{b}$ and $\mathbf{c}$.
3. The volume of a unit cell in real space is

$$V_c = \mathbf{c} \cdot (\mathbf{a} \times \mathbf{b}) = \mathbf{b} \cdot (\mathbf{c} \times \mathbf{a}) = \mathbf{a} \cdot (\mathbf{b} \times \mathbf{c}) \tag{3.13}$$

   The volume of a unit cell in resp is

$$V_c^* = \mathbf{c}^* \cdot (\mathbf{a}^* \times \mathbf{b}^*) = \mathbf{b}^* \cdot (\mathbf{c}^* \times \mathbf{a}^*) = \mathbf{a}^* \cdot (\mathbf{b}^* \times \mathbf{c}^*) \tag{3.14}$$

4. The relations between real and reciprocal space vectors are given by

$$\mathbf{a}^* = \frac{\mathbf{b} \times \mathbf{c}}{V_c} \qquad \mathbf{b}^* = \frac{\mathbf{c} \times \mathbf{a}}{V_c} \qquad \mathbf{c}^* = \frac{\mathbf{a} \times \mathbf{b}}{V_c} \tag{3.15}$$

$$\mathbf{a} = \frac{\mathbf{b}^* \times \mathbf{c}}{V_c} \qquad \mathbf{b} = \frac{\mathbf{c}^* \times \mathbf{a}^*}{V_c} \qquad \mathbf{c} = \frac{\mathbf{a}^* \times \mathbf{b}^*}{V_c} \tag{3.16}$$

   Using $\mathbf{a} \cdot (\mathbf{b} \times \mathbf{c}/V_c) = 1$ from (3.13) and $\mathbf{a} \cdot \mathbf{a}^* = 1$ from the definition of the reciprocal lattice (3.10), the first equation of (3.15), follows immediately. The rest of (3.15) and (3.16) is proved similarly.
5. Any reciprocal lattice vector $\mathbf{r}^*_{hkl} = h\mathbf{a}^* + k\mathbf{b}^* + l\mathbf{c}^*$ is perpendicular to the hkl plane in real space. Consider Fig. 3.5*a*. According to the definition of Miller indices, the intercepts of hkl with the principal real space axes are $\mathbf{a}/h$, $\mathbf{b}/k$, $\mathbf{c}/l$. Figure 3.5*b* shows three vectors in

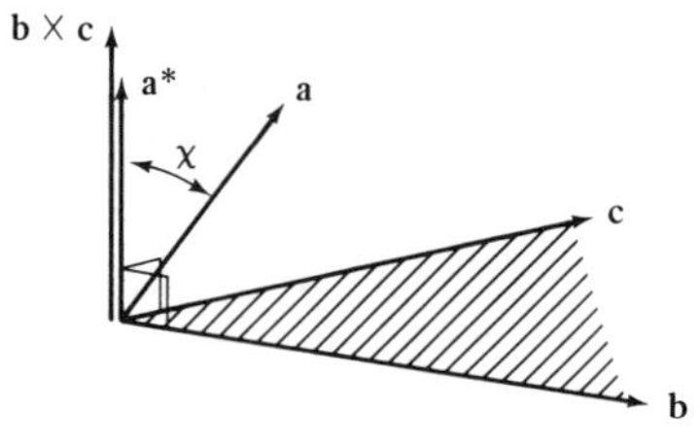

Figure 3.4. The relation among **a, b, c,** and **a***. **a*** lies normal to **b** and **c** and at an angle χ to **a.**

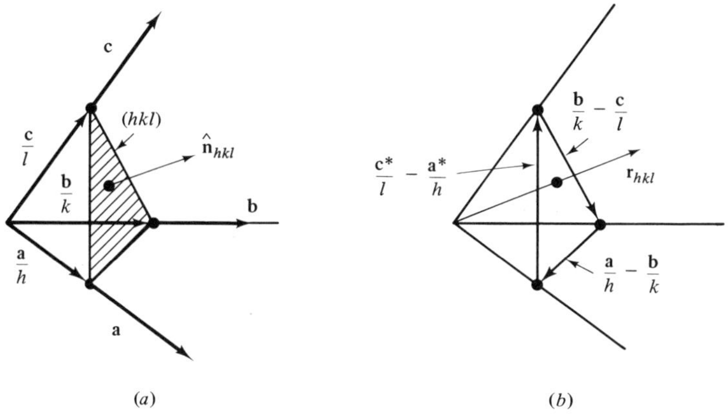

Figure 3.5. Reciprocal vector geometries used to show that $\mathbf{r}^*_{hkl}$ is normal to *hkl*.

hkl and also the rel vector $\mathbf{r}^*_{hkl}$. If we can now show that $\mathbf{r}^*_{hkl}$ is perpendicular to any two nonparallel vectors in *hkl*, then $\mathbf{r}^*_{hkl}$ must be perpendicular to *hkl*. This is shown by carrying out the dot product of $\mathbf{r}^*_{hkl}$ with two of the vectors shown

$$\begin{aligned} \mathbf{r}^*_{hkl} \cdot \left(\frac{\mathbf{c}}{l} - \frac{\mathbf{a}}{h}\right) &= (h\mathbf{a}^* + k\mathbf{b}^* + l\mathbf{c}^*) \cdot \left(\frac{\mathbf{c}}{l} - \frac{\mathbf{a}}{h}\right) = 1 - 1 = 0 \\ \mathbf{r}^*_{hkl} \cdot \left(\frac{\mathbf{b}}{k} - \frac{\mathbf{c}}{l}\right) &= (h\mathbf{a}^* + r\mathbf{b}^* + l\mathbf{c}^*) \cdot \left(\frac{\mathbf{b}}{h} - \frac{\mathbf{c}}{l}\right) = 1 - 1 = 0 \end{aligned} \tag{3.17}$$

6. $r^*_{hkl} = d^{-1}_{hkl}$, where d_{hkl} is the spacing of *hkl* planes. Consider again Fig. 3.5. Here the dot product of $\mathbf{a}/h$ with the unit vector $\hat{n}_{hkl}$ normal to *hkl* is, by definition of an interplanar distance

$$\frac{\mathbf{a}}{h} \cdot \hat{n}_{hkl} = d_{hkl} \tag{3.18}$$

Also, since $\hat{n}_{hkl}$ and $\mathbf{r}^*_{hkl}$ are parallel

$$\frac{\mathbf{a}}{h} \cdot \mathbf{r}^*_{hkl} = \frac{\mathbf{a}}{h} \cdot (r^*_{hkl}\,\hat{\mathbf{n}}) = r^*_{hkl}\, d_{hkl} \tag{3.19}$$

But, algebraically

$$\frac{\mathbf{a}}{h} \cdot \mathbf{r}^* = \frac{\mathbf{a}}{h} \cdot (h\mathbf{a}^* + r\mathbf{b}^* + l\mathbf{c}^*) = 1 \tag{3.20}$$

Equating (3.19) and (3.20) we have the desired result.

7. The angle ϕ_{hkl} between two lattice planes *hkl* and $h'\,k'\,l'$ is given by

$$\cos \phi_{hkl} = \frac{\mathbf{r}^*_{hkl} \cdot \mathbf{r}^*_{h'k'l'}}{r^*_{hkl}\, r^*_{h'k'l'}} \tag{3.21}$$

This result follows immediately from the definition of the dot product. As examples of (3.21), we look at cubic and tetragonal systems. For the cubic system, $|\mathbf{a}^*| = |\mathbf{b}^*| = |\mathbf{c}^*| = a^{-1}$ and the angles between $\mathbf{a}^*$, $\mathbf{b}^*$, and $\mathbf{c}^*$ are all 90°. Thus

$$\cos \phi_{hkl}^{h'k'l'} = \frac{hh' + kk' + ll'}{\sqrt{h^2 + k^2 + l^2}\sqrt{h'^2 + k'^2 + l'^2}} \tag{3.22}$$

For the tetragonal system $|\mathbf{a}^*| = |\mathbf{b}^*| = a^{-1} \neq |\mathbf{c}^*| = c^{-1}$ and again the angles between reciprocal axes are all 90°. Thus

$$\cos \phi_{hkl}^{h'k'l'} = \frac{hh' + kk' + (a/c)^2 ll'}{\sqrt{h^2 + k^2 + (a/c)^2 l^2}\sqrt{h'^2 + k'^2 + (a/c)^2 l'^2}} \tag{3.23}$$

Relations similar to (3.22) and (3.23) for all seven crystal system are found in Table 3.1.

8. The spacing d_{hkl} between neighboring hkl planes is given by

$$\frac{1}{d_{hkl}^2} = r_{hkl}^{*2} \tag{3.24}$$

This is merely item six rewritten for purposes shown below. In terms of the rel vectors $\mathbf{a}^*$, $\mathbf{b}^*$, and $\mathbf{c}^*$ and the Miller indices h, k, and l, we have

$$r_{hkl}^{*2} = \mathbf{r}_{hkl}^* \cdot \mathbf{r}_{hkl}^* = \frac{1}{d_{hkl}^2} = (h\mathbf{a}^* + k\mathbf{b}^* + l\mathbf{c}^*) \cdot (h\mathbf{a}^* + k\mathbf{b}^* + l\mathbf{c}^*) \tag{3.25}$$

Carrying out the scalar multiplication on the right of (3.25) gives (3.24). The explicit forms of (3.24) for the seven crystal classes (see Chap. 5) are

Cubic $$\frac{1}{d_{hkl}^2} = \frac{h^2 + k^2 + l^2}{a^2}$$

Tetragonal $$\frac{1}{d_{hkl}^2} = \frac{h^2 + k^2}{a^2} + \frac{l^2}{c^2}$$

Orthorhombic $$\frac{1}{d_{hkl}^2} = \frac{h^2}{a^2} + \frac{k^2}{b^2} + \frac{l^2}{c^2}$$

Hexagonal $$\frac{1}{d_{hkl}^2} = \frac{4}{3}\frac{h^2 + hk + k^2}{a^2} + \frac{l^2}{c^2} \tag{3.26}$$

Rhombohedral $$\frac{1}{d_{hkl}^2} = (h^2 + k^2 + l^2)a^{*2} + 2(hk + kl + lh)a^{*2}\cos\alpha^*$$

Monoclinic $$\frac{1}{d_{hkl}^2} = h^2a^{*2} + h^2b^{*2} + l^2c^{*2} + 2lha^*c^*\cos\beta^*$$

Table 3.1
Interplanar Angles

Crystal system	*Cosine of angle between* hkl *and* h′k′l′
Cubic	$\cos\phi = \dfrac{h_1h_2 + k_1k_2 + l_1l_2}{\sqrt{h_1^2 + k_1^2 + l_1^2}\sqrt{h_2^2 + k_2^2 + l_2^2}}$
Tetragonal	$\cos\phi = \dfrac{h_1h_2 + k_1k_2 + (a/c)^2 l_1l_2}{\sqrt{h_1^2 + k_1^2 + (a/c)^2 l_1^2}\sqrt{h_2^2 + k_2^2 + (a/c)^2 l_2^2}}$
Orthorhombic	$\cos\phi = \dfrac{\dfrac{h_1h_2}{a^2} + \dfrac{k_1k_2}{b^2} + \dfrac{l_1l_2}{c^2}}{\sqrt{\left(\dfrac{h_1}{a}\right)^2 + \left(\dfrac{k_1}{b}\right)^2 + \left(\dfrac{l_1}{c}\right)^2}\sqrt{\left(\dfrac{h_2}{a}\right)^2 + \left(\dfrac{k_2}{b}\right)^2 + \left(\dfrac{l_2}{c}\right)^2}}$
Hexagonal	$\cos\phi = \dfrac{h_1h_2 + k_1k_2 + \frac{1}{2}(h_1k_2 + h_2k_1) + \frac{3}{4}(a/c)^2 l_1l_2}{\sqrt{h_1^2 + k_1^2 + h_1k_1 + \frac{3}{4}(a/c)^2 l_1^2}\sqrt{h_2^2 + k_2^2 + h_2k_2 + \frac{3}{4}(a/c)^2 l_2^2}}$
Rhombohedral	$\cos\phi = \dfrac{h_1h_2 + k_1k_2 + l_1l_2 + [(h_1k_2 + h_2k_1) + (k_1l_2 + k_2l_1) + (l_1h_2 + l_2h_1)]\cos\alpha^*}{\sqrt{(h_1^2 + k_1^2 + l_1^2) + 2(h_1k_1 + k_1l_1 + l_1h_1)\cos\alpha^*}\sqrt{h_2^2 + k_2^2 + l_2^2) + 2(h_2k_2 + k_2l_2 + l_2h_2)\cos\alpha^*}}$
Monoclinic	$\cos\phi = \dfrac{h_1h_2a^{*2} + k_1k_2b^{*2} + l_1l_2c^{*2} + (l_1h_2 + l_2h_1)a^*c^*\cos\beta^*}{\sqrt{h_1^2a^{*2} + k_1^2b^{*2} + l_1^2c^{*2} + 2h_1l_1a^*c^*\cos\beta^*}\sqrt{h_2^2a^* + k_2^2b^{*2} + l_2^2c^{*2} + 2h_2l_2a^*c^*\cos\beta^*}}$
Triclinic	$\cos\phi = \dfrac{h_1h_2a^{*2} + k_1k_2b^{*2} + l_1l_2c^{*2} + (h_1k_2 + h_2k_1)a^*b^*\cos\gamma^* + (k_1l_2 + k_2l_1)b^*c^*\cos\alpha^* + (l_1h_2 + l_2h_1)c^*a^*\cos\beta^*}{(h_1^2a^{*2} + k_1^2b^{*2} + l_1^2c^{*2} + 2h_1k_1a^*b^*\cos\gamma^* + 2k_1l_1b^*c^*\cos\alpha^* + 2l_1h_1c^*a^*\cos\beta^*)^{1/2}\,(h_2^2a^{*2} + k_2^2b^{*2} + l_2^2c^{*2} + 2h_2k_2a^*b^*\cos\gamma^* + 2k_2l_2h_2c^*a^* + 2k_2l_2b^*c^*\cos\alpha^* + 2l_2h_2c^*a^*\cos\beta^*)^{1/2}}$

Triclinic
$$\frac{1}{d^2_{hkl}} = h^2a^{*2} + k^2b^{*2} + l^2c^{*2} + 2hka^*b^*\cos\gamma^* + 2klb^*c^*\cos\alpha^* + 2lhc^*a^*\cos\beta^*$$

In these relationships, α^*, β^*, and γ^* are the angles between **b*** and **c***, **c*** and **a***, and **a*** and **b***, respectively.

3.3 THE EWALD CONSTRUCTION AND DIFFRACTION EXPERIMENTS

3.3.1 Basic Construction

The Ewald construction is a very useful graphic tool for describing diffraction phenomena in crystals. To obtain the construction we proceed in the following way. We first draw the reciprocal lattice for our problem, as in Fig. 3.6. Next we draw an arrow representing the vector $\hat{\mathbf{s}}_o/\lambda$ to the origin. Using the *base* of the arrow as an origin we describe a sphere. From the center of the sphere we now draw any other radius vector and note that it is $\hat{\mathbf{s}}/\lambda$. The vector from the rel origin to the intersection of $\hat{\mathbf{s}}/\lambda$ with the sphere is **s**. Now recall that for nonzero intensity, $\mathbf{s} = \mathbf{r}^*_{hkl}$ and the arrow **s** must have its head on a rel point. Thus the appropriate rel point must lie on the Ewald sphere to record a diffracted intensity. We shall now see how the results of several types of diffraction experiments are visualized in reciprocal space.

3.3.2 Rotating Crystal Method

We first consider the rotating crystal experiment. In this experiment a single crystal is placed on a goniometer base such that one of its principal real lattice directions is along the axis of rotation. Monochromatic x-radiation is then input normal to this axis and the specimen is rotated in the beam. The physical

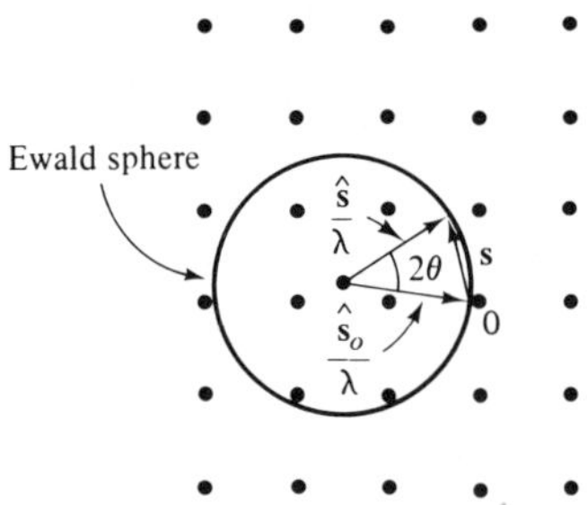

$$s = \frac{2\sin\theta_{hkl}}{\lambda} = r^*_{hkl} = \frac{1}{d_{hkl}}$$

$$\lambda = 2\,d_{hkl}\sin\theta_{hkl}$$

Figure 3.6. The Ewald construction. The origin of the Ewald sphere is located s_o/λ from the origin of the rel.

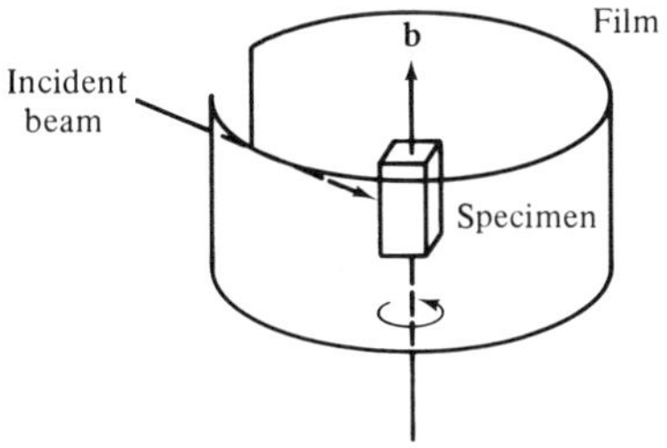

Figure 3.7. Schematic representation of the rotating crystal experiment. The specimen is oriented so that a rel vector is parallel to the axis of rotation. A film (or the detector path) is concentric about the specimen.

situation is as pictured in Fig. 3.7. The diffraction data are recorded on a film wrapped concentrically about the specimen or by a counter moving in a similar arc.

In resp the situation is as follows. The orientation of the rel is dictated by the orientation of the crystal. Thus **b** is along the rotation axis, as in Fig. 3.8.

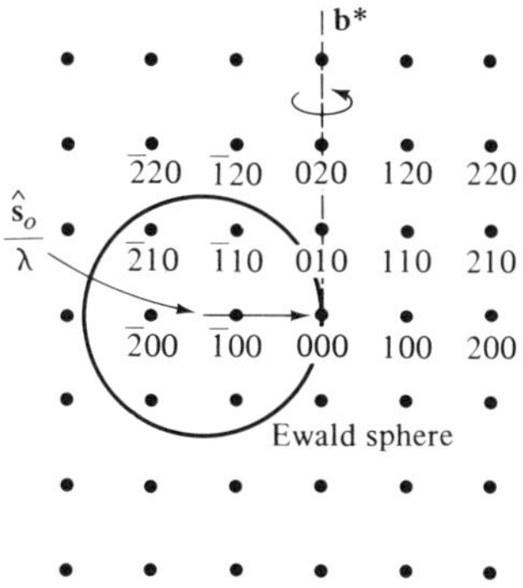

Figure 3.8. The rotating crystal experiment in resp. The reciprocal is rotated about **b***, causing relnodes to intercept the Ewald sphere.

We now place $\hat{\mathbf{s}}_o/\lambda$ normal to the axis and draw the Ewald sphere. As the real crystal rotates, the rel rotates with it and the rel points (which can) pass through the Ewald sphere during the rotation. Note that all the $h0l$ rel points pass through the Ewald sphere at its equator, while the $h1l$'s pass through on a longitude further north. The $h2l$ come even farther north. Thus, as shown in Fig. 3.9, the $h1l$ reflections all lie on a cone of diffracted rays. This cone will intersect the film in a horizontal line, called the first layer line. There will likewise be other layer lines from the other planes in the rel. This situation is shown in Fig. 3.10.

Using a Bernal chart, it is possible to index the spots of the rotating crystal

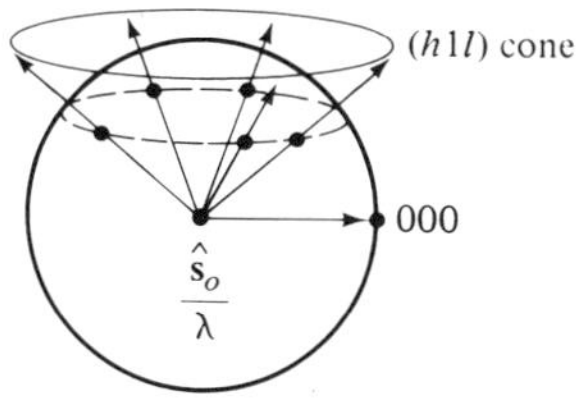

Figure 3.9. Formation of the $h1l$ cone. The $h1l$ relplane points have intercepted the Ewald sphere only in that plane. Hence the scattering vectors **s** lie on the cone shown.

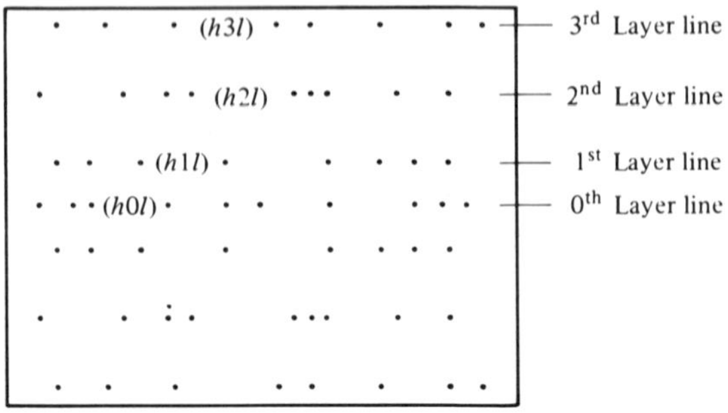

Figure 3.10. Sketch of a rotating crystal photograph, showing layer lines and Miller indices of the families of spots on each line.

pattern. As seen in Fig. 3.10, the relnodes of the kth plane form the kth layer line, parallel to the equator, in the rotation pattern. If the film is X cm from the specimen and if the height of the kth layer above the equator is Y cm, then, from simple trigonometry (Fig. 3.11), the reciprocal distance ζ of the kth relplane from the origin is given by

$$\zeta\lambda = \frac{Y}{\sqrt{X^2 + Y^2}} \tag{3.27}$$

As seen in Fig. 3.12, the vector $\mathbf{r}^*_{hkl}$ to any reciprocal lattice point can be viewed as the sum of two vector components: ζ, parallel to the rotation axis; and ξ, perpendicular to the axis.

$$r^{*2}_{hkl} = \zeta^2 + \xi^2 \tag{3.28}$$

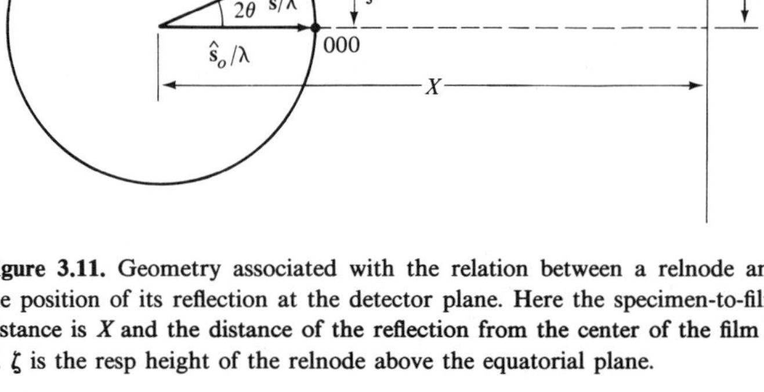

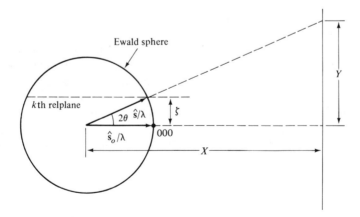

Figure 3.11. Geometry associated with the relation between a relnode and the position of its reflection at the detector plane. Here the specimen-to-film distance is X and the distance of the reflection from the center of the film is Y. ζ is the resp height of the relnode above the equatorial plane.

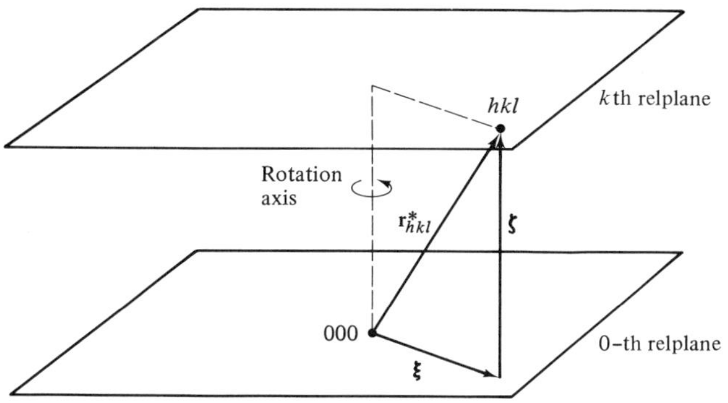

Fig. 3.12. ξ and ζ components of the rel vector $\mathbf{r}^*_{hkl}$.

We have just learned how to find ζ from the height of the layer line. The position of ξ is obtained from the position on the layer line. The trigonometry is a little bit more involved in this case. But the trigonometric results are conveniently plotted as Bernal charts. Figure 3.13 shows a Bernal chart. As seen in (3.27) the value assigned to ζ depends on the specimen-to-film distance X. Thus any chart used to find ζ must depend on the camera radius. The Bernal chart of Fig. 3.13 is scaled for a 3-cm camera radius. To use the chart, the rotation photograph is superposed, with its center at the center of the chart. The horizontal lines are read directly in $\lambda\zeta$. The closed figures give $\lambda\xi$. Using (3.28), the magnitude of the reciprocal vector $\mathbf{r}^*_{hkl}$ is found. Using (3.25) and (3.26) and known lattice constants, the values $h, k,$ and l can be found. These last steps can also be done graphically by first plotting the reciprocal lattice for a plane perpendicular to the rotation axis and then drawing circles, at the same scale, using the measured values of ξ.

The result is, however, not unambiguous. Consider, for example, the reflections (500) and (340) for a cubic material. According to (3.26), these both have the same value of r^*_{hkl} and will superimpose on a c-axis rotation photograph. For crystal structure determination, it is essential to remove this ambiguity. Weissenberg cameras and precession cameras have been developed to eliminate the superimposition of spots. In the Weissenberg method, the film is continuously translated along the rotation axis as the specimen rotates. In this way, no two spots superpose. The precession camera moves both specimen and film in such a way that an undistorted picture of a plane in resp is obtained.

An interesting and useful situation occurs in the case of electron diffraction. Electron diffraction is most often performed by transmission through thin films, with a thickness of the order of 100 Å, in the electron microscope. Such microscopes conventionally use accelerating voltages of the order of 100 kV. According to Table 1–1, the wavelength of 100-kV electrons is about 0.04 Å. The radius $1/\lambda$ of the Ewald sphere would then be 25 Å^{-1}. The spacing within the rel, for lattice spacings near 4 Å, is two orders of magnitude smaller than the

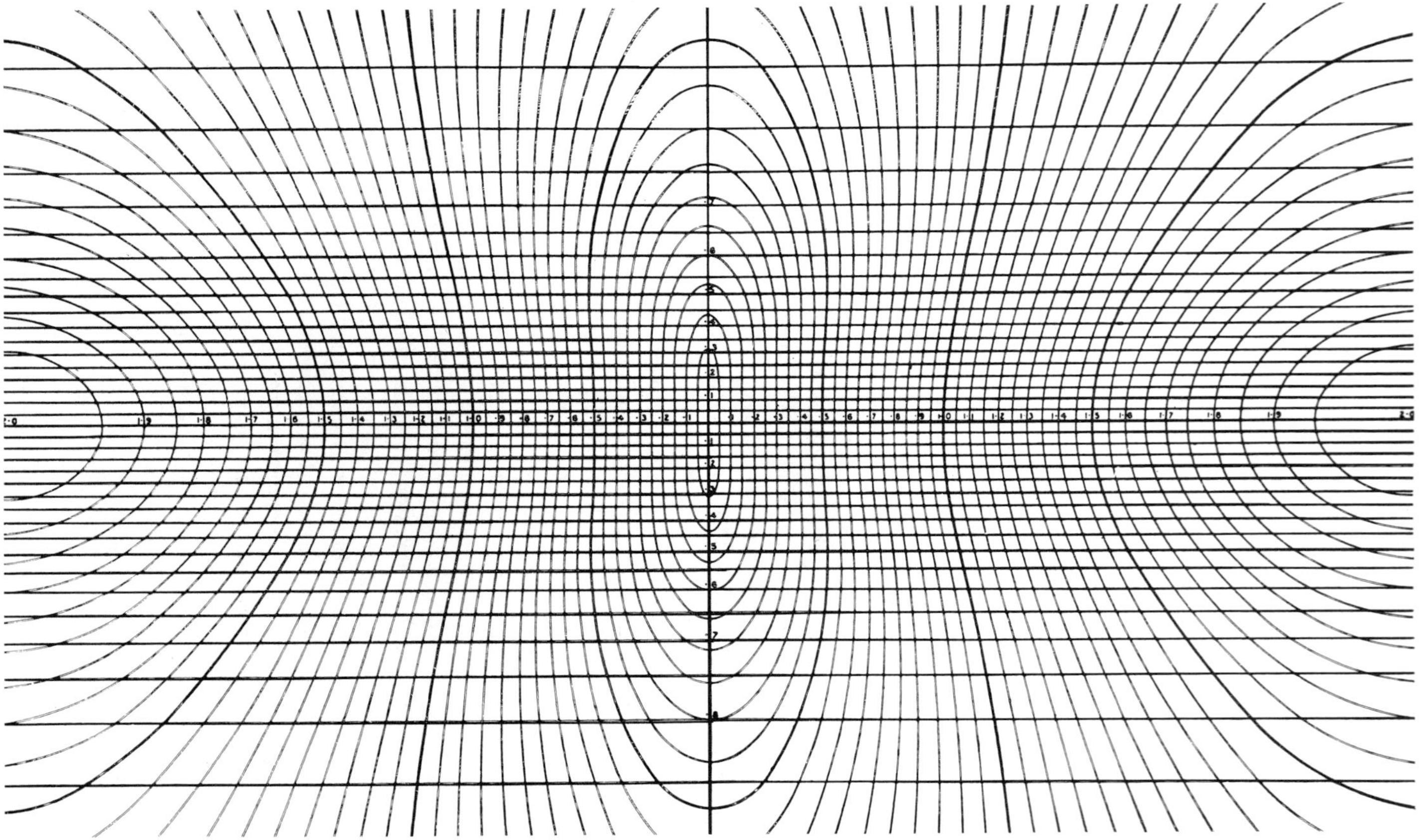

Figure 3.13. Bernal chart, giving $\lambda\xi$ and $\lambda\zeta$ for a cylindrical camera of radius 3.0 cm.

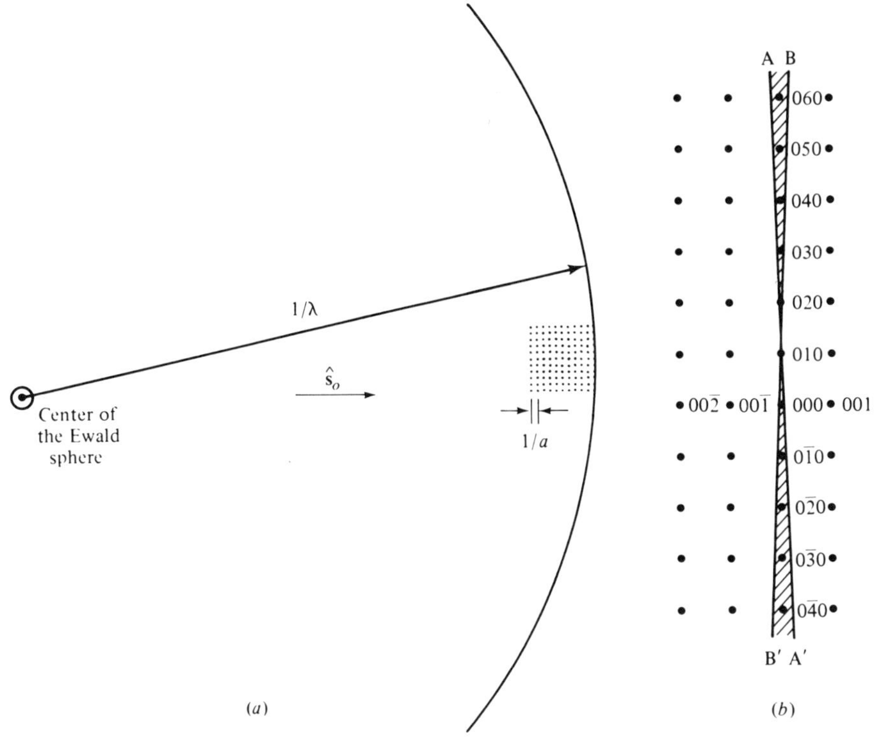

Figure 3.14. Ewald sphere geometry for electron diffraction: *(a)* relative dimensions of the rel and the Ewald sphere; *(b)* magnification of the neighborhood of the origin.

Ewald sphere radius. The situation is as depicted in Fig. 3.14*a*. The surface of the Ewald sphere passes very near several spots in the rel plane perpendicular to the incident beam. Since the Ewald sphere passes so close to those spots, the Bragg condition can be met by using a slightly divergent electron beam. The situation in resp is then as depicted in Fig. 3.14*b*. Here AA' and BB' are the local Ewald sphere surfaces for rays entering the crystal at slightly different angles. If all the intermediate entry angles are also available, then any relnodes lying in the shaded region, between AA' and BB' will lie on an Ewald sphere and will produce a diffracted beam. The result is similar to that of the rotating crystal experiment using x-rays or neutrons. In the electron microscope geometry, a film is placed normal to the incident beam and behind the specimen film. For the case of Fig. 3.14, where the incident beam is along [001], the film will capture the diffracted rays from the (hk0) rel plane. These diffracted rays form a relatively undistorted image of that rel plane. In general, an image of whatever rel plane is perpendicular to the incident beam is produced in this type of experiment. An example of an electron diffraction pattern—here for the (111) rel plane of cubic η-Al_2O_3—is shown in Fig. 3.15.

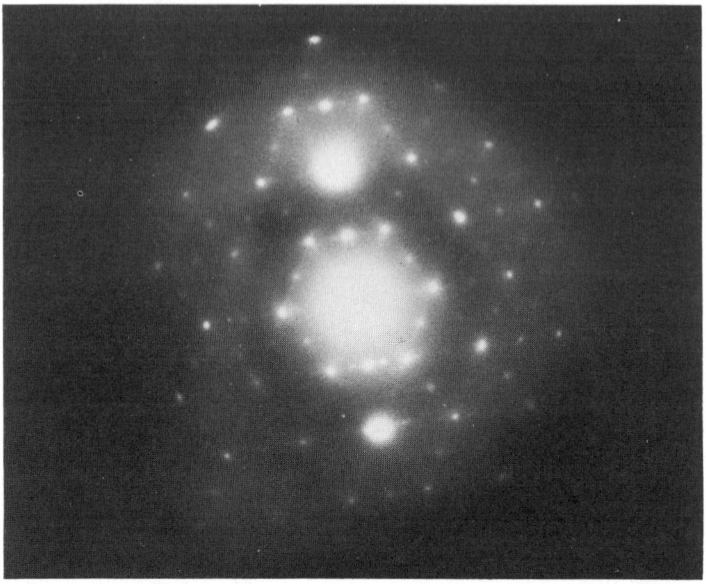

Figure 3.15. 100-kV electron diffraction pattern of cubic η-Al_2O_3 taken with the electron beam approximately normal to (111). Note the sixfold symmetry.

3.3.3 Powder Method

Here the sample is a powder or polycrystal of fine crystallites. The experiment is described in Fig. 3.16*a* in both real and reciprocal space. The Ewald sphere and the locus sphere for the *hkl* relpoint intersect at a circle, as in Fig. 3.16*b*. The vectors **s**, originating at the center of the Ewald sphere, terminate on that circle. Thus in this method a cone of rays is again generated, this time coaxial

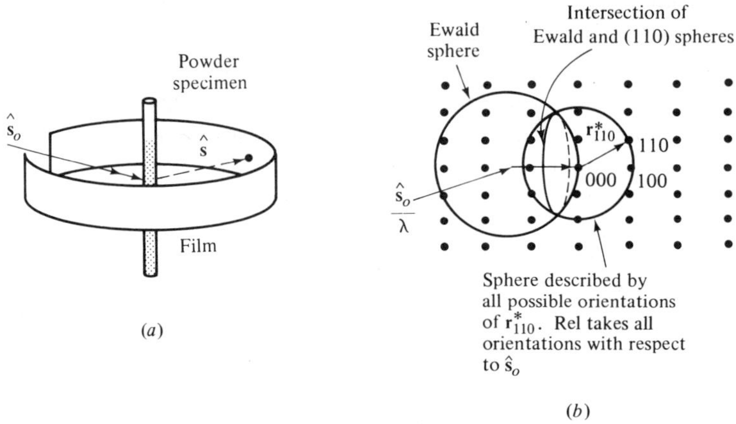

Figure 3.16. The powder experiment in *(a)* real and *(b)* reciprocal space.

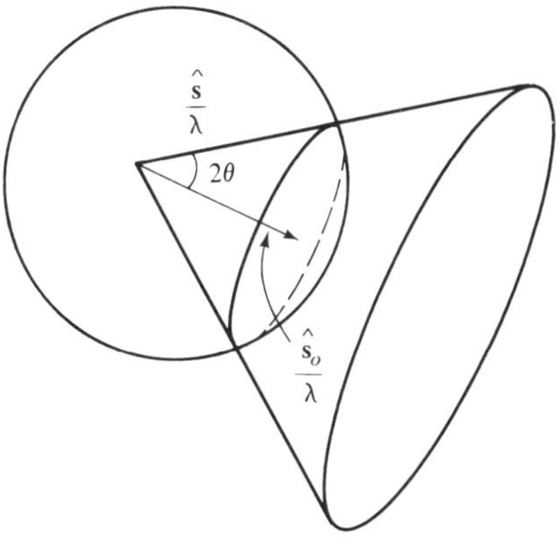

Figure 3.17. Generation of 110 cone, after the geometry of Fig. 3.16*b*.

with the main beam and having apex angle 4θ, as shown in Fig. 3.17. These cones intersect the film or counter in arcs, as shown in Fig. 3.18. The d_{hkl} spacings are determined from the θ_{hkl} angles. This method is widely used to identify unknown materials and to determine precise lattice constants. Using this method one can follow changes in composition or perfection in some detail.

In practice, the film is sometimes replaced by an electronic radiation detector which moves along the circle which would have been described by the film. Such an instrument is called a powder diffractometer. The output of such an instrument will then be directly a plot of intensity against scattering angle 2θ. An example of such an output is shown in Fig. 3.19.

The identification of unknown materials is carried out using the JCPDS (Joint Committee on Powder Diffraction Standards) file of powder diffraction data. A typical data set, for one material, is given in Fig. 3.20. The name and file number of the material are found in the upper left. At the lower right is a listing of d_{hkl} values, as found by powder diffraction, and the associated intensities. This list is specific to each material and can be thought of as a sort of "fingerprint." Search manuals are also available from the JCPDS. The search

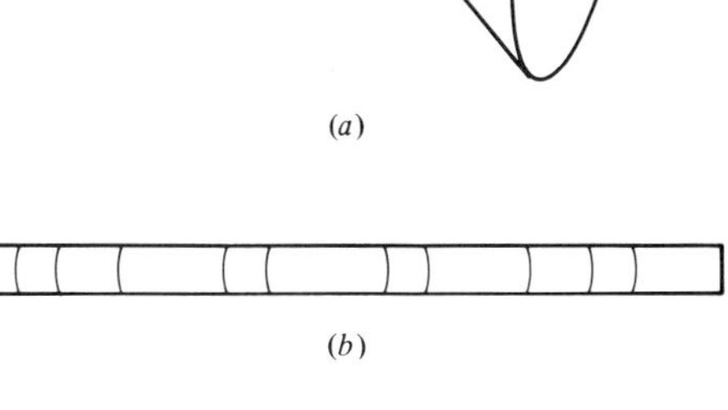

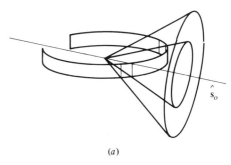

Figure 3.18. Formation of diffraction arcs on Debye-Scherrer film: *(a)* arc formation, *(b)* appearance of Debye-Scherrer film.

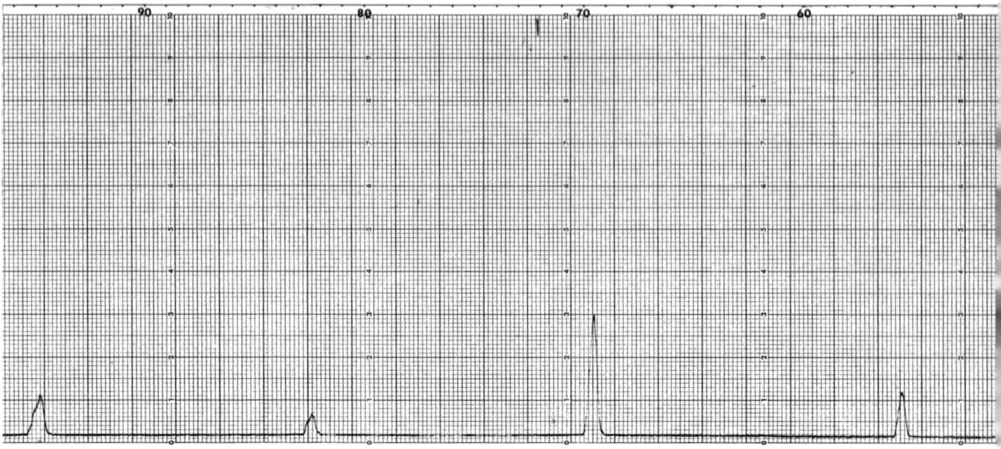

Figure 3.19. X-ray diffractometer scan of tantalum powder. Note that the scattering angle 2θ is read from right to left (an artifact of the standard recorder output).

manuals list materials by their three strongest diffraction lines. Using the search manual (and, usually, some knowledge of the composition), one can narrow the possibilities immediately to a few data sets out of the approximately 70,000 included in the diffraction file. Final identification is complete, of course, only when the d_{hkl} values and associated intensities for a listed data set match those measured experimentally.

d	2.24	3.99	2.71	3.99	UB_2	★
I/I_1	100	85	80	85	URANIUM BORIDE	

Rad. CuKα λ 1.5418 Filter Dia. GUINIER CAMERA
Cut off I/I_1 VISUAL (CALIBRATED STRIPS)
Ref. THE PLESSEY COMPANY LTD., CASWELL, TOWCESTER, NORTHANTS, ENGLAND

Sys. HEXAGONAL S.G. *P6/MMM (191)
a_0 3.131 b_0 c_0 3.987 A C 1.274
α β γ Z*1 Dx
Ref. IBID. (LATTICE PARAMETERS CALCULATED FROM DEBYE-SCHERRER PHOTOGRAPH)

εα nωβ εγ Sign
2V D mp Color
Ref.

*BREWER ET AL., U.S. A.E.C. PUBL. A.E.C.D. 2823, (1950).
THE THREE STRONGEST LINES IN THIS PHOTOGRAPH ARE COMPLETELY DIFFERENT FROM THOSE ON A PHOTOGRAPH IN A DEBYE-SCHERRER CAMERA, BECAUSE NO PRECAUTIONS WERE TAKEN WITH THE LATTER EXPOSURE TO DILUTE THE SPECIMEN, AND THE VERY HIGH ABSORPTION OF THE MATERIAL GAVE RISE VERY REDUCED LOW-ANGLE INTENSITIES.

d Å	I/I_1	hkl	d Å	I/I_1	hkl
3.99	85	001			
2.71	80	100			
2.24	100	101			
1.99	15	002			
1.607	30	102			
1.565	15	110			
1.457	25	111			
1.355	10	200			
1.329	5	003			
1.284	10	201			
1.231	10	112			
1.194	10	103			
1.121	10	202			

Figure 3.20. JCPDS card for uranium boride (UB_2). Diffraction data is found in the lower right quadrant. At the upper left are found the d values of the four most intense lines; these data are used in the search procedure for locating candidate JCPDS cards. (Reprinted by permission of the International Centre for Diffraction Data, Swarthmore, PA, U.S.A.)

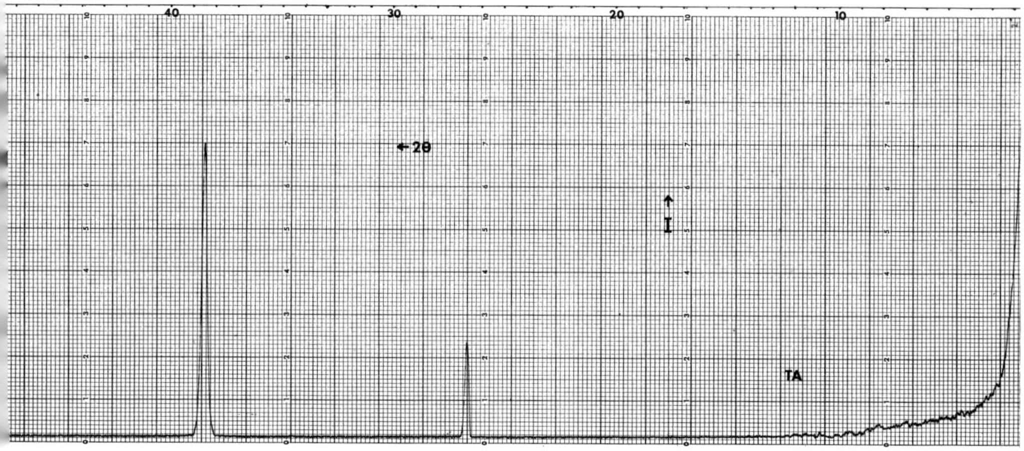

Figure 3.19. (Continued).

3.3.4 Laue Method

The third common geometry is that of the Laue method. In this method a polychromatic beam is incident on a stationary single crystal. The experiment in real and reciprocal space is shown in Figs. 3.21 and 3.22. A resulting film is shown in Fig. 3.23. All relpoints between the spheres representing λ_{max} and λ_{min} will show up in the diffraction pattern. The geometry is better seen if we draw the rel with lattice vectors $\lambda\mathbf{a}^*$, $\lambda\mathbf{b}^*$, and $\lambda\mathbf{c}^*$. An Ewald geometry still holds good, as we see in Fig. 3.24. Here the $\mathbf{r}^*_{hkl}\lambda$ are represented by lines instead of points, the length of the lines being given by the spread in λ. The advantage of this geometry is that all vectors $\hat{\mathbf{s}}$ have a common origin and the direction of each reflection can be easily obtained. In the geometry of Fig. 3.22, each wavelength gives an Ewald sphere with a different origin. One thus

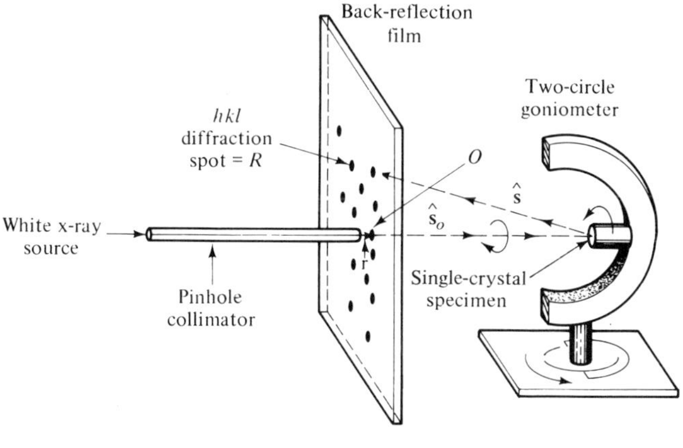

Figure 3.21. Real space geometry for the Laue back-reflection method.

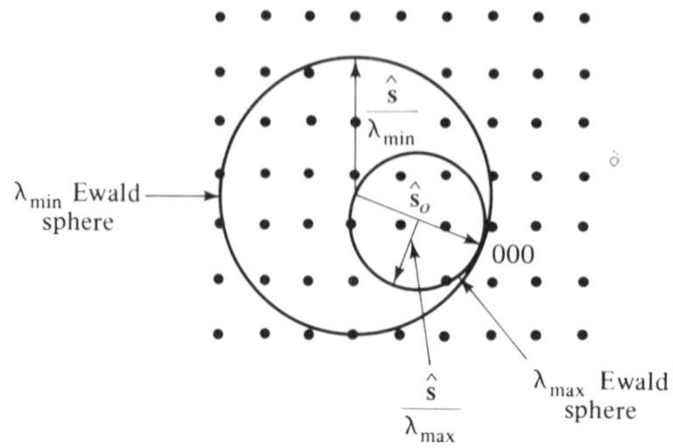

Figure 3.22. Reciprocal space geometry for the Laue back-reflection method.

Figure 3.23. Laue back-reflection film from a copper specimen of arbitrary orientation. Specimen-to-film distance: 3.0 cm.

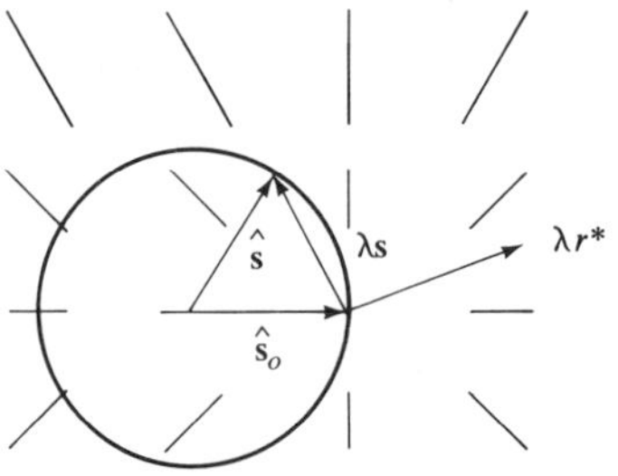

Figure 3.24. Modified resp geometry for the Laue back-reflection method. Here distances are dimensionless λr^*, rather than r^*.

would have to draw the Ewald sphere connected with each possible *hkl* reflection in order to determine the diffracted beam direction.

The Laue method is widely used throughout the world to set the axes of a crystal in any given orientation relative to the laboratory coordinate system—that is, "to orient the crystal." As we have already seen, the properties of crystals are quite anisotropic. Thus, to achieve a given behavior, the axes of a crystal must be set in a known way relative to the direction of application of driving forces (e.g., mechanical force, electrical field, magnetic field, flux of atoms). The ability to orient a crystal thus has both industrial and research significance.

The orientation of crystals by the Laue method is based on two principles:

(1) *If x-rays with a broad, continuous range of wavelengths are played upon a set of crystal planes, each set of planes acts as a mirror for some component of the x-ray beam that has just the right wavelength for constructive interference.* To see how these conditions are satisfied consider Fig. 3.25 in which the *hkl* planes are set at an arbitrary angle to the polychromatic incident beam. Recall Bragg's law, $\lambda = 2d_{hkl} \sin \theta$. In this case θ is fixed by the crystal position. The *hkl* planes diffract that wavelength λ_{hkl} which satisfies Bragg's law for θ as fixed by the crystal orientation. The *hkl* planes act effectively as a mirror for the λ_{hkl} rays, as indicated.

(2) The angle between the normals to *hkl* and *h' k' l'* are fixed by the geometry of the crystal. For instance, in the cubic system, the angle between the normal to (100) and the normal to (110) is 45°.

In Fig. 3.26, the *hkl* and *h' k' l'* planes are shown to diffract portions of the beam at different output angles. Both planes act as mirrors for appropriate wavelengths among the x-rays input to them. Notice that the angle between the ray reflected from *(hkl)* and the ray reflected from *h' k' l'* is related to the dihedral angle $\phi_{hkl-h'k'l'}$ between those planes, hence to the angle between

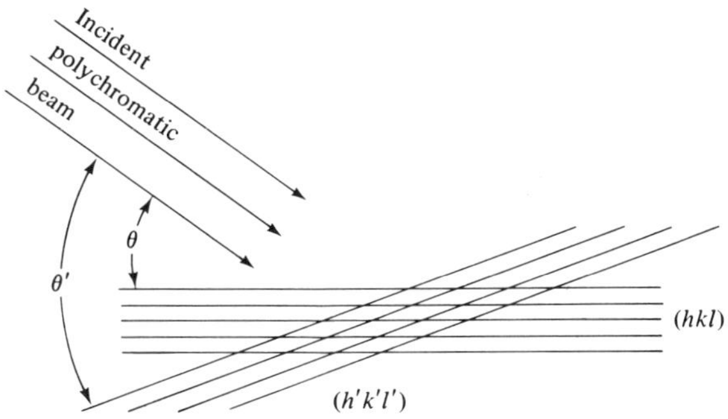

Figure 3.25. *(hkl)* and *(h'k'l')* planes of a single crystal in a polychromatic x-ray beam.

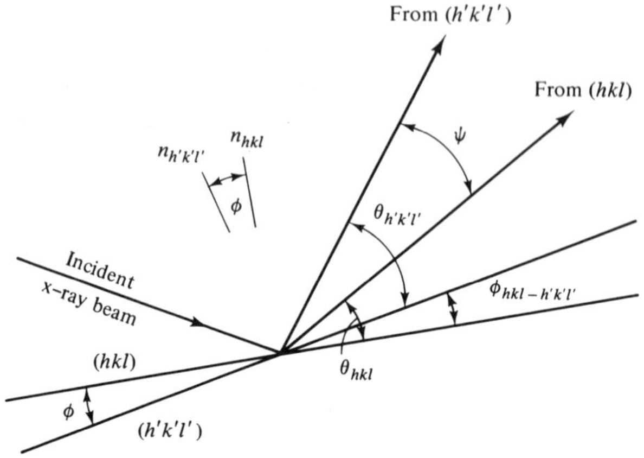

Figure 3.26. Relation of diffracted beams to the incident beam and crystal planes. θ_{hkl} and $\theta_{h'k'l'}$ are the reflection angles for *(hkl)* and *(h' k' l')*. $\phi_{hkl\text{-}h'k'l'}$ is the angle between *(hkl)* and *(h' k' l')*.

the normals. In fact, ψ, the angle between the two reflected rays is $\theta_{hkl} - \theta_{h'k'l'} + \phi_{hkl-h'k'l'}$, where θ_{hkl} and $\theta_{h'k'l'}$ are the angles between incident beam and *hkl* and *h' k' l'*, respectively.

The angles between all sets *hkl* and *h' k' l'* in the cubic system are known. Thus the determination of orientation consists essentially of (a) measurement of angles ψ between sets of diffraction spots and (b) the determination by a systematic approach of how the lattice planes must have been set so as to produce the given sets of $\psi_{ij} = \theta_i - \theta_j + \phi_{ij}$. Once this initial orientation is found, the crystal can be reoriented in any desired way.

The Laue spots are observed to form along hyperbolic arcs or straight lines (if the spots go through the center of the film). Those spots belonging to a given arc are associated with planes of a given zone. Note that some of the spots belong to more than one zone. These are usually some of the more important planes in the crystal.

If the sequence of spots go through the center of the film the zone axis associated with the spots is perpendicular to the incoming x-ray beam and the array forms a straight line. Zone axes that are not perpendicular to the x-ray beam form the various hyperbolic arcs. Measurement of the deviation of the zone axis in terms of both a tilt angle ϕ and twist angle α can be accomplished by means of a Greninger Chart. A Greninger chart is given in Fig. 3.27 (pp. 70–71). This chart has been computed for a specimen-to-film distance of 3 cm. It can be used not only to measure tilt and twist angles, as in Fig. 3.28, but also to measure the angle between spots on the film, to compare with known angles between planes.

Standard procedures for obtaining and conveying orientation information from Laue photographs are found in the following sources: C. S. Barrett and

T. B. Massalski, *Structure of Metals,* 3rd ed., McGraw-Hill, NY, 1966, pp 211–7; D. B. Cullity, *Elements of X-ray Diffraction,* Addison-Wesley, Reading, MA, 1966, pp 215–29; and J. B. Cohen, *Diffraction Methods in Materials Science,* Macmillan, 1966, pp 152–68.

3.4 THE STRUCTURE FACTOR

We have now seen that the amplitude function for a large, perfect crystal is given by

$$\frac{A(\mathbf{s})}{A_e(s)} = N\left[\sum \sum \sum \delta(\mathbf{s} - \mathbf{r}^*_{hkl})\right] \cdot F(\mathbf{s}) \tag{3.29}$$

We also noted that the positions of the intensity nodes are controlled by the (Bravais) lattice vectors in the real lattice. The amplitudes related to those relnodes are given, as we see in (3.5), by $F(\mathbf{s})$, the structure factor.

Let us now write the structure factor in its most useful form. Our definition of $F(\mathbf{s})$ gives

$$F(\mathbf{s}) = \int_{\text{cell}} \rho(\mathbf{r}) e^{2\pi i \mathbf{s}\cdot\mathbf{r}}\, dv_{\mathbf{r}} \tag{3.30}$$

Converting to atom-centered coordinates in the unit cell, we have

$$F(\mathbf{s}) = \sum_{n=1}^{N_c} f_n e^{2\pi i \mathbf{s}\cdot\mathbf{r}_n} \tag{3.31}$$

where f_n is the structure factor of the nth atom and N_c is the number of atoms in the unit cell. We have already seen that a necessary condition on $\mathbf{s}$ is that it be a rel vector $\mathbf{r}^*_{hkl}$. We then substitute $\mathbf{r}^*_{hkl}$ for $\mathbf{s}$. We also write $\mathbf{r}_n$ in terms of the real space lattice vectors

$$\mathbf{r}_n = x_n\,\mathbf{a} + y_n\,\mathbf{b} + z_n\,\mathbf{c} \tag{3.32}$$

Thus

$$F(\mathbf{s}) = \sum_{n=1}^{N_c} f_n \exp\left[2\pi i(h\mathbf{a}^* + k\mathbf{b}^* + l\mathbf{c}^*) \cdot (x_n\mathbf{a} + y_n\mathbf{b} + z_n\mathbf{c})\right]$$

$$= \sum_{1}^{N_c} f_n\, e^{2\pi i(hx_n + ky_n + lz_n)} \tag{3.33}$$

Let us look at a few examples of the application of the structure factor equation to simple solids.

First, let us take any crystal system with one atom per unit cell. Let us place that atom at position $x,y,z = 0,0,0$. Thus $F(\mathbf{s}) = f$. That is, each rel point will have amplitude f.

Figure 3.27. Greninger chart for the back-reflection Laue method.

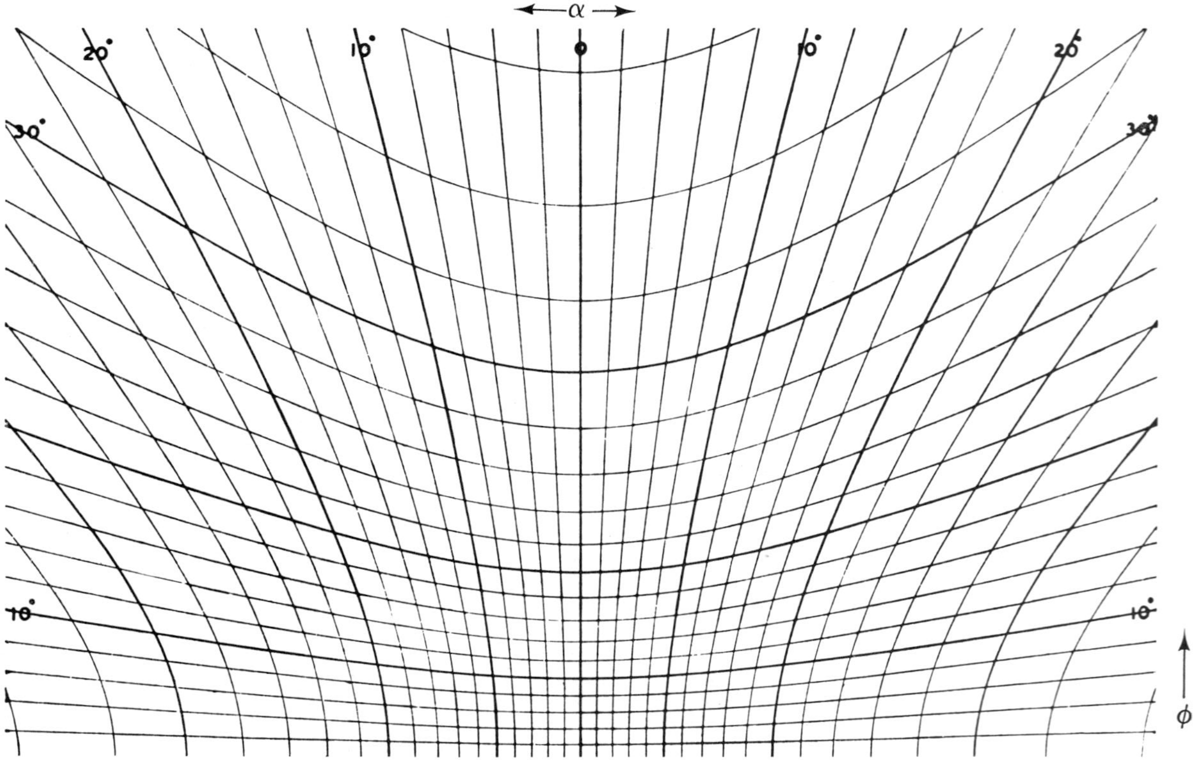

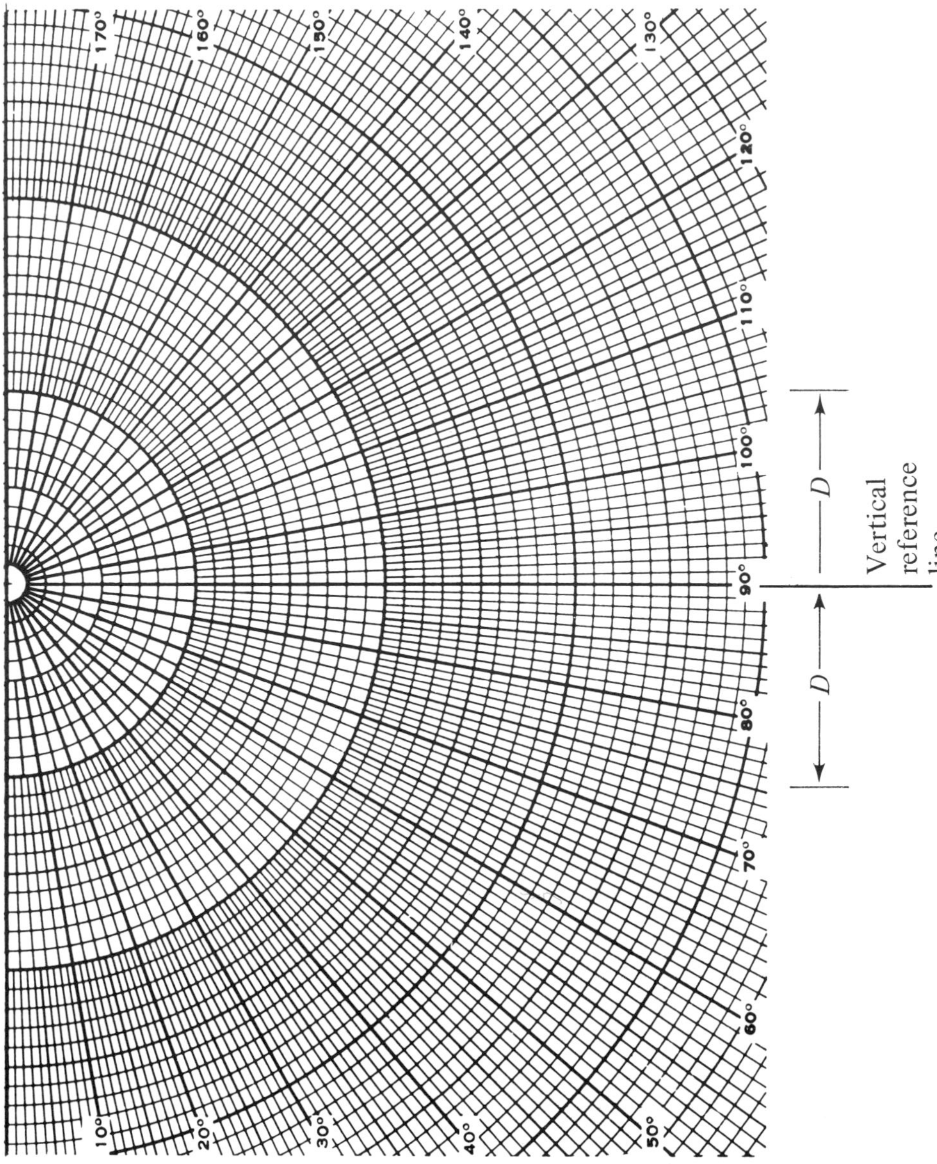
10°
20°
30°
40°
50°
60°
70°
80°
90°
100°
110°
120°
130°
140°
150°
160°
170°
D
D
Vertical reference line

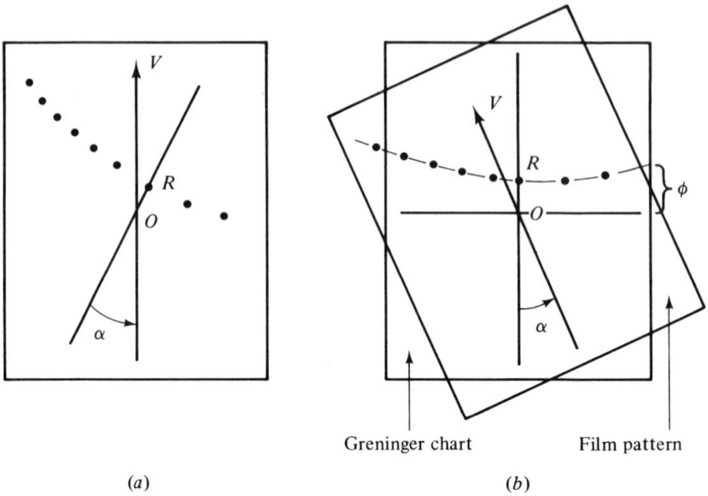

Figure 3.28. *(a)* Laue back-reflection film, here showing only one diffraction hyperbola. *(b)* Rotation of the pattern, to align diffraction hyperbola with a hyperbola on the Greninger chart.

Second, let us examine any system with a centered cell of one atomic type. Let us then place the origin of coordinates such that the two atoms in the cell occur at 0,0,0 and ½, ½, ½. Thus

$$F(\mathbf{s}) = F_{hkl} = f[1 + e^{\pi i(h+k+l)}] = \begin{cases} 0 \text{ if } h+k+l = \text{odd} \\ 2f \text{ if } h+k+l = \text{even} \end{cases} \tag{3.34}$$

Third, let us look at any face-centered system of one atomic type. Here the atoms are taken at (0,0,0), (½, ½, 0), (0, ½, ½), and (½, 0, ½). Thus

$$F_{hkl} = f[1 + e^{\pi i(h+k)} + e^{\pi i(k+l)} + e^{\pi i(h+l)}] = \begin{cases} 0 \text{ if } h,k,l \text{ are mixed} \\ \text{odd and even} \\ 4f \text{ if } h,k,l \text{ are all} \\ \text{odd or all even} \end{cases} \tag{3.35}$$

These simple examples are instructive by implication. The absent reflections allow us to determine the symmetry of the atomic arrangement within the unit cell. Such symmetry is described in Chap. 5. The three unit cells chosen represent all simple, body-centered, and face-centered Bravais lattices. If, in addition, we know the macroscopic unit cell symmetry (e.g., cubic, hexagonal, monoclinic, etc.) we will have defined the space group. In practice, the macroscopic symmetry is easily obtained. Thus for these simple monatomic systems, we note that determination of macroscopic symmetry plus analysis of the *systematic extinction* has defined the space group. This occurs because of the specific systematic symmetry of placement of atoms in the unit cell for each space group. By implication, then, and indeed in fact, each space group has its own specific

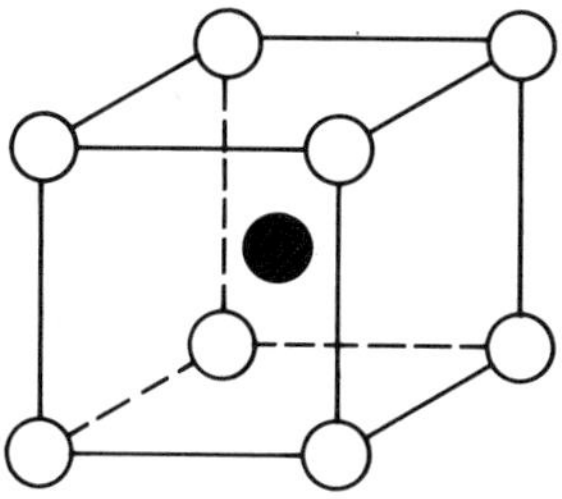

Figure 3.29. The CsCl structure.

set of macroscopic symmetry plus systematic extinctions and may be identified thereby. The business of establishing the space group symmetry by determination of the systematic extinction rules is one of the first steps in crystal structure analysis.

A last example is that of the CsCl structure. The unit cell for CsCl is shown in Fig. 3.29. The Cs atom is located at 0,0,0 and the Cl at ½, ½, ½. The structure factor is

$$F_{hkl} = f_{Cs} + f_{Cl}\, e^{\pi i(h+k+l)} = \begin{cases} f_{Cs} + f_{Cl} \text{ for } h+k+l = \text{even} \\ f_{Cs} - f_{Cl} \text{ for } h+k+l = \text{odd} \end{cases} \tag{3.36}$$

We learn two things by implication here. First, it was the *space group* (simple cubic, $P\frac{4}{m}\bar{3}\frac{2}{m}$) which determined the systematic extinctions. Here there were no extinctions, as for any primitive lattice. Second, the amplitude F_{hkl} associated with each peak *is* determined by the placement of atoms acting as the basis for the unit cell. In general, it is a study of the intensity of hkl intensities which enables one to deduce the exact placement of atoms within the unit cell. We shall return to this again in Chap. 5.

3.5 INTEGRATED REFLECTING POWER

3.5.1 General

In this section we derive the intensity formulas which pertain to actual powder and rotating crystal experiments. Basically, the problem evolves to one of translating geometry from **s**-space to scans over 2θ.

To begin we need to return to a general expression for the amplitude of scattering from a crystalline material. Using (3.5) for the amplitude, we write

$$\frac{I(\mathbf{s})}{I_e(s)} = |F(s)|^2 |Z(s)|^2 = |F(s)|^2 \left(\frac{\sin^2 \pi N_1 s_1 a}{\sin^2 \pi s_1 a}\right) \cdot \left(\frac{\sin^2 \pi N_2 s_2 b}{\sin^2 \pi s_2 b}\right) \cdot \left(\frac{\sin^2 \pi N_3 s_3 c}{\sin^2 \pi s_3 c}\right) \tag{3.37}$$

Considering one of the factors on the right, we plot $\sin^2\pi N_1 s_1 a$, $\sin^2\pi s_1 a$, and $(\sin^2\pi N_1 s_1 a)/(\sin^2\pi s_1 a)$ against s_1 in Fig. 3.30, using values of 5 and 10 for N_1. We see from these curves that

1. $(\sin^2\pi N_1 s_1 a)/(\sin^2\pi s_1 a)$ peaks strongly whenever $s_1 = n_1/a$, n_1 being an integer.
2. The sharpness of the peak at $s_1 = n_1/a$ increases as N_1 increases.
3. Except for sign, all n_1/a peaks are identical.

Furthermore, the height of each peak is N_1^2. This is seen by taking the limit of the quotient

$$\lim_{s_1\to 0} \frac{\sin^2\pi N_1 s_1 a}{\sin^2\pi s_1 a} = \lim_{s_1\to 0} N_1^2 \left(\frac{\sin \pi N_1 s_1 a}{\pi N_1 s_1 a}\right)^2 \left(\frac{\pi s_1 a}{\sin \pi s_1 a}\right)^2 = N_1^2 \qquad (3.38)$$

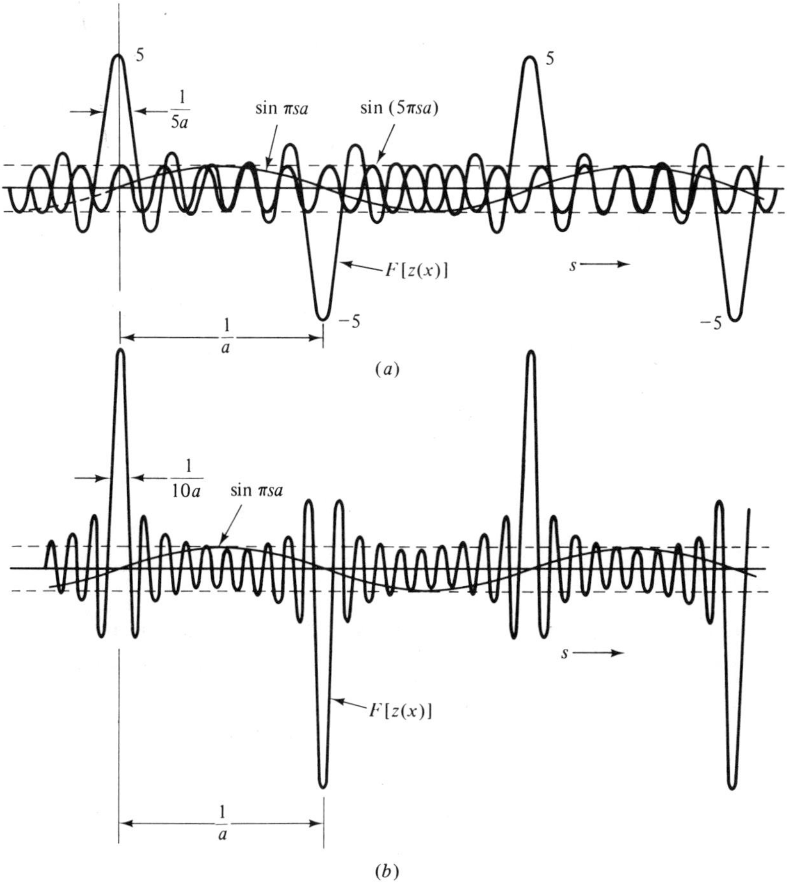

Figure 3.30. Plots of $x = \sin(\pi s_1 a)$, $y = \sin(N_1 \pi s_1 a)$, and $\psi = y/x$ for line lattices of *(a)* 5 points and *(b)* 10 points.

the last step utilizing (four times) the zero limit of (sin x)/x. Lastly, we note that the value of s_1 at half maxima values of $(\sin^2\pi N_1 s_1 a)/(\sin^2\pi s_1 a)$ is given by

$$\frac{\sin^2\pi N_1(s_1)_{1/2}a}{\sin^2\pi(s_1)_{1/2}a} \simeq \frac{\sin^2\pi N_1(s_1)_{1/2}a}{(\pi s_1 a)^2} = \frac{N_1^2}{2} \tag{3.39}$$

or

$$\frac{\sin x}{x} = \sqrt{½} = 0.707 \tag{3.40}$$

In this case $x = 1.39$ radian. Solving for $(s_1)_{1/2}$, we have

$$(s_1)_{1/2} = \frac{1.39}{\pi N_1 a} \tag{3.41}$$

We note that the breadth of the reflection is inversely proportional to N_1. This *particle-size broadening* (to which we shall return later) becomes a nice tool by which the dimensions of small crystallites may be determined. As in Chap. 1, we note again that the area under the peak can be estimated by replacing the true peak by a step function of height N_1^2 and breadth $2(s_1)_{1/2}$. In this way, the total area under the peak is given approximately by $(N_1^2)(2)(0.983/\pi N_1 a) = (1.966/\pi)(N_1/a)$. If this area had been rigorously obtained we would have found

$$\int_{-\infty}^{\infty} \frac{\sin^2\pi N_1 s_1 a}{\sin^2\pi s_1 a}\, ds_1 = \frac{N_1}{a} = \frac{L_1}{a^2} \tag{3.42}$$

where $N_1 a = L$ is the length of the crystal in the direction parallel to a.

We note from the above paragraph that the height and breadth of a diffraction pattern line may vary, but that its integral remains constant. Thus, if one wishes to derive quantitative structural data beyond unit cell size from a diffraction pattern, one should work with an integrated intensity.

3.5.2 Integrated Reflecting Power in the Powder Experiment

Armed with these results, let us observe what a powder pattern represents, in resp. One can visualize the *hkl* line of a powder pattern as being produced by the rotation of the *hkl* relnode about 000. This is illustrated in Fig. 3.31. The relnode is rotated about the entire 4π solid angle. Consider now a plane in the relnode such that that plane is perpendicular to $\mathbf{r}^*_{hkl}$, as in Fig. 3.32*a*. The vector from the center of the relnode to that plane is w. Consider also an increment of area $du\,dv$ in that plane (Fig. 3.32*b*). Recall now that all possible orientations about 000 are possible. The fraction of all these orientations occupied by our crystal, with $du\,dv$ positioned exactly as shown is given by the solid

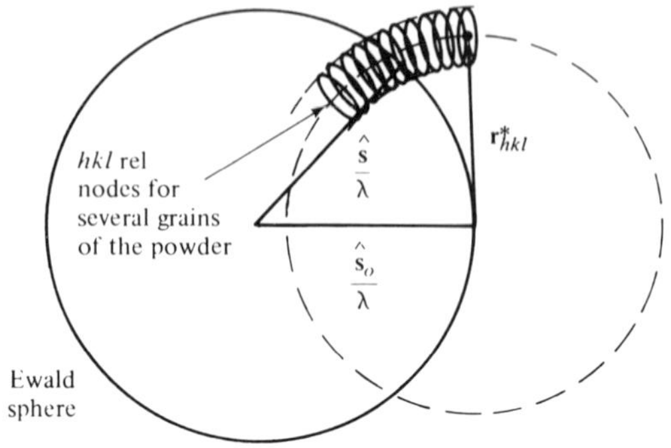

Figure 3.31. Relnode loci for a powder pattern.

angle $d\Omega$ subtended by $du\ dv$, divided by the total solid angle, 4π. This fraction is

$$\frac{du\ dv}{\left(\frac{2}{\lambda}\sin\theta\right)^2}/4\pi$$

Now observe that the intensity associated with a specific position on the Ewald sphere is given by an integration over all the relnodes that pass through that point. If the distance of the Ewald sphere point is $\mathbf{r}^*_{hkl} + \mathbf{w}$ from 000, then the integration would be simply over the relnode plane sketched in Fig. 3.32. We must remember to associate with each element of intensity $i(u,v,w)du$

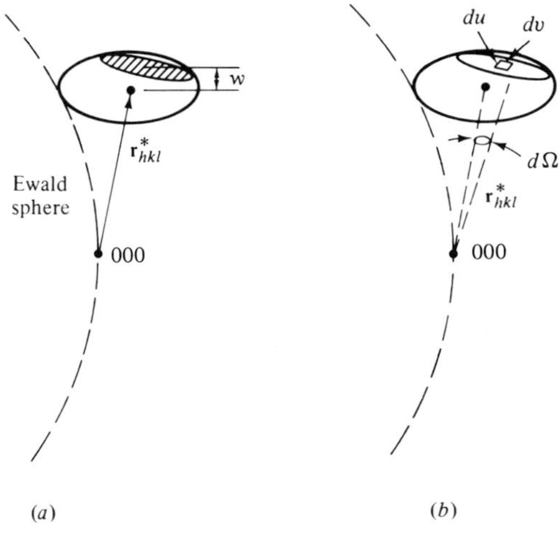

Figure 3.32. Geometry associated with the intersection of a relnode with the Ewald sphere. In *(a)* a relnode plane perpendicular to $\mathbf{r}^*_{hkl}$ is shaded. The plane lies a reciprocal distance w above the centroid of the relnode. In *(b)* an area element $du\ dv$ on the chosen plane is shown. Corresponding to the area element is a solid angle element $d\Omega$.

dv in that plane the appropriate weight $\left[4\pi\left(\frac{2}{\lambda}\sin\theta\right)^2\right]^{-1}$, as described above. Thus the intensity I^* associated with the Ewald sphere point is

$$I^*(2\theta) = \int_{\substack{\text{one}\\\text{node}}} \frac{i(u,v,w)du\,dv}{\left[4\pi\left(\frac{2}{\lambda}\sin\theta\right)^2\right]} \tag{3.43}$$

This $I^*(2\theta)$ is constant over a ring on the Ewald sphere, as shown in Fig. 3.33*a*. Now relnode planes of other heights w above $\mathbf{r}^*_{hkl}$ will give different values of $I^*(2\theta)$, at different 2θ positions. The intensity scattered in the spherical ring between 2θ and $2\theta + d(2\theta)$, relative to its intensity scattered over all solid angle is

$$I^*(2\theta)\frac{2\pi(\sin 2\theta)d(2\theta)}{4\pi} \tag{3.44}$$

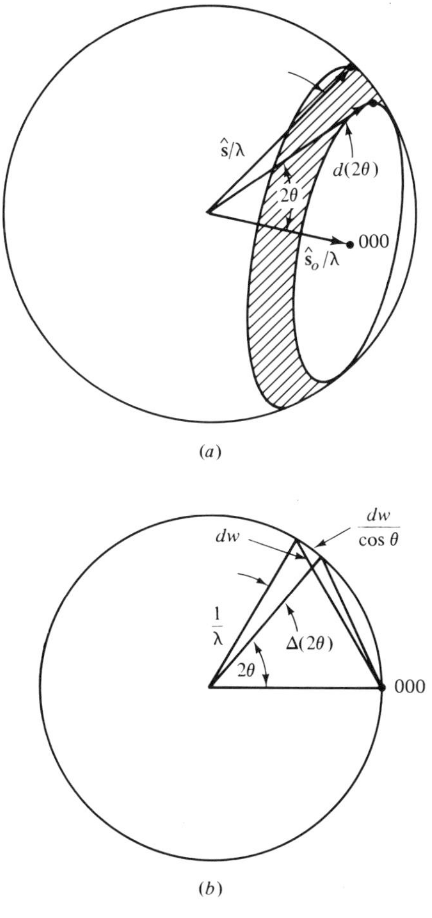

Figure 3.33. Geometry associated with ring area elements on the Ewald sphere. *(a)* shows the basic geometry of a ring of angular breadth $\Delta(2\theta)$. *(b)* shows the relationships leading to (3.45).

According to Fig. 3.33*b*

$$\frac{1}{\lambda}\,d(2\theta) = \frac{dw}{\cos\theta} \tag{3.45}$$

Thus the relative scattering, integrated over all 2θ associated with *hkl* is

$$\mathrm{I}_{hkl}^{\mathrm{int}} = \int_{\substack{\text{one}\\ \text{node}}} I^*(2\theta)\,\frac{\sin 2\theta}{2}\,\frac{\lambda dw}{\cos\theta} \tag{3.46}$$

This intensity manifests itself as a Debye ring. If a Debye-Scherrer camera of radius R were used, the relative intensity *per unit length of diffraction arc* would be

$$I_{\mathrm{INT}}^{hkl} = \frac{I_{hkl}^{\mathrm{int}}}{2\pi R \sin 2\theta} \tag{3.47}$$

Inserting (3.43) and (3.46) into (3.47), we have

$$\begin{aligned} I_{\mathrm{INT}}^{hkl} &= \frac{1}{2\pi R \sin 2\theta} \int_{\substack{\text{one}\\ \text{node}}} \frac{i(u,v,w)(\sin 2\theta)\lambda du\, dv\, dw}{16\pi \left(\dfrac{\sin\theta}{\lambda}\right)^2 2\cos\theta} \\ &= \frac{\lambda^3}{64\pi^2 R \sin 2\theta} \int \frac{i(u,v,w)\sin 2\theta}{\sin^2\theta\cos\theta}\, du\, dv\, dw \end{aligned} \tag{3.48}$$

We have previously seen, however, that the angular breadth associated with one relnode is very small. Thus the integration of (3.48) is sensibly constant in θ, and the trigonometric terms can be removed from the integral. Therefore,

$$I_{\mathrm{INT}}^{hkl} = \frac{1}{64\pi^2 R}\,\frac{\lambda^3}{\sin^2\theta\cos\theta} \int i(u,v,w) du\, dv\, dw \tag{3.49}$$

The integrated intensity of one relnode is, from (3.37) and (3.42)

$$\int_{\substack{\text{one}\\ \text{node}}} i(u,v,w) du\, dv\, dw = \frac{\mathrm{V}}{\mathrm{V}_c^2}\, I_e(s)|F_{hkl}|^2 \tag{3.50}$$

where V is the irradiated volume and V_c is the volume of one unit cell. As is shown in Appendix A, the Thompson scattering $I_e(s)$ can be written

$$I_e(s) = (r_e^2/R^2)\left(\frac{1+\cos^2 2\theta}{2}\right) I_o \tag{3.51}$$

where r_e is a universal constant. Inserting (3.50) and (3.51) in (3.49) we have

$$I_{\mathrm{INT}}^{hkl} = I_o\,\frac{r_e^2 \quad V \quad \lambda^3}{64\pi^2 R^3\; V_c^2 \sin\theta \sin 2\theta}\left(\frac{1+\cos^2 2\theta}{2}\right)|F_{hkl}|^2 \tag{3.52}$$

In (3.52) the term $\lambda^3/\sin\theta\sin(2\theta)$ is referred to as the *Lorentz factor.* The term $\left(\frac{1+\cos^2 2\theta}{2}\right)$ is the *polarization factor* (as we shall see).

One other intensity determiner also needs to be taken into account. This deals with the number of planes of type *hkl* available in the unit cell—i.e., how many different orientations of the crystal may give rise to an *hkl* plane at the appropriate orientation. Considering a cubic structure, there are six orientations which provide (100) at the favorable orientation, 12 for (110), eight for (111), and so on. Table 3.2 gives a full listing of such *multiplicity factors* for cubic materials.

TABLE 3.2
MULTIPLICITY FACTORS FOR THE POWER METHOD
(Cubic Symmetry)

hkl	*n*
*h*00	6
hhh	8
*hh*0	12
hhk	24
*hk*0	24
hkl	48

Finally using the multiplicity factor *n*, the integrated intensity per unit length of power pattern ring, $\mathscr{I}^{hkl}_{\mathrm{INT}}$, becomes

$$\mathscr{I}^{hkl}_{\mathrm{INT}} = I_o \frac{r_e^2}{64\pi^2 R^3} \frac{\mathrm{V}}{\mathrm{V}_c^2} \frac{\lambda^3}{\sin\theta\sin 2\theta}\left(\frac{1+\cos^2 2\theta}{2}\right) \mathrm{n}_{hkl}|F_{hkl}|^2 \qquad (3.53)$$

In general, we use intensity data to obtain relative $|F_{hkl}|$ values. Equation (3.53) gives us the geometric factors by which I^{hkl}_{INT} must be multiplied—$\frac{\sin\theta\sin 2\theta}{1+\cos^2 2\theta}$—to produce a relative $|F_{hkl}|$.

Let us test the predictive ability of (3.53) by computing the relative intensities of the *hkl* lines of an α-Fe powder, using CuK_α ($\lambda = 1.54$Å) x-rays. These intensities can then be compared to the experimental relative intensities found in the JCPDS file. We first note that, for a given material, the quantity $(I_o r_e^2 V\lambda^3)/(64\pi^2 R^3 V_c^2)$ is a constant, regardless of which *hkl* reflection we consider. Thus, the relative intensities are proportional to $[(1+\cos^2 2\theta)/(\sin\theta\cdot\sin 2\theta)]\, n_{hkl}|F_{hkl}|^2$. For BCC α-iron, $|F_{hkl}|^2 = 4(f^x_{\mathrm{Fe}})^2$. In order to find the $f_{\mathrm{Fe}}(\theta)$, we use the table of Appendix Table B.2. Table 3.3 shows, for all Fe reflections, the computed factors and the computed relative intensities. The last column shows the experimentally determined relative intensities. We see that the match

TABLE 3.3
COMPUTED RELATIVE INTENSITIES FOR AN IRON POWDER, USING CUK_α RADIATION

hkl	θ,°	$\frac{1+\cos^2 2\theta}{\sin\theta \sin 2\theta}$	n	f^2	$\left(\frac{I_{hkl}}{I_{110}}\right)_{calc}$	$\left(\frac{I_{hkl}}{I_{110}}\right)_{exp}$
110	22.30	5.66	12	343	1.00	1.00
200	32.41	2.44	6	236	0.15	0.19
211	41.03	1.56	24	175	0.28	0.30
220	49.29	1.36	12	136	0.10	0.09
310	57.94	1.56	24	114	0.18	0.12
222	68.18	2.38	8	94	0.08	0.06

is qualitatively good, but that the experimental values lie significantly below the computed values as higher scattering angles are approached. This latter effect is due to thermal vibrations within the crystals.

3.5.3 Integrated Reflecting Power in the Rotating Crystal Experiment

In the rotating crystal experiment, the specimen rotates at constant angular velocity ω. Here the hkl relnode moves at the same constant angular velocity through the Ewald sphere. In this case we need to integrate the intensity—more properly, the reflecting power—over time. Here we let $R(\theta)$, in units of energy/(area × time), be the reflecting power. Then we define the integrated reflecting power P by

$$P \equiv \int_{\theta_0-\epsilon}^{\theta_0+\epsilon} \frac{R(\theta)}{I_o}\, d\theta = \frac{\omega}{I_o}\int R(\theta)dt \tag{3.54}$$

Here $\int R(\theta)dt$ is the total flux collected as the specimen goes through the Bragg angle.

We need now to calculate P from microscopic principles. The geometry used is shown in Fig. 3.34. We choose a coordinate system within the relnode, such that $\mathbf{s}_1$ lies parallel to the rotation axis, $\mathbf{s}_2$ lies parallel to the Ewald sphere and normal to $\mathbf{s}_1$, and $\mathbf{s}_3$ lies normal to the Eward sphere. The volume element swept out during rotation of the relnode is $ds_1 ds_2 ds_3$. During the rotation time t, an angle $d\alpha$ is swept by the crystal

$$\omega dt = d\alpha \tag{3.55}$$

The angle $d\alpha$ is related to resp dimensions by

$$d\alpha = \frac{ds_3}{\left(\frac{2\sin\theta}{\lambda}\right)\cos\theta} = \frac{\lambda ds_3}{\sin 2\theta} \tag{3.56}$$

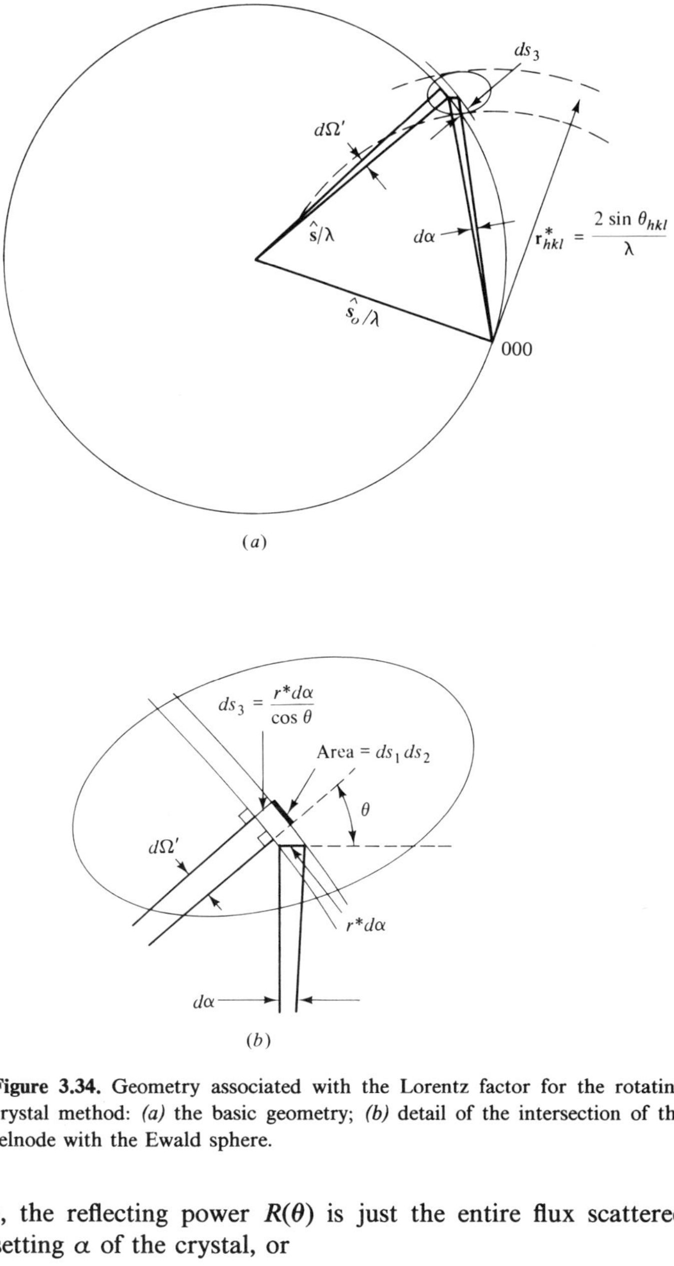

Figure 3.34. Geometry associated with the Lorentz factor for the rotating crystal method: *(a)* the basic geometry; *(b)* detail of the intersection of the relnode with the Ewald sphere.

In (3.54), the reflecting power $R(\theta)$ is just the entire flux scattered at each angular setting α of the crystal, or

$$R(\theta) = \int_{\substack{\text{one}\\ \text{node}}} i(s_1, s_2, s_3) d\Omega' = \int i(s_1, s_2, s_3) \frac{ds_1 ds_2}{\left(\frac{1}{\lambda}\right)^2} \tag{3.57}$$

Inserting (3.55) to (3.57) into (3.54), we have

$$P=\frac{\omega}{I_o}\int i(s_1,s_2,s_3)\frac{\lambda^3}{\omega\sin 2\theta}\,ds_1ds_2ds_3 \tag{3.58}$$

Again, the relnode is so small (usually) that $\sin 2\theta$ is effectively constant as the relnode moves through the Ewald sphere. Thus

$$P=\frac{1}{I_o}\frac{\lambda^3}{\sin 2\theta}\int_{\substack{\text{one}\\\text{node}}} i(s_1,s_2,s_3)ds_1ds_2ds_3=\frac{1}{I_o}\frac{\lambda^3}{\sin 2\theta}\frac{V}{V_c^2}I_e(s)|F_{hkl}|^2 \tag{3.59}$$

Inserting our expression for $I_e(s)$, we find

$$P=\frac{E_{hkl}\omega}{I_o}=\frac{r_e^2}{R_o^2}\frac{V}{V_c^2}\frac{\lambda^3}{2\sin 2\theta}m_{hkl}\left(\frac{1+\cos^2 2\theta}{2}\right)|F_{hkl}|^2 \tag{3.60}$$

In this expression m is a multiplicity factor. For a diffractometer or a Weissenberg camera experiment, in which each node is collected independently, m is unity. However for a simple film method, spots from different hkl may overlap. For instance, consider a cubic crystal rotating about (001) with the incident beam normal to that axis. In this case, (100) and (010) will satisfy Bragg's law under the identical diffracting conditions. Accounting for looking at these planes from both sides, $m_{100} = 4$. Clearly, the value of m will depend not only on the crystal system, but will also depend on the axis of rotation.

3.6 THE EFFECT OF TEMPERATURE

As we saw in the example computation of the intensity of scattering from an iron powder, (3.46) puts us in the right ballpark, but there is a systematic deviation between measured and computed values. More explicitly, the ratio of measured to computed intensity decreases as 2θ increases. The principal cause of this deviation is the vibration of the constituent atoms due to heat. What follows here is a simple treatment which shows that the effect of temperature T is to multiply the right side of (3.53) or (3.60) by a term $e^{-k^2s^2}$. This is the *Debye-Waller factor.*

Suppose that the atoms of the crystal are in thermal vibration in such a way that the deviation of each unit cell from its equilibrium site is independent of every other unit cell. This is the Einstein approximation. Now let the deviation of the nth unit cell be $\Delta\mathbf{x}_n$. That is, every atom in the nth cell is rigidly displaced by $\Delta\mathbf{x}_n$ from its equilibrium position $\mathbf{r}_j$. The structure factor F_n' of this cell is then

$$F_n'=\sum f_je^{2\pi i\mathbf{s}\cdot(\mathbf{r}_j+\Delta\mathbf{x}_n)}=\left[\sum f_je^{2\pi i\mathbf{s}\cdot\mathbf{r}_j}\right]e^{2\pi i\mathbf{s}\cdot\Delta\mathbf{x}_n}=F_{hkl}e^{2\pi i\mathbf{s}\cdot\Delta\mathbf{x}_n} \tag{3.61}$$

where F_{hkl} is the structure factor of the perfect crystal. Averaging over the positions of all the (independent) unit cells, we have

$$\overline{F'_{hkl}} = F_{hkl}\,\overline{e^{2\pi i\mathbf{s}\cdot\Delta\mathbf{x}_n}} \tag{3.62}$$

Recall that, in general

$$e^{ia} = 1 - ia + \frac{(ia)^2}{2} - \cdots \tag{3.63}$$

In our case

$$\overline{e^{2\pi i\mathbf{s}\cdot\Delta\mathbf{x}_n}} = 1 - 2\pi i\overline{\mathbf{s}\cdot\Delta\mathbf{x}_n} - 2\pi^2\overline{(\mathbf{s}\cdot\Delta\mathbf{x}_n)^2} \tag{3.64}$$

Since we have defined the equilibrium lattice site at $\Delta\mathbf{x}_n = 0$, the second term on the right of (3.64) is zero. Thus, eliminating that term and reexponentiating

$$\overline{e^{-2\pi i\mathbf{s}\cdot\Delta\mathbf{x}_n}} \simeq e^{-2\pi^2\overline{(\mathbf{s}\cdot\Delta\mathbf{x}_n)^2}} \tag{3.65}$$

Expanding the dot product and using l_1, l_2, and l_3 to denote direction cosines in the x_1, x_2, and x_3 directions, we have

$$\overline{e^{-2\pi i\mathbf{s}\cdot\Delta\mathbf{x}_n}} \simeq e^{-2\pi^2 s^2\overline{\Delta x^2_{n_1}}\,l_1^2}\,e^{-2\pi^2 s^2\overline{\Delta x^2_{n_2}}\,l_2^2}\,e^{-2\pi^2 s^2\overline{\Delta x^2_{n_3}}\,l_3^2} \tag{3.66}$$

Now using an isotropy condition

$$\overline{(\Delta x)^2} = \overline{\Delta x^2_{n_1}} + \overline{\Delta x^2_{n_2}} + \overline{\Delta x^2_{n_3}} = \overline{3\Delta x^2_{n_1}} = \overline{3\Delta x^2_{n_2}} = \overline{3\Delta x^2_{n_3}} \tag{3.67}$$

and the general property that $l_1^2 + l_2^2 + l_3^2 = 1$, we find

$$\overline{e^{-2\pi i\mathbf{s}\cdot\Delta\mathbf{x}_n}} \simeq e^{-(2/3)\pi^2 s^2\overline{(\Delta x)^2}} \tag{3.68}$$

Substituting (3.68) into our expression for the structure factor, (3.62), we have

$$\overline{F'_{hkl}} = F_{hkl}e^{(-2\pi^2 s^2\overline{(\Delta x)^2})/3} \tag{3.69}$$

In the intensity or power formulas, (3.53) for a powder pattern or (3.60) for a rotating crystal, we must now substitute $|\overline{F'_{hkl}}|^2$ for $|F_{hkl}|^2$. This operation is tantamount to introducing a new factor

$$D = e^{\overline{\frac{-4\pi^2 s^2(\Delta x)^2}{3}}} = e^{-2M} \tag{3.70}$$

to account for the temperature effect.

The effect of temperature clearly decreases as the scattering angle decreases, and vanishes as 2θ approaches zero. Further, we intuitively expect that $\overline{(\Delta x)^2}$ will increase as temperature increases and that the effect of temperature would then be increasingly great as the temperature is raised. We shall see in Chap. 6 that the Debye-Waller factor is related to fundamental lattice quantities through an analysis of $\overline{(\Delta x)^2}$. In fact, analysis of D can lead to information regarding the Debye temperature and the elastic constants of the material.

As we shall see in Chap. 6, the expression relating $\overline{(\Delta x)^2}$ to the absolute temperature T is

$$\overline{(\Delta x)^2} = \frac{9h^2}{4\pi^2 mk\,\Theta_D}\left[\frac{1}{4} + \frac{T}{\Theta_D}\,\phi\left(\frac{\Theta_D}{T}\right)\right] \tag{3.71}$$

where m is the atomic mass, k is Boltzmann's constant and

$$\phi\left(\frac{\Theta_D}{T}\right) = \int_0^{\Theta_D/T} \frac{y\,dy}{e^y - 1} \tag{3.72}$$

and Θ_D is the Debye temperature of the solid. Figure 3.35 is a plot of $\frac{1}{4} + \frac{T}{\Theta_D}\phi\left(\frac{\Theta_D}{T}\right)$ versus temperature. The effect of thermal vibrations can in some cases be quite large, reducing the intensity to a small fraction of its ideal crystal value. Table 3.4 shows Debye-Waller factors, e^{-2M}, for the (111) reflection from selected fcc metals. We observe that the harder the metal, the less important (i.e., closer to unity) the Debye-Waller factor.

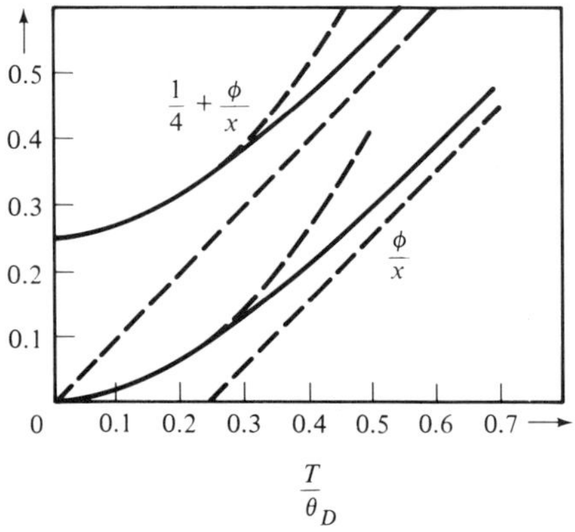

Figure 3.35. The Debye functions, ϕ/x and $\frac{1}{4} + (\phi/x)$, plotted against the reduced temperature, T/Θ_D. Here $x = \Theta_D/T$.

We must note again here that (3.67) has expressed a condition of isotropy for the crystal. For cubic metals, this is a good approximation. For molecular crystals, the approximation is not good at all. In such cases it is necessary to give each atom in the unit cell its own Δx_j value. In this general case the perfect crystal structure factor is replaced by

$$\overline{F'_{hkl}} = \sum_j f_j\, e^{-(4\pi^2 s^2 \overline{(\Delta x_j)^2_s})/3} e^{2\pi i(hx_j + ky_j + lz_j)} \tag{3.73}$$

TABLE 3.4
DEBYE-WALLER FACTORS FOR SELECTED FACE-CENTERED CUBIC (FCC) METALS

Element	Θ_DK	T/Θ_D(*at RT*)	$\sqrt{\overline{(\Delta x)^2}}$, Å	e^{-2M}_{111}
Pb	94.5	3.15	0.78	0.52
Au	165	1.81	0.46	0.71
Pd	275	1.08	0.37	0.79
Al	418	0.71	0.41	0.76
Ni	456	0.65	0.25	0.88

where $(\Delta x_j)_s$ is the component of $(\Delta x_j)^2$ normal to the *hkl* plane. Clearly this analysis becomes more complex, since the temperature correction is carried in the summation.

3.7 THE EFFECT OF ABSORPTION

In general, the absorption of an electromagnetic wave by a crystal may be related to its path length l in the crystal by Beer's law

$$\frac{I}{I_o} = e^{-\mu' l} \tag{3.74}$$

where I_o is the incident intensity, I is the intensity after passing through the medium and μ' is a *linear absorption coefficient*. Most often, in order to obviate tabulation for different states of aggregation of the same element, the mass absorption coefficient $\mu = \mu'/\rho$ is tabulated (Appendix Table B.5). In this case

$$\frac{I}{I_o} = e^{-\mu \rho l} \tag{3.75}$$

Mass absorption coefficients of compounds or alloys are related to the mass absorption coefficients μ_j and weight fractions w_j of the components through

$$\mu = \Sigma w_j \mu_j \tag{3.76}$$

The effect of absorption depends on the geometry of the experiment. Three common experimental geometries are shown in Fig. 3.36. Since it is easily han-

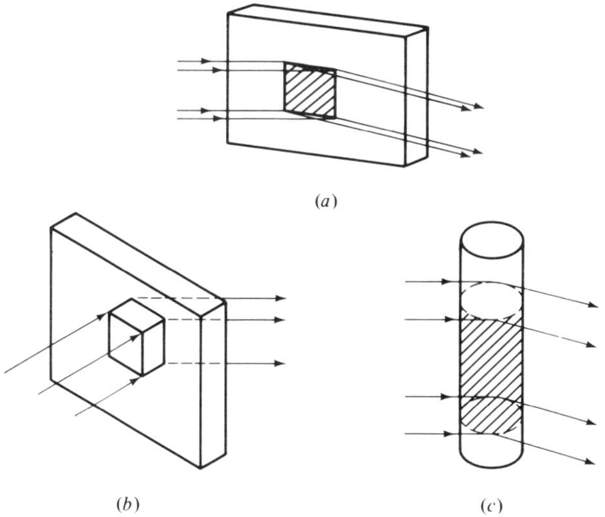

Figure 3.36. Common experimental geometries: *(a)* back reflection from a slab; *(b)* transmission through a plate or film; *(c)* transmission or back reflection for a cylinder.

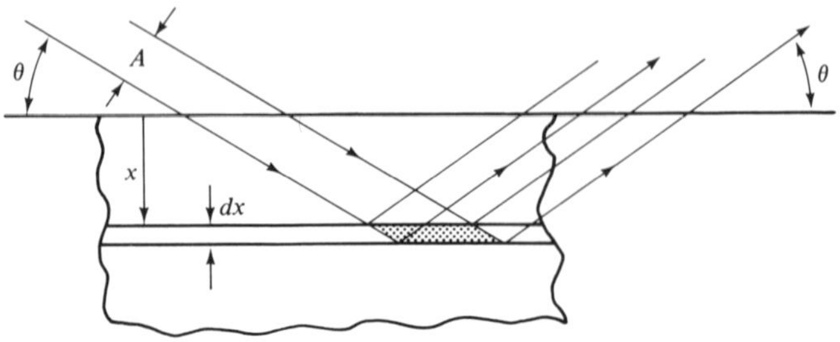

Figure 3.37. Symmetrical back reflection from a flat slab. The shaded element of irradiated volume of thickness dx lies a distance x below the entrant surface. The cross-sectional area of the incident beam is A.

dled, let us look at the effect of absorption on the symmetrical reflection from a slab. The geometry of our problem is set in Fig. 3.37.

For either a powder pattern or a rotating crystal experiment, the intensity of the hkl peak may be given by

$$I_{hkl} = Q_{hkl}\ V \tag{3.77}$$

where V is the irradiated volume. Better, let us look at each element δV of volume and the intensity δI_{hkl} contributed by it. Thus

$$\delta I_{hkl} = Q_{hkl}\ \delta V \tag{3.78}$$

In Fig. 3.37 the element of volume located a distance x below the surface is given by

$$\delta V = \frac{A}{\sin\theta}\, dx \tag{3.79}$$

where A is the cross-sectional area of the incident beam. The absorption path for the wave reaching this volume element and then diffracted through the crystal face again is $2x/\sin\theta$. Thus, integrating over all volume elements in the beam path, for an infinitely thick crystal

$$I_{hkl} = \int \delta I_{hkl} = Q_{hkl} A \int_0^{\infty} e^{-2\mu\rho x/\sin\theta}\, \frac{dx}{\sin\theta} = \frac{Q_{hkl} A}{2\mu\rho} \tag{3.80}$$

Thus in (3.53) or (3.60) we replace V by $A/(2\mu\rho)$. In this particular geometry there is no dependence of the effect of absorption on the angle of diffraction 2θ. Clearly for other geometries there must be a 2θ dependence.

For symmetrical transmission through a thin film of thickness t for which the angle of incidence is α, the volume V in (3.53) or (3.60) is replaced by

$$\frac{A}{\mu\rho\left[1 - \dfrac{\cos\alpha}{\cos(2\theta-\alpha)}\right]}\left[\exp\left(-\frac{\mu\rho t}{\cos(2\theta-\alpha)}\right) - \exp\left(\frac{-\mu\rho t}{\cos\alpha}\right)\right] \tag{3.81}$$

The geometry of a cylindrical specimen has also been solved. Tables of absorption corrections for this case are given in the *International Tables for Crystallography.*

PROBLEMS

3.1. Find the reciprocal lattice vector **a*** for cubic and for hexagonal crystals, both having **a** = 3.0 Å.

3.2. What is the angle between (110) and $(3\bar{1}2)$ in the cubic system?

3.3. What is the angle between $(3\bar{2}1)$ and $(\bar{1}11)$ for a tetragonal crystal with $c/a = 1.5$?

3.4. Derive the relationship between d_{hkl} and the crystal axes for the monoclinic system.

3.5. Figure P3.5 is a rotating crystal pattern of an orthorhombic crystal. The rotation is about (100). Camera radius = 3 cm; λ = 1.54 Å. A few diffraction spots are indexed.
(a) Find lattice spacings *a* and *b*.
(b) Find the indices of the circled spot. (Lattice spacing *c* is 8.33 Å).

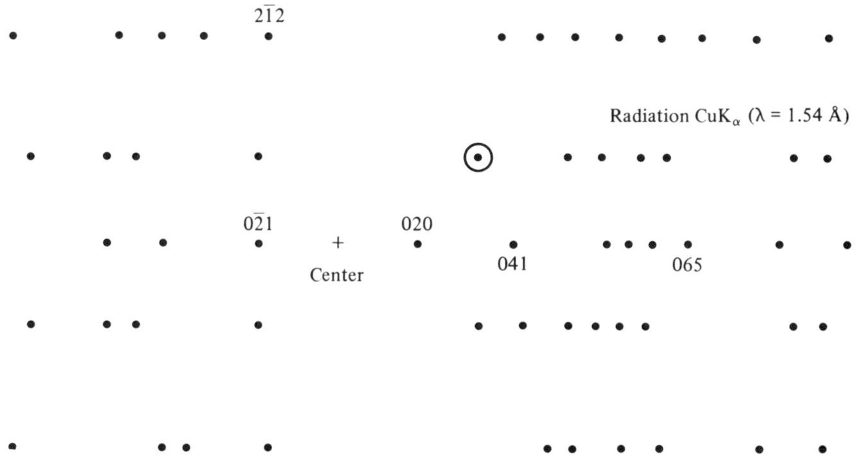

Figure P3.5.

3.6. A single crystal of copper (face-centered cubic; $a = 3.61$ Å) is set up for a rotating crystal experiment. In the experiment the specimen rotates about [001] and the FeK$_\alpha$ radiation (λ = 1.94 Å) is incident normal to that axis. Using the Ewald construction, list all the reflections which will appear on the film.

3.7. Figure P3.7 shows a rotating crystal pattern from a supposed single crystal of polyoxymethylene. The camera radius is 3.0 cm and CuK$_\alpha$ radiation (λ = 1.54 Å) was used.
(a) Prove by layer line analysis that this pattern cannot be from a single crystal.
(b) Find the *d* spacings of the points marked *A* and *B*.

Figure P3.7.

(c) How would you go about proving whether this pattern is from two different crystal types or from two different orientations of the same crystal type?

3.8. In general, if a crystalline polymer is drawn out into a fine thread, the chain axes will become aligned with the tensile axis. The structure of polymer chains and polymer crystals is generally taken from studies of such drawn fibers.

Demonstrate that a Debye-Scherrer x-ray pattern from such a fiber is identical to a rotating crystal pattern of a single crystal. (In this experiment the film and specimen are located just as in a Debye-Scherrer camera, but the film is much wider, so that nearly entire diffraction cones are recorded). For purposes of your discussion let the c-axis of the unit cell be parallel to the polymer chain axis.

3.9. The output of a certain x-ray tube is adjusted so that the effective limits of the white radiation emitted are $\lambda_{min} = 1.00$ Å and $_{max} = 3.00$ Å. A single crystal of FCC structure and $a = 2.50$ Å is irradiated along [010] by this white radiation. List the lattice planes which can potentially contribute to a back reflection Laue photograph in this case.

3.10. Why is it not possible to use visible light to study crystals by diffraction?

3.11. X-ray diffraction is often used to measure elastic strains in crystalline solids. In our case, we have a polycrystalline molybdenum slab (with random orientation) held under tension as shown in Fig. P3.11. The tensile stress σ is 50,000 psi. Below are listed some of the physical parameters for molybdenum.

Young's modulus:	4.74×10^7 psi
Crystal structure:	body-centered cubic (BCC)
Lattice parameter:	$a = 3.15$ Å
Poisson's ratio:	$\nu = 0.292$
Yield stress:	70,000 psi

X-ray diffraction data are taken, using CuK_α radiation ($\lambda = 1.54$ Å), according to AA. The plane of AA contains the tensile axis.

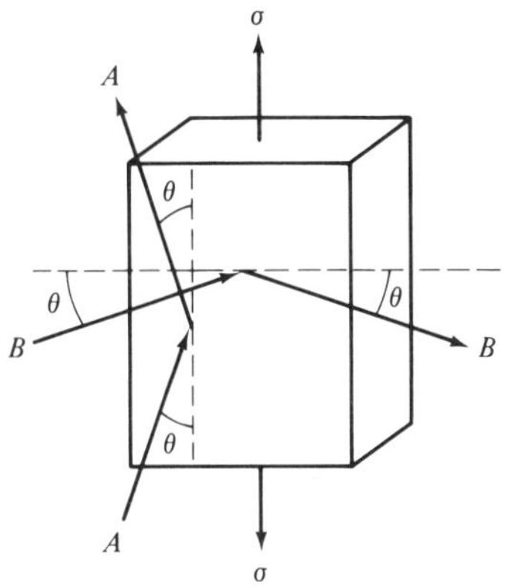

Figure P3.11.

(a) At what angular position does the (200) peak occur under no applied load?
(b) At what angular position does (200) occur during application of the applied stress?
(c) If data were taken according to *BB*, would one find the same Bragg angle as in (b), for the specimen under stress? If not, what would be the Bragg angle? The plane of *BB* contains the normal to the tensile axis.
(d) Show that the relative sensitivity of measurement of the change $\Delta(2\theta)$ in the Bragg angle, for a given variation $\Delta a/a$ in the lattice spacing, varies with Bragg angle according to

$$\Delta(2\theta) = -(\tan\theta)\frac{\Delta a}{a}$$

(e) According to (d), exactly which reflection would, for molybdenum, provide the most sensitive measurement of $\Delta a/a$?

3.12. Write out and simplify to a concise set of rules the structure factors for:
(a) face-centered orthorhombic crystals
(b) the NaCl structure
(c) diamond cubic crystals

3.13. Given a set of intensity data, how could you distinguish NaCl from KCl, which also has the NaCl arrangement of atoms?

3.14. The structure of ordered Cu_3Au is as depicted in Fig. P3.14.
(a) Compute the structure factor for Cu_3Au and determine the extinction rules.

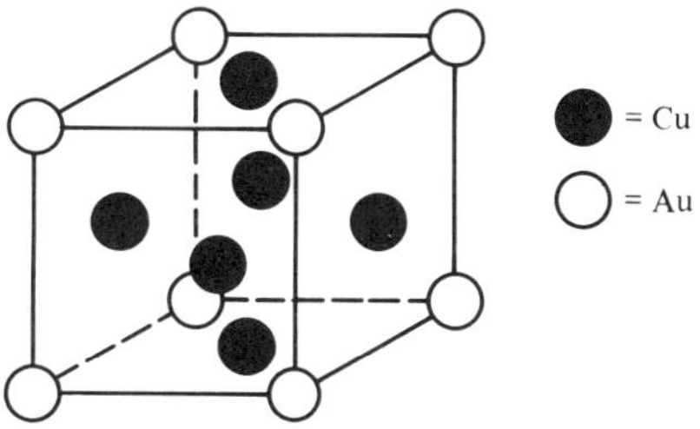

Figure P3.14.

(b) The lattice parameter of this material is $a = 3.73$ Å. Compute the positions of the first three powder pattern peaks, using CuK_α (1.54 Å) radiation.

3.15. β-brass (50% Cu, 50% Zn) can exist at room temperature in either an ordered (equilibrium) or a disordered (nonequilibrium) form. The cubic unit cell has atoms at (0, 0, 0) and at (½, ½, ½). The lattice constant, a, is 2.94 Å.

(a) List the three lowest angle reflections for both forms. Find the Bragg angles for these three reflections. Use CuK_α radiation: $\lambda = 1.54$ Å.

(b) For the ordered alloy, some of the reflections have intensities which are proportional to $(f_{Zn} - f_{Cu})^2$ and some which are proportional to $(f_{Zn} + f_{Cu})^2$, where the f's are the atomic scattering factors. Which planes are associated with which intensity formulas? Can you find, by induction, a general rule relating hkl to $(f_{Cu} + f_{Zn})$ and $(f_{Cu} - f_{Zn})$?

3.16. In drawing or discussing the rel, one includes only those relnodes which have nonzero intensity value. Using the structure factors and a sketch of resp, show that the rel for FCC is BCC and that the rel for BCC is FCC.

3.17. Sketch the $hk0$ and hkl reciprocal lattice planes for diamond and for zincblende (ZnS). The unit cells are shown in Fig. P3.17.

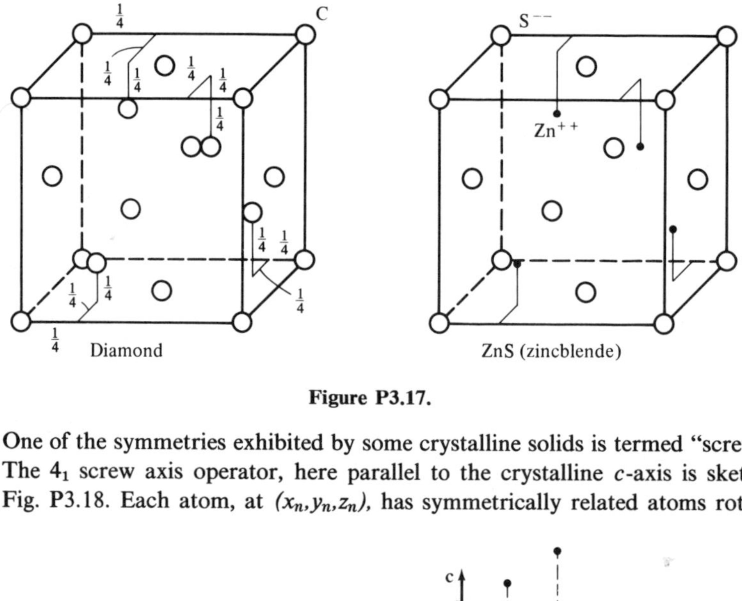

Figure P3.17.

3.18. One of the symmetries exhibited by some crystalline solids is termed "screw axis." The 4_1 screw axis operator, here parallel to the crystalline c-axis is sketched in Fig. P3.18. Each atom, at (x_n, y_n, z_n), has symmetrically related atoms rotated se-

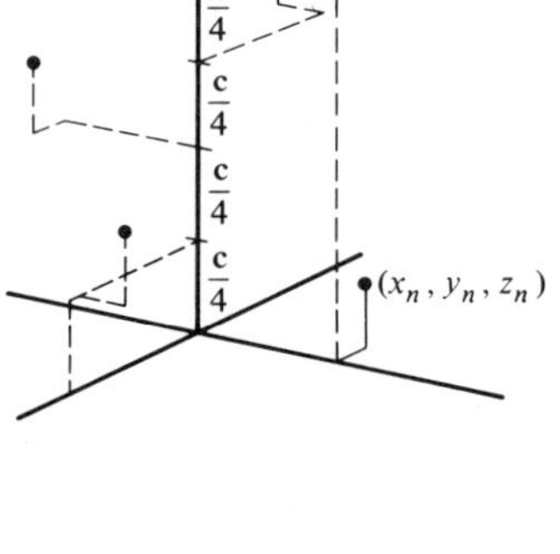

Figure P3.18.

quentially 90° about **c** and elevated in steps of ¼ **c**. Show, by examination of the structure factor, how you determine whether or not a 4_1 screw axis is present in a crystal. That is, what are the systematic absences for a 4_1 screw axis. In your work remember that all atoms, of all elements present, have screw-related partners.

3.19. Figure P3.19 shows the unit cell of CaF_2.

(a) Derive the structure factor and show that it is equivalent to

- $4f_{Ca} + 8f_F$ if h, k, and l are all even and have values such that either
  - (a) two of the three h, k, l are 2,6,10,14,18, . . . and the other is 0,4,8,12,16, . . .
  - (b) one of the three h, k, l are 2,6,10,14,18, . . . and the others are 0,4,8,12, 16 . . .
- $4f_{Ca} + 8f_F$ if h, k, and l are all even and have values such that all three are either 2,6,10,14, . . . or 0,4,8,12, . . .
- $4f_{Ca}$ if h, k, and l are all odd
- 0 if h, k, and l are mixed

(b) The lattice spacing of CaF_2 is 5.4646 Å. What is the Bragg angle of its second powder pattern ring? What is the Bragg angle of the first equatorial (zeroth layer line) spot of a rotating crystal pattern of CaF_2?

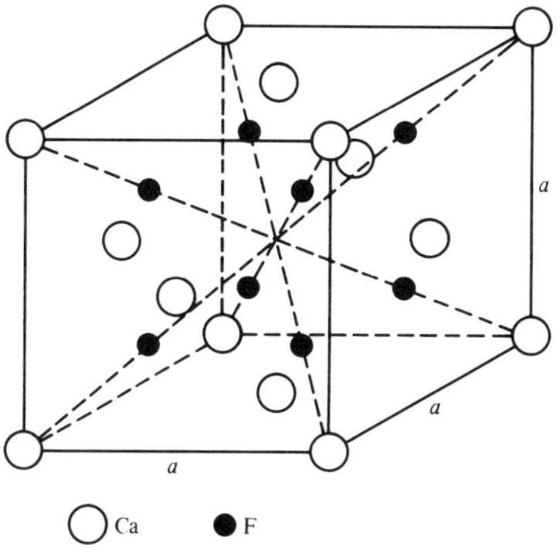

Figure P3.19.

3.20. (a) What would be the multiplicity factor for powder diffraction from (211) for cubic, tetragonal, and orthorhombic systems? (b) What would be the multiplicity factors for those three systems for a rotating crystal experiment with [001] parallel to the rotation axis?

3.21. A small spherical FCC crystal has been mounted in a rotating crystal camera in two different ways: (a) with [001] vertical and (b) with [111] vertical. Should the relative intensity of [111] spots for case (a) relative to case (b) be greater, the same, or less? Why?

3.22. In a Debye-Scherrer pattern of molybdenum (BCC, $a = 3.15$ Å), will the (111) or the (220) reflection appear more intense if CuK_α radiation ($\lambda = 1.54$ Å) is used? Do not consider the temperature factor.

3.23. Without worrying about the temperature correction, compute the relative intensities of the first three powder pattern peaks for a random polycrystal of lead (FCC, $a = 4.95$ Å) using CuK_α radiation ($\lambda = 1.54$ Å) and a symmetrical reflective geometry.

3.24. Repeat problem 3.23, but now using 100-kV electrons. (Note that electrons have no polarization factor). Are the ratios in problems 3.23 and 3.24 much different?

3.25. Which reflection, (111) or (200), will have greater intensity in a FCC copper ($a = 3.61$ Å) powder pattern taken using CuK_α x-rays ($\lambda = 1.54$ Å)?

3.26. It is found that the intensities of the (211) and (220) powder pattern lines in sodium (BCC, $a = 4.30$ Å) at room temperature are in the ratio 20:5 when CuK_α radiation is used ($\lambda = 1.54$ Å). Which contributes most greatly to the deviation of this ratio from unity: (a) the structure factor, (b) the Lorentz polarization factor, or (c) the temperature factor? Do not forget to take multiplicity into account in your answer.

3.27. Compute the relative intensities for the first five diffraction lines of a powder (Debye-Scherrer) pattern from BCC iron ($a = 287$ Å). Use FeK_α radiation (1.94 Å). Use the following assumptions:

(a) No absorption, no temperature correction

(b) No absorption, but temperature corrected for 298 K. The Debye temperature for iron is 467 K.

(c) As in (b), but also corrected for absorption by the cylindrically shaped powder. The absorption correction is $A = a(\mu R)^{-1} + b(\mu R)^{-2} + \cdots$, where μ is the linear absorption coefficient $\rho(\mu/\rho)$ and R is the specimen radius. In our case $R = 0.02$ cm. A tabulation of $a, b, c, \ldots$ is given on p. 93.

3.28. (a) Using the graph of $\left[\frac{1}{4} + \frac{1}{X}\phi(X)\right]$ versus X, where $X = \Theta_D/T$, determine the room temperature value of the Debye-Waller factor at $\sin\theta = 1$ (near the last observable reflection) for the following crystals. Use the values of Θ_D given.

Crystal	Θ_DK
Pb	88
Ag	215
W	280
Cu	315
Al	398
Fe	453
Diamond	1860

(b) Using the same graph, recompute the intensities of the *hkl* lines for an iron powder, using CuK_α ($\lambda = 1.54$ Å) radiation. Is the fit to measurement significantly better now than in column 6 of Table 3.3?

$$A = (A^*)^{-1} = a(\mu R)^{-1} + b(\mu R)^{-2} + \cdots$$

θ \ i	a	b	c	d	e	f	g
0	0	0	0.3183	0	0.955	0	7.1
5	0.001615	0.0314	0.0648	0.835	0.14		
10	0.00642	0.053	0.179	0.057			
15	0.01442	0.086	0.061	0.33			
20	0.0254	0.105	0.080	0.13			
25	0.0394	0.123	0.073	−0.16			
30	0.0560	0.139	0				
35	0.0752	0.150	−0.022				
40	0.0966	0.159	−0.10				
45	0.1199	0.164	−0.16				
50	0.1448	0.16	−0.20				
55	0.1741	0.13	−0.30				
60	0.1984	0.11	−0.21				
65	0.226	0.10	−0.33				
70	0.250	0.105	−0.48				
75	0.274	0.08	−0.52				
80	0.295	0.064	−0.64				
85	0.3107	0.03	−0.5				
90	0.3183	0	−0.53				

3.29. According to the ASTM card file, the (111) and (222) reflections from Pb powder at room temperature have an intensity ratio of 100 : 9, using CuK$_\alpha$ radiation (λ = 1.54 Å). What is the rms amplitude of vibration of the Pb atoms at room temperature? Some useful data are given below. Lattice parameter of Pb at RT = 4.9502 Å

$10^{-8} \times \frac{(\sin\theta)}{\lambda}$	0.0	0.1	0.2	0.3	0.4	0.5	0.6	0.7	0.8	0.9	1.0	1.1	1.2
Tl	81	75.5	66.7	58.7	51.2	45.0	41.1	37.4	34.1	31.1	28.3	26.0	24.1
Pb	82	76.5	67.5	59.5	51.9	45.7	41.6	37.9	34.6	31.5	28.8	26.4	24.5
Bi	83	77.5	68.4	60.4	52.7	46.4	42.2	38.5	35.1	32.0	29.2	26.8	24.8

3.30. Compute the relative intensities of the first two lowest angle powder diffraction lines of CsCl (a = 4.11 Å). A diffractometer with flat specimen is used, so no absorption correction is necessary. Take CuK$_\alpha$ (λ = 1.54 Å) radiation. Ignore the temperature correction. Data for the atomic scattering factors Cs^+ and Cl^- are shown on p. 94.

$(\sin\theta)/\lambda(A^{-1})$		0.0	0.1	0.2	0.3	0.4	0.5	0.6	0.7
Cl	17	17.0	15.33	12.00	9.44	8.07	7.29	6.64	5.96
Cl^-	17	18.0	16.02	12.20	9.40	8.03	7.28	6.64	5.97
Xe	54	54.0	50.3	43.7	37.9	33.1	29.2	25.9	23.2
Cs	55	55.0	51.3	44.5	38.7	33.8	29.8	26.5	23.7

$(\sin\theta)/\lambda(A^{-1})$		0.8	0.9	1.0	1.1	1.2	1.3	1.4	1.5
Cl	17	5.27	4.60	4.00	3.47	3.02	2.65		
Cl^-	17	5.27	4.61	4.00	3.47	3.03	2.65	2.35	2.11
Xe	54	20.9	18.9	17.2	15.8	14.5	13.4	12.5	11.6
Cs	55	21.3	19.4	17.7	16.2	14.9	13.8	12.8	11.9

Use xenon data for Cs^+.

3.31. The table shown is a listing of all d values obtained in a Debye-Scherrer pattern of a certain cubic material. Determine the lattice parameter a and the lattice type (simple, body-centered, or face-centered).

d Å
3.751
2.294
1.956
1.622
1.489
1.325
1.249
1.1470
1.0968
1.0260
0.9895
0.9365
0.9087
0.8671
0.8450

3.32. A powder pattern from a cubic material exhibits the following lines upon CuK_α radiation ($\lambda = 1.54$ Å):

2θ, degrees
38.2
43.3
44.4
50.4

64.7
74.1
77.5
81.7
89.9
95.1
98.1
110.8
115.2
116.9
135.4
136.5
144.7

(a) Index the lines.
(b) Find the lattice parameter *a*. You need not use an extrapolative method.

3.33. Copper and gold exhibit solid solubility at all compositions. Both are FCC. The lattice constant, a_{Cu}, of copper is 3.615 Å; the lattice constant, a_{Au}, of gold is 4.079 Å.

(a) The powder pattern of a Cu-Au alloy of unknown composition is obtained, using CuK_α ($\lambda = 1.54$ Å). The pattern shows lines at the following Bragg angles θ

20.86°
24.28°
35.56°
42.99°
45.42°
55.33°

What is the approximate composition of the alloy? Assume random distribution of Cu and Au on the *fcc* lattice.
(b) Ordered 50% Cu/50% Au alloys exhibit the CuAuI structure (100 layers alternately all Cu or all Au). What is the Bravais lattice for this material?
(c) Derive, and simplify, the structure factor formula for the ordered 0.50 Cu/0.50 Au crystal.

BIBLIOGRAPHY

A. General

1. L. V. AZAROFF, *Elements of X-ray Crystallography,* Chs. 7, 9, 10, 13–20, McGraw-Hill Book Company, N.Y. (1968).

2. L. V. AZAROFF, R. KAPLOW, N. KATO, R. J. WEISS, A. J. C. WILSON, and R. A. YOUNG, *X-ray Diffraction,* Chs. 2, 6, and 7, McGraw-Hill Book Company, N.Y. (1974).

3. J. M. COWLEY, *Diffraction Physics,* Ch. 6, North-Holland, Amsterdam (1975).

4. B. D. CULLITY, *Elements of X-ray Diffraction, 2nd Ed.,* Addison-Wesley Publishing Co., Inc., Reading, MA (1976).

5. A. GUINIER, *X-ray Crystallographic Technology,* Chs. 2–4, Hilger and Watts, London (1952).

6. A. GUINIER, *X-ray Diffraction in Crystals, Imperfect Crystals, and Amorphous Bodies,* Ch. 4, W. H. Freeman & Company Publishers, San Francisco (1963).

7. N. F. M. HENRY, H. LIPSON, and W. A. WOOSTER, *The Interpretation of X-ray Diffraction Photographs,* St. Martin's Press, Inc., N.Y. (1960).

8. R. W. JAMES, *The Optical Principles of the Diffraction of X-rays,* Chs. 1 and 2, G. Bell & Sons, Ltd., London (1954).

9. M. KAKUDO and N. KASAI, *X-ray Diffraction in Polymers,* Elsevier North-Holland, Inc., N.Y. (1972).

10. J. S. KASPER and KATHLEEN LONSDALE, eds., *International Table for X-ray Crystallography, 2nd Ed., Vol. II,* The Kynoch Press, Birmingham, England (1965).

11. KATHLEEN LONSDALE, *Crystals and X-rays,* Ch. 3, Van Nostrand Reinhold Company, N.Y. (1949).

12. D. MCKIE and CHRISTINE MCKIE, *Crystalline Solids,* Chs. 6–9, Wiley-Halsted, N.Y. (1974).

13. L. H. SCHWARTZ and J. B. COHEN, *Diffraction from Materials,* Chs. 3 and 5, Academic Press, Inc., N.Y. (1977).

14. B. E. WARREN, *X-ray Diffraction,* Chs. 3–7, Addison-Wesley Publishing Co., Inc., Reading, MA (1969).

15. A. J. C. WILSON, *Elements of X-ray Crystallography,* Chs. 2–8, Addison-Wesley Publishing Co., Inc., Reading, MA (1970).

B. Specific Methods

1. J. L. AMOROS and M. L. CANUT DE AMOROS, *La Difraccion Difusa de los Cristales Moleculares,* C.S.I.C., Madrid (1965).

2. L. V. AZAROFF, *The Powder Method in X-ray Crystallography,* McGraw-Hill Book Company, N.Y. (1958).

3. M. J. BUERGER, *The Precession Method in X-ray Crystallography,* John Wesley & Sons, Inc., N.Y. (1964).

4. R. W. M. D'EYE and E. WAIT, *X-ray Powder Photography in Inorganic Chemistry,* Butterworths, London (1960).

5. J. R. WORMALD, *Diffraction Method,* Clarendon, Oxford (1973).

6. A. GUINIER, *X-ray Diffraction in Crystals, Imperfect Crystals, and Amorphous Bodies,* W. H. Freeman & Company Publishers, San Francisco (1963).

7. A. J. C. WILSON, *Mathematical Theory of X-ray Powder Diffractometry,* Centrex, Eindhoven (1963).

C. Electron and Neutron Diffraction

1. J. W. EDINGTON, *Electron Diffraction in the Electron Microscope,* Macmillan Inc., London (1975).

2. P. A. EGELSTAFF, *Thermal Neutron Scattering,* Chs. 4 and 8, Academic Press, Inc., N.Y. (1965).

3. P. B. HIRSCH, A. HOWIE, R. B. NICHOLSON, D. W. PASHLEY, and M. J. WHELAN, *Electron Microscopy of Thin Crystals,* Chs. 4 and 5, Plenum Publishing Corporation, N.Y. (1967).

4. M. A. KRIVOGLAZ, *Theory of X-ray and Thermal-Neutron Scattering by Crystals,* Plenum Publishing Corporation, N.Y. (1969).

5. T. RYMER, *Electron Diffraction,* Ch. 4, Methuen, London (1970).

X-ray Scattering Physics and Foundations of the Dynamical Theory

4

4.1 THE SCATTERING OF ELECTROMAGNETIC RADIATION BY ONE ELECTRON

In this section we are interested in viewing the origin of the Thomson scattering term—the intensity of scattering of x-rays by one electron

$$I_e(2\theta) = \frac{r_e^2}{R_0^2} I_0 \left(\frac{1 + \cos^2 2\theta}{2}\right) \tag{4.1}$$

To understand what happens, let us first develop a simplified physical model. Consider, as in Fig. 4.1, an electron acted upon by a wave traveling in the

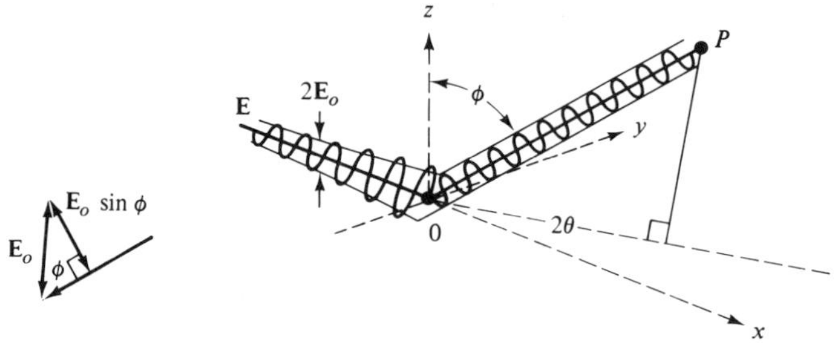

Figure 4.1. Propagation of incident and scattered rays. An incident ray, polarized in the *xz*-plane, propagates in the *x*-direction. A scattered ray originates at point *O* and is detected at the position *P*. The angles ϕ and θ define the direction of the scattered ray. The inset indicates the vector component of the field strength for a wave scattered at azimuthal angle ϕ.

x-direction. If the incident wave is polarized in the xz plane, then the electron responds to the periodic field $\mathbf{E} = \mathbf{E}_0 \cos \omega t$ by oscillating back and forth along z. As we know from elementary physics, the effect of accelerating an electric charge (here the electron) is to produce a magnetic field. The electron has also associated with it a magnetic field (due to its spin); hence an electric field is also created. The period of this newly created electromagnetic field clearly must replicate that of the incident beam. Further, it must be polarized in the xz plane, just as was the incident wave. This new wave is termed the "scattered wave." Let us allow it, for now, to have a maximum amplitude $\beta\mathbf{E}_0$ in the forward $(+x)$ direction. Thus, in the forward direction the electric field strength is $\beta\mathbf{E}_0 \cos \omega t$. In all scattering directions in the xy plane, the electric vector $\beta\mathbf{E}$ lies normal to the direction of propagation. Hence, anywhere in xy, the maximum field strength is βE_0. For directions of propagation lying out of the xy plane—as for propagation toward P in Fig. 4.1—the maximum strength will be $\beta E_0 \sin \phi$.

Consider now the case of an unpolarized incident beam propagating again along x. This wave can be broken down into E-field equal components lying in xz and in yz. It is, however, more convenient to choose new axes, x, y', and z', as in Fig. 4.2, such that one of the components of $\mathbf{E}$ lies perpendicular to OP; i.e., OP now lies in the xz' plane. In this case the xz'-plane component of $\mathbf{E}$, call it $E_{\parallel}$ yields a maximum field strength $\beta E_0 \sin \phi$. Since P now lies in the xz' plane, the component of scattered field due to the xy' plane component of $\mathbf{E}$ (call it $E_{\perp}$) must be βE_0. Supposing that we know the value of the constant of proportionality β, we are now in position to compute the intensity of the scattered wave.

The relation between the electric intensity of the scattered ray and its electric field strength $\mathbf{E}^*$ is shown as follows. Suppose that there is some region of

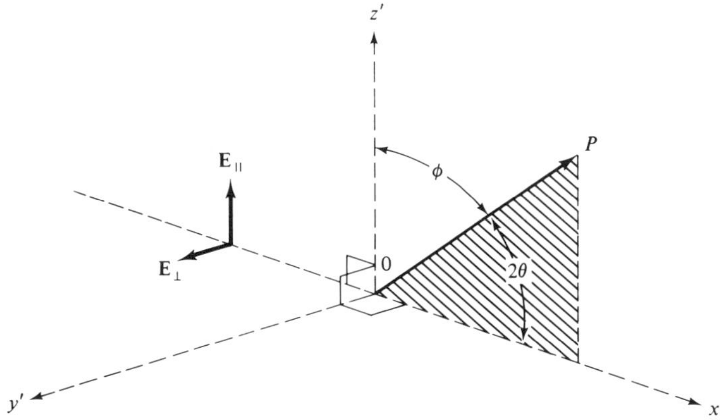

Figure 4.2. New coordinate system, with OP now in the xz-plane. The new coordinates are formed by rotation of y and z about x. Note also the components $\mathbf{E}_{\parallel}$ and $\mathbf{E}_{\perp}$ of the electric vector $\mathbf{E}$.

space where the electric field **E*** causes a current flow of density **j**. The total electric power dissipated will then be given by

$$P_E = \int \mathbf{E}^* \cdot \mathbf{j}\, dv \tag{4.2}$$

where the integral is taken over the volume V of the region of space. We now make use of one of Maxwell's equations

$$\nabla \times \mathbf{H}^* = \mathbf{j} + \frac{\partial \mathbf{D}^*}{\partial t} = \mathbf{j} + \epsilon_0 \epsilon \left(\frac{\partial \mathbf{E}^*}{\partial t} \right) \tag{4.3}$$

where **D*** is the electric diplacement, ϵ_0 is a constant, and ϵ is the dielectric constant of the medium. Substituting (4.3) into (4.2), we have

$$P_E = \int \mathbf{E}^* \cdot (\nabla \times \mathbf{H}^*) dv - \int \mathbf{E}^* \cdot \left(\frac{\partial \mathbf{D}^*}{\partial t} \right) dv \tag{4.4}$$

where **H*** is the magnetic field associated with the scattered wave. It can be shown that the first term on the right of (4.4) represents the net rate at which energy flows across the boundary between our region of space and the remainder of the universe; as such it drops out in a steady state system. Thus the rate of electric energy flow within the wave is given by

$$P_E = \frac{1}{dt} \int E^* \cdot D^*\, dv \tag{4.5}$$

In this formulation, we see that the energy density within the wave must be **E*** · **D***. For a wave propagating in vacuum, **D*** = **E***, and the energy flux is just $|\mathbf{E}^*|^2$. An intensity is an energy crossing unit area in unit time; hence, if $\mathscr{E}$ is an electric field strength per unit area (i.e., $\mathscr{E} = \mathbf{E}/\text{area}$), then the electric energy flux W, the intensity I, and $\mathscr{E}^*$ are related through

$$W = \int I dt = |\mathscr{E}^*|^2 \tag{4.6}$$

In the laboratory, the electric flux W is measured; the average intensity is derived from it by dividing total flux by time of measurement.

For the case of scattering by one electron, we have two components to the energy flux W, call these $W_\parallel$ and $W_\perp$, due, respectively, to incident wave electric components in the xz and yz planes. Thus

$$\begin{aligned} W_\parallel &= \beta^2 \mathscr{E}_0^2 \sin^2 \phi \\ W_\perp &= \beta^2 \mathscr{E}_0^2 \end{aligned} \tag{4.7}$$

Since the energy of the incident beam is taken to be equally partitioned over all polarization directions, the total energy flux of the scattered wave must be

$$W = \tfrac{1}{2} W_\perp + \tfrac{1}{2} W_\parallel = \beta^2 \mathscr{E}_0^2 \left(\frac{1 + \sin^2 \phi}{2} \right) \tag{4.8}$$

according to Fig. 4.2, $\sin\phi = \cos 2\theta$. Hence

$$W = \beta^2 \mathscr{E}_0^2 \left(\frac{1 + \cos^2 2\theta}{2}\right) \tag{4.9}$$

The intensity of the scattered wave is then

$$I_e(2\theta) = \beta^2 \left(\frac{1 + \cos^2 2\theta}{2}\right) \frac{\partial \mathscr{E}_0^2}{\partial t} = I_0 \beta^2 \left(\frac{1 + \cos^2 2\theta}{2}\right) \tag{4.10}$$

where $I_0 = \partial \mathscr{E}_0^2/\partial t$ is the intensity of the incident wave.[1]

We lack now only an expression for β in order to evaluate the Thomson scattering term. This conversion efficiency is derived by a straightforward electrodynamic analysis. The analysis, found as Appendix A, is, however, quite tedious, and no insights are gained which may be useful to our further work in this text. The results of the analysis yield, in rationalized mks. units

$$\beta^2 = \frac{\mu\mu_0 e^2}{4\pi m R_0^2} \tag{4.11}$$

where μ is the magnetic permeability of the medium in which the wave propagates, μ_0 is a constant, e is the charge of the electron, m_e is the electronic mass, and R_0 is the distance to the point of detection. Numerically the universal constant r_e^2 has the value

$$r_e^2 = \beta^2 R_0^2 = \frac{\mu\mu_0 e^2}{4\pi m_e} = 7.94 \times 10^{-30}\ \text{m}^2 \tag{4.12}$$

We learn immediately from (4.12) that the probability of a given incident photon being scattered is very small (of the order of 10^{-28}, considering the entire 4π solid angle of scattering). Thus for many cases, we may quite nicely assume that each photon is scattered at most one time. In large, perfect crystals this assumption is no longer valid and interesting properties ensue, as we shall see in Sects. 4.4 and 4.5. For electrons, the parameter corresponding to β is much larger and the assumption of single-scattering breaks down at relatively small crystal diameters.

4.2 INCOHERENT RADIATION

We have already mentioned, in broad terms, that some portion of the beam incident on an atom is scattered as incoherent radiation. We ask here (1) the

[1] What we have learned in the above regards the origin of the polarization factor $(1 + \cos^2 2\theta)/2$. We have seen that this arises through the effect of the plane of oscillation of the excited electron relative to a specific direction of propagation of the scattered wave. It should be obvious that when the incident beam is not entirely depolarized, the factor $(1 + \cos^2 2\theta)/2$ will not correctly account for the effect of polarization. If the incident beam were preferentially polarized along z and if the ratio of E_z to $E_y + E_z$ were q, then

$$I = (1 - q)^2 I_y^2 + q^2 I_z^2 = I_0 \beta^2 \left[\frac{(1 - q)^2 + q^2 \cos^2 2\theta}{2}\right]$$

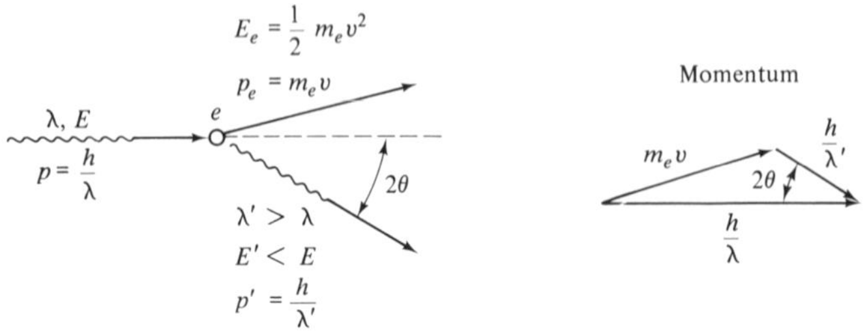

Figure 4.3. Conservation of energy and momentum in Compton (incoherent) scattering. At the left are the scalar quantities used in a conservation of energy expression. Here an x-ray of wavelength λ (energy E) strikes an electron. A portion E_e of the energy of the x-ray is imparted to the electron (kinetic energy). The final wavelength and energy of the x-ray are λ′ and E'. On the right is shown the vector momentum balance for this process.

nature of the incoherent scattering and (2) its relative strength. We seek an answer to (1) first.

Although an idealization, we consider first a free electron with a wave of wavelength λ impinging upon it, as sketched in Fig. 4.3. As a result of the interaction with the incident x-ray photon, the electron can have been caused to move. The photon will correspondingly lose energy. Energy and momentum balances for this process are written

$$\text{Conservation of energy:} \qquad \frac{hc}{\lambda} - \frac{hc}{\lambda'} = \frac{1}{2} m_e v^2 \tag{4.13}$$

$$\text{Conservation of momentum:} \quad \frac{h^2}{\lambda^2} + \frac{h^2}{\lambda'^2} - 2\frac{h^2}{\lambda\lambda'} \cos 2\theta = m_e^2 v^2 \tag{4.14}$$

Eliminating v^2 between (4.13) and (4.14), we have

$$\frac{2hc}{m_e\lambda} - \frac{2hc}{m_e\lambda'} = \frac{h^2}{m_e^2\lambda^2} + \frac{h^2}{m_e^2\lambda'^2} - \frac{2h^2}{\lambda\lambda' m_e^2} \cos 2\theta \tag{4.15}$$

Rearranging terms

$$\frac{2c}{\lambda} - \frac{2c}{\lambda'} = \frac{h}{m_e\lambda^2} + \frac{h}{m_e\lambda'^2} - \frac{2h}{m_e\lambda\lambda'} \cos 2\theta \tag{4.16}$$

Assuming that the change in wavelength is small $\left(\frac{\lambda' - \lambda}{\lambda} << 1\right)$

$$\frac{2c(\lambda' - \lambda)}{\lambda^2} \simeq \frac{2h}{m_e\lambda^2}(1 - \cos 2\theta) \tag{4.17}$$

Finally, solving for the shift in wavelength, we find

$$\lambda' - \lambda \simeq \frac{h}{m_e c}(1 - \cos 2\theta) = 0.024(1 - \cos 2\theta) \tag{4.18}$$

Note in particular that the wavelength shift exhibits a maximum at $2\theta = 90°$. At either very high or very low scattering angle, the wavelength shift is very small.

For a bound electron, however, not every interaction will give rise to a release of energy to the electron. In such a case only certain quantized energies are allowed. The question which must be asked in this case is, "How is the total energy partitioned between the coherent and incoherent scattering?" The answer to the question is approached in the following way. Consider an atom containing one electron whose equilibrium wave function is $\psi_n e^{-i W_n t/\hbar}$, where W_n denotes the nth energy level and whose other energy levels have the set of wave functions $\psi_m e^{-i W_m t/\hbar}$, where $m \neq n$. We then consider the effect of an incident x-ray wave on Schrödinger's equation.

The problem of the apportionment of energy between coherent and incoherent scattered radiation is somewhat involved. In outline, however, it is as follows. The time-dependent Schrödinger equation is written

$$-\frac{\hbar^2}{2m_e}\nabla^2\left(\psi_n e^{-i\frac{W_n t}{\hbar}}\right) + V\left(\psi_n e^{-i\frac{W_n t}{\hbar}}\right) = ih\frac{d\left(\psi_n e^{-i\frac{W_n t}{\hbar}}\right)}{dt} \tag{4.19}$$

Here V is the potential energy of the electron. An electron may experience both the potential energy V_0 due to its usual environment and an additional potential eA due to the electromagnetic wave. Here

$$A = \frac{E_0}{\omega}\sin\left(\omega t - \frac{2\pi}{\lambda}\hat{\mathbf{g}}_0 \cdot \mathbf{R}\right) \tag{4.20}$$

where E_0 is the electromagnetic potential, ω is the frequency, $\hat{\mathbf{g}}_0$ is a unit vector transverse to the propagation direction of the wave, and $\mathbf{R}$ is a real space vector. In this case (4.19) becomes, after cancelling $e^{-i W_n t/\hbar}$ terms

$$-\frac{\hbar^2}{2m_e}\nabla^2\psi_n + (V_0 + eA)\psi_n = W_n\psi_n \tag{4.21}$$

From (4.21), we can, after some tedious arithmetic, obtain a solution for ψ_n. This solution obviously must contain elements from the impressed wave (through the eA term in (4.21)).

Simultaneously, we can consider a "Schrödinger current density." From classical electrodynamics we have

$$\frac{d\rho}{dt} + \nabla \cdot \mathbf{j} = 0 \tag{4.22}$$

where ρ is a charge density and $\mathbf{j}$ a current density. In quantum mechanical terms, the charge density becomes

$$\rho = e\left|\psi_n e^{-i\frac{W_n t}{\hbar}}\right|^2 \tag{4.23}$$

The ψ_n previously found can be inserted into (4.23) and (4.22). In this way, a value of the *current density* is found. Using classical electrodynamics we can also relate this current density to the electric field created by it. And this is just what we need.

Two different cases must be considered in determining this electric field: that in which the state function ψ_n is unchanged by interaction with the electromagnetic wave and that in which ψ_n is transformed to some other state function ψ_m. The former case is coherent scattering; the latter is incoherent scattering. The solution for coherent scattering yields

$$\mathrm{I}^n_{\mathrm{coh}}(s) = \beta^2\left(\frac{1+\cos^2 2\theta}{2}\right) I_0 \left|\int \left|\psi_n\right|^2 e^{-2\pi i \mathbf{s}\cdot\mathbf{r}_n} dv_{\mathbf{r}}\right|^2 \tag{4.24}$$

where β is defined in (4.11). The integral on the right of (4.24) is similar to an atomic scattering factor and will exhibit its same general form, but is for only one electron (the nth one). We may refer to this integral as an electron scattering factor f^e_n. For the incoherent scattering representing transition from the nth to the mth electron state, we would find

$$[I^n_{\mathrm{inc}}(s)]_{n,m} = \beta^2\left(\frac{1+\cos^2 2\theta}{2}\right) I_0 \left|\int \psi_n\psi^*_m\, e^{-2\pi i \mathbf{s}\cdot\mathbf{r}}\, dv_{\mathbf{r}}\right|^2 \tag{4.25}$$

The total incoherent scattering is a sum of such terms

$$I^n_{\mathrm{inc}}(s) = \beta^2\left(\frac{1+\cos^2 2\theta}{2}\right) I_0 \sum \left|\int \psi_n\psi^*_m\, e^{-2\pi i \mathbf{s}\cdot\mathbf{r}}\, dv_{\mathbf{r}}\right|^2 \tag{4.26}$$

The total scattering by this electron is

$$I^n_{\mathrm{tot}}(s) = \beta^2\left(\frac{1+\cos^2 2\theta}{2}\right) I_0 \sum_{\text{all } m} \left|\int \psi_n\psi^*_m\, e^{-2\pi i \mathbf{s}\cdot\mathbf{r}}\, dv_{\mathbf{r}}\right|^2 \equiv I_e(s) \tag{4.27}$$

The term $I_e(s)$, called the Thomson intensity, is the total intensity scattered by one electron. Combining (4.24), (4.26), and (4.27), and using (4.9), we have the following relationship among I^n_{coh}, I^n_{inc}, and I^n_{tot}

$$I^n_{\mathrm{tot}} = I_e(s) = I^n_{\mathrm{coh}} + I^n_{\mathrm{inc}} = I_e(s)(f^e_n)^2 + I^n_{\mathrm{inc}} \tag{4.28}$$

Rearranging terms

$$I^n_{\mathrm{inc}} = I_e(s)[1-(f^e_n)^2] \tag{4.29}$$

Now let the atom have many (z) electrons. Then

$$I_{\mathrm{inc}} = \sum_{n=1}^{z} I^n_{\mathrm{inc}} = I_e(s) \sum_{n=1}^{z} [1-(f^e_n)^2] \tag{4.30}$$

As s approaches zero, f^e_n approaches 1 and $1-(f^e_n)^2$ approaches zero. As s becomes large, f^e_n moves toward zero and $1-(f^e_n)$ approaches 1. Thus the righthand side of (4.30) becomes large.

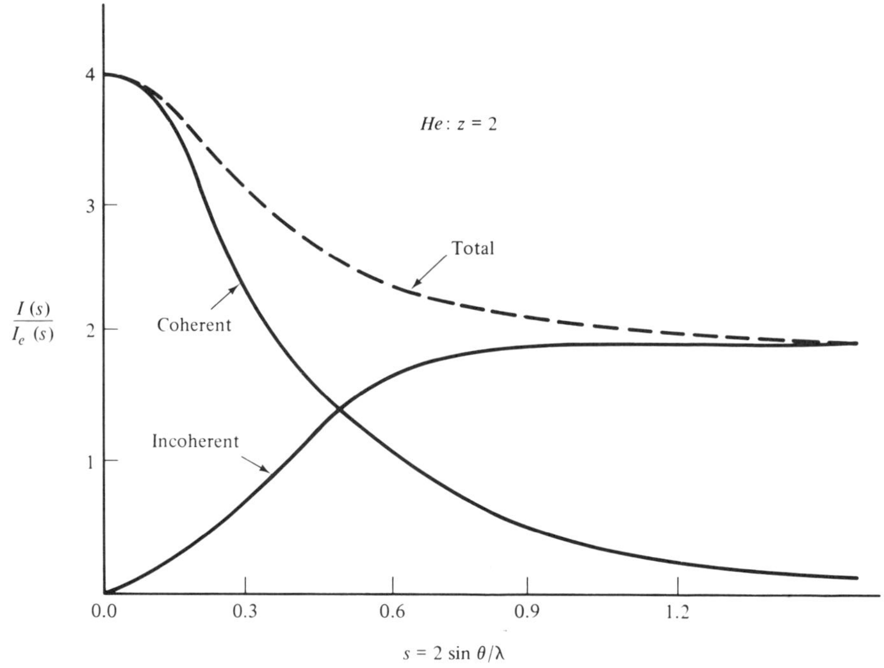

Figure 4.4. Coherent, incoherent, and total x-ray scattering intensities for helium.

On the other hand, the coherent scattering I_{coh} depends on the phase relations between all electrons simultaneously and is written

$$I_{\text{coh}} = I_e(s)\left(\sum_{n=1}^{z}\int|\psi_n|^2 e^{2\pi i\mathbf{s}\cdot\mathbf{r}}dv_{\mathbf{r}}\right)^2 = I_e(s)\left(\sum_{n=1}^{z} f_n^e\right)^2 \tag{4.31}$$

As we have already seen, this quantity is $Z^2 I_e$ at $s = 0$ and becomes small as s becomes large. To a first approximation, the maximum value of I_{coh} goes as Z^2 and the maximum value of I_{inc} goes as Z. This is exactly so for He, as shown in Fig. 4.4. As we increase Z, the relative contribution of incoherent scattering decreases. This is predicted by (4.30) and (4.31). The behavior of the Ca atom ($Z = 20$) is shown in Fig. 4.5.

4.3 DYNAMICAL INTERACTIONS: INTRODUCTION

When the x-irradiated crystal is sufficiently large and perfect it is possible for the x-ray to be multiple diffracted within the crystal, as indicated in Fig. 4.6. In this figure we see possible diffraction events for a ray incident at the upper

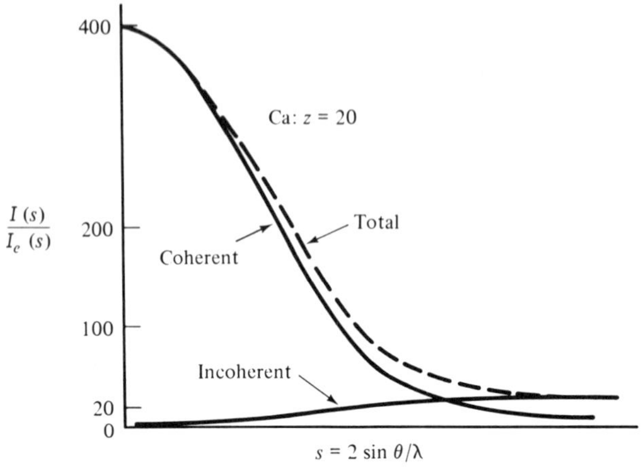

Figure 4.5. Coherent, incoherent, and total x-ray scattering intensities for calcium.

left. We note that so long as the crystal remains perfect over large distances the diffracted rays will continue to meet lattice planes, approaching them from top *and* bottom, at the Bragg angle, θ. For this perfect, semiinfinite crystal in the geometry shown, the net effect of the multiple interactions must be to divert energy from the once-reflected ray into the incident direction. One should thus observe a decrease of intensity per unit volume, relative to a smaller or less perfect crystal (where multiple events are not possible). This phenomenon is actually found and is termed *primary extinction.* Clearly primary extinction must be identified and taken into account when one wishes to analyze intensity data in terms of a structure factor.

Primary extinction can be defeated locally by the presence of lattice defects which distort the lattice and disrupt the sequence of sequential Bragg interactions. This effect is schematically shown in Fig. 4.7 for a crystal containing a single edge dislocation whose Burgers' vector lies normal to the diffracting planes. In this illustration regions A and C are perfect and permit multiple scattering, whereas the lattice planes are distorted in region B, such that the Bragg conditions

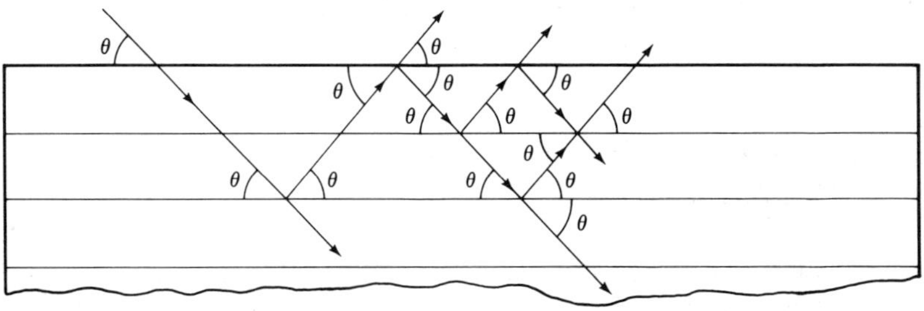

Figure 4.6. Multiple diffraction within a perfect crystal.

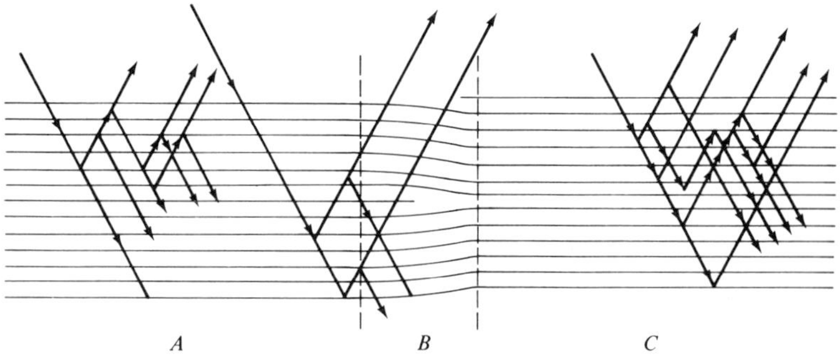

Figure 4.7. Effect of a dislocation in locally interrupting primary extinction. Here regions *A* and *C* are perfect, while *B* contains the core and strain field of an edge dislocation.

are not met locally by the once-diffracted ray, and multiple scattering cannot occur. Thus, rays diffracted first in *A* will not be attenuated by primary extinction as they pass through *B*. These very intense diffracted rays then add to the rays weakly once-diffracted from region *B* to give a net quite intense local diffraction contribution. In this way, we see that in principle one should be able to observe images of dislocations directly in a magnified diffraction spot. Indeed this is observed. Fig. 4.8 is a magnified section of a zinc $(11\bar{2}0)$ diffraction

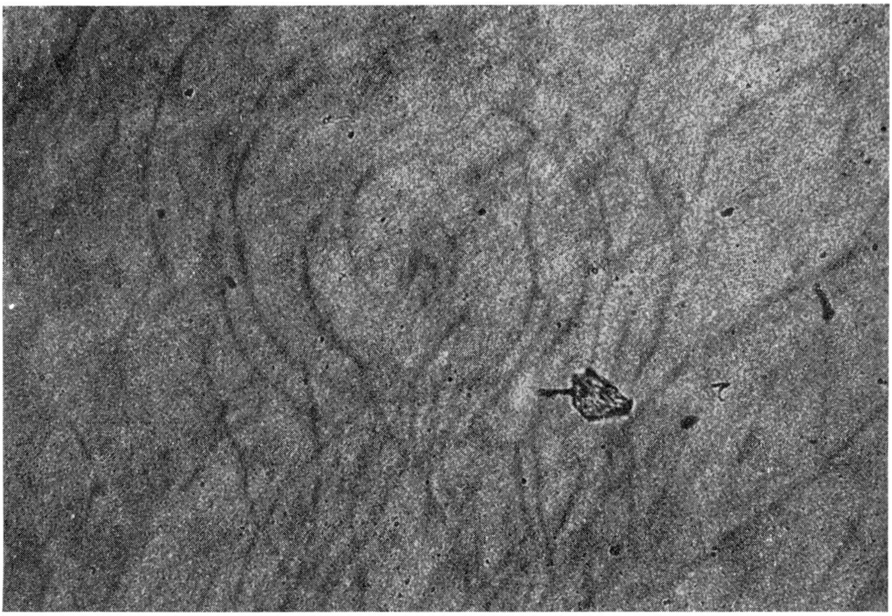

Figure 4.8. Berg-Barrett $<11\bar{2}3>$ topograph of a nonplanar Frank-Read source in a zinc single crystal. (Reprinted with permission from *Acta Met.*, **14,** 436, J. M. Schultz and R. W. Armstrong, Copyright 1966, Pergamon Press, Ltd.)

spot, showing a nonplanar Frank-Read source originating at an inclusion particle. There are several geometrical arrangements for obtaining such defect images. The study of diffraction spot detail is referred to as *x-ray topography* and will be treated in further detail in Sect. 4.5. The analogous phenomenon in electron diffraction is termed *dark-field electron microscopy.* In this case, electrons diffracted through a thin film, using a specific lattice plane, are imaged.

Two other effects which are due to multiple interactions in very perfect crystals are the Borrmann effect and the observation of *pendellösung* fringes. The Borrmann effect is a very low x-ray absorption of a beam incident at the Bragg angle. *Pendellösung* fringes are a regular set of light and dark lines in a beam diffracted from a wedge-shaped perfect crystal. These effects are treated later in this section.

The inverse effect is found when waves are transmitted through a solid. If the crystal is precisely at the Bragg angle and is not thicker than the extinction distance (see Sect. 4.5), then rays transmitted through defective regions will appear less intense than will rays which have passed through the more perfect regions. This is simply because more wave energy is transmitted to the diffracted rays in the more defective areas. This is the principle of bright field electron microscopy (magnification of the forward-directed beam) and of Lang-geometry (transmitted beam, thin specimen) x-ray topography. Figure 4.9 is a bright-field electron micrograph of a mechanically induced dislocation array in copper. The contrast of this micrograph has been photographically reversed.

All these phenomena can be explained and quantified using formalisms developed by Ewald[2] in 1917 and von Laue[3] in 1931. Both analyses treat the dynamics of interacting waves in a periodic medium. These treatments are therefore termed the *dynamical theory.* The Ewald and von Laue approaches treat the same physical problem using slightly different points of view. Wagenfeld has shown the formal identity of the two approaches. In the next section we shall predominantly follow the von Laue approach.

4.4 THE DYNAMICAL THEORY

In this section we shall derive the relations between incident wave and waves scattered by periodically spaced (atomic) scattering centers. We will ignore effects of absorption; absorptive results will be considered later.

We begin by deriving a fundamental equation which relates all the internal wave fields. Consider Fig. 4.10. Here an incident wave $\boldsymbol{\kappa}$ ($|\kappa| = 2\pi/\lambda$) whose electric field $\mathbf{E}_0^o$ is incident on a crystal. Internally the primary beam is slightly refracted, giving the new wave vector $\mathbf{k}_0$ ($|\mathbf{k}_0| = 2\pi/\lambda_0$) with electric field $\mathbf{E}_0$ or displacement $\mathbf{D}_0$. In addition, any number of diffracted waves $\mathbf{k}_H$, with electric

[2] P. P. Ewald, *Ann. d. Physik,* **49,** 117 (1916); **49,** 1 (1916); **54,** 519 (1917).

[3] M. v. Laue, *Ergeb. der exakt, Naturwiss.,* **10,** 133 (1931).

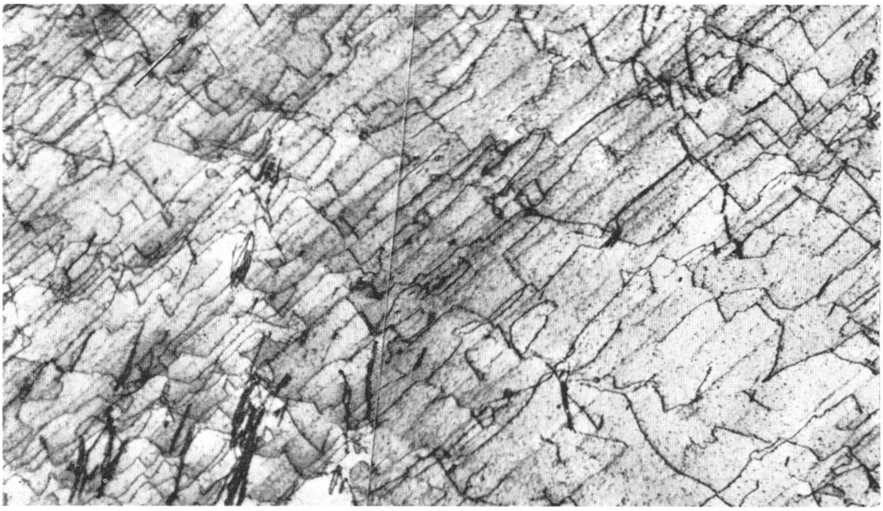

Figure 4.9. Mechanically induced dislocation array in a thin copper foil. [U. Essmann, *Phys. Stat. Sol.*, **12,** 723 (1965)]

displacement $\mathbf{D}_H$ may be created, depending upon the experimental geometry. The displacement fields of the several reflected rays are all of the form $\mathbf{D}_H \exp(i\omega t + i\mathbf{k}_H \cdot \mathbf{r})$. The total electrical displacement field $\mathbf{D}$ within the crystal is then

$$\mathbf{D} = \sum_H \mathbf{D}_H e^{i(\omega t + \mathbf{k}_H \cdot \mathbf{r})} \tag{4.32}$$

The amplitude D_H are what we seek. The intensities I_H and I_O of the forward-directed and diffracted waves are D_0^2 and D_H^2. In fact, it is precisely the ratio D_H^2/D_0^2 that we must find. This ratio can be found by application of the classical electric field theory.

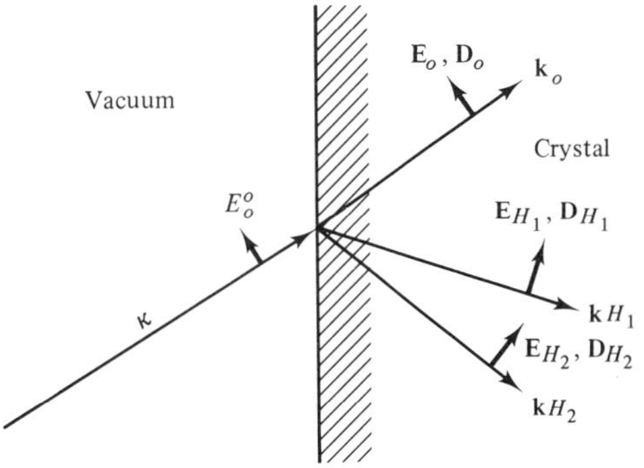

Figure 4.10. Incident and internal wave fields.

In order to relate all the internal waves we apply the wave equation

$$\nabla \times (\nabla \times \mathbf{E}) = -\frac{1}{c^2}\frac{\partial^2 \mathbf{D}}{\partial t^2} \tag{4.33}$$

which is obtained by combining two of Maxwell's equations. We further specify the polarizability ψ by

$$\mathbf{E} = (1 - \psi)\mathbf{D} \tag{4.34}$$

We now use the lattice periodicity. ψ relates to the structure of the medium and must exhibit the periodicity of that structure. In fact, it is convenient to write ψ as a Fourier series, using the rel. In this way, we write

$$\psi(r) = \sum_H \psi_H \, e^{2\pi i \mathbf{r}^*_H \cdot \mathbf{r}} \tag{4.35}$$

Written this way ψ may be viewed as a superposition of several polarizabilities, relating to different lattice directions and each exhibiting a lattice periodicity through $\mathbf{r}^*_H$.

Insertion of (4.32), (4.34), and (4.35) into (4.33) yields a relationship among the several electrical displacement fields

$$\sum_L \{\psi_{H-L}(\mathbf{k}_H \cdot \mathbf{D}_L)\mathbf{k}_H - \psi_{H-L}k_H^2\mathbf{D}_L\} = (\kappa^2 - k_H^2)\mathbf{D}_H \tag{4.36}$$

Here κ is the wave vector of the beam incident on the crystal. This is the *fundamental equation of the dynamical theory.* The mathematical steps leading from (4.33) to (4.36) are given in Appendix C.

Most often we deal with only a transmitted wave and one diffracted wave. Let us look then at this two wave case. There within the crystal only the displacements $\mathbf{D}_0$ and one $\mathbf{D}_H$ have significant amplitude. That is, in (4.36)

$$\begin{aligned} \mathbf{D}_0 &\neq 0 \\ \mathbf{D}_H &\neq 0 \\ \mathbf{D}_{L\neq 0} &= 0 \end{aligned} \tag{4.37}$$

In this case (4.36) becomes the following. Taking $H = H$ and using $L = 0$, H in the sum, we have

$$\psi_H(\mathbf{k}_H \cdot \mathbf{D}_0)\mathbf{k}_H - \psi_H k_H^2 \mathbf{D}_0 + \psi_0(\mathbf{k}_H \cdot \mathbf{D}_H)\mathbf{k}_H - \psi_0 k_H^2 \mathbf{D}_H = (\kappa^2 - k_H^2)\mathbf{D}_H \tag{4.38}$$

Taking $H = 0$ and using $L = 0$, H in the sum, we have

$$\psi_0(\mathbf{k}_0\mathbf{D}_0)\mathbf{k}_0 - \psi_0 k_0^2 \mathbf{D}_0 + \psi_{-H}(\mathbf{k}_0 \cdot \mathbf{D}_H)\mathbf{k}_0 - \psi_{-H} k_0^2 \mathbf{D}_H = (\kappa^2 - k_0^2)\mathbf{D}_0 \tag{4.39}$$

Noting that $\mathbf{k}_0 \cdot \mathbf{D}_0 = 0$ and $\mathbf{k}_H \cdot \mathbf{D}_H = 0$—i.e., the waves are tranverse—and rearranging terms, the two expressions become

$$\psi_H(\mathbf{k}_H \cdot \mathbf{D}_0)\mathbf{k}_H - \psi_H k_H^2 \mathbf{D}_0 = [\kappa^2 - (1 - \psi_0)k_H^2]\mathbf{D}_H \tag{4.40a}$$

$$\psi_{-H}(\mathbf{k}_0 \cdot \mathbf{D}_H)\mathbf{k}_0 - \psi_{-H} k_0^2 \mathbf{D}_H = [\kappa^2 - (1 - \psi_0)k_0^2]\mathbf{D}_0 \tag{4.41a}$$

Equation (4.40a) and (4.41a) can be simplified still further, by scalar multiplication of (4.40a) by $\mathbf{D}_H$ and (4.41a) by $\mathbf{D}_0$. We then have

$$\psi_H k_H^2 D_0 D_H \sin \xi + [\kappa^2 - (1 - \psi_0) k_H^2] D_H^2 = 0 \tag{4.40b}$$

$$\psi_{-H} k_0^2 D_0 D_H \sin \xi + [\kappa^2 - (1 - \psi_0) k_0^2] D_0^2 = 0 \tag{4.41b}$$

where ξ is the angle between $\mathbf{D}_0$ and $\mathbf{D}_H$.

Equations (4.40b) and (4.41b) can now be solved for the ratio D_H/D_0. This is, after all, what we seek—the apportionment of energy between forward and diffracted waves. Such an expression can be obtained for either (4.40b) or (4.41b). The ratio is expressed

$$\frac{D_H}{D_0} = \frac{\psi_H k_H^2 \sin \xi}{(1 - \psi_0) k_H^2 - \kappa^2} = \frac{(1 - \psi_0) k_0^2 - \kappa^2}{\psi_{-H} k_0^2 \sin \xi} \tag{4.42}$$

This expression relates the wave field ratios to the wave vectors and the internal polarizabilities. Clearly this solution exists only so long as

$$\psi_H \psi_{-H} k_0^2 k_H^2 \sin^2 \xi = [(1 - \psi_0) k_0^2 - \kappa^2][(1 - \psi_0) k_H^2 - \kappa^2] \tag{4.43}$$

This last relation is known as the *dispersion equation.*

It turns out to be convenient at this juncture to define the new functions

$$\zeta_0 = \frac{1}{2\kappa}[(1 - \psi_0) k_0^2 - \kappa^2] \tag{4.44a}$$

$$\zeta_H = \frac{1}{2\kappa}[(1 - \psi_0) k_H^2 - \kappa^2] \tag{4.44b}$$

Inserting (4.44) into (4.43) we have

$$\zeta_0 \zeta_H = \frac{1}{4} \frac{k_0^2 k_H^2}{k^2} \psi_H \psi_{-H} \sin^2 \xi \tag{4.45}$$

We shall see that the index of refraction of x-rays is very nearly unity. Hence $k_0 \simeq k_H \simeq \kappa$, and, to a good approximation, our dispersion equation becomes

$$\zeta_0 \zeta_H \equiv \frac{\kappa^2}{4} \psi_H \psi_{-H} \sin^2 \xi \tag{4.46}$$

The right side of (4.46) is constant for a given reflection H. The left side contains the exact incident angle, as a deviation from the exact Bragg angle. In the same terms

$$\frac{D_H}{D_0} \simeq -\frac{2\zeta_0}{\kappa \psi_{-H} \sin \xi} = -\frac{\kappa \psi_H \sin \xi}{2\zeta_H} \tag{4.47}$$

The dispersion equation affords us a nice conceptual representation which aids in our interpretation of dynamical problems. Clearly the dispersion equation represents somehow the locus of all diffraction conditions which permit $\mathbf{D}_H$ and $\mathbf{D}_0$ to exist simultaneously. Let us first see in what sense the quantities ζ_0

and ζ_H can be interpreted in terms of an experiment. Consider for the moment (4.44a). Using the facts (to be shown) that ψ_0 is small and $(k_0 - \kappa)/\kappa \ll 1$, we write (4.44a) as

$$2\kappa\zeta_0 \simeq \left[\left(1 - \frac{\psi_0}{2}\right) k_0 + \kappa\right]\left[\left(1 - \frac{\psi_0}{2}\right) k_0 - \kappa\right] \tag{4.48}$$
$$\zeta_0 \simeq \left(1 - \frac{\psi_0}{2}\right) k_0 - \kappa \simeq k_0 - \left(1 + \frac{\psi_0}{2}\right)\kappa$$

Now the term $\left(1 + \frac{\psi_0}{2}\right)$ is just the average refractive index of the crystal. That is seen in the following way. Consider the fundamental equation for the case in which all $\mathbf{D}_H$ are zero, except, of course, $\mathbf{D}_0$. This is the case where no relnodes are near the Ewald sphere. In this case (4.36) is written

$$-\psi_0 k_0^{*2}\mathbf{D}_0 = (\kappa^2 - k_0^{*2})\mathbf{D}_0 \tag{4.49}$$

and

$$(1 - \psi_0)k_0^{*2} = \kappa^2 \tag{4.50}$$
$$k_0^* \simeq \left(1 + \frac{\psi_0}{2}\right)\kappa = \epsilon^*\kappa$$

where, by definition, ϵ^* is the refractive index in the case of no diffraction. Thus we see that ζ_0 is the difference in $1/\lambda$ between the actual forward-directed wave and the forward wave under the condition of no excitation. Similarly $\zeta_H = k_H - (1 + \psi_0/2)\kappa$ is the difference between the diffracted reciprocal wavelength and that of the forward wave under no excitation. The dispersion relation, (4.46), tells us that the loci of the allowable values of ζ_0 and ζ_H are found on the surface of a hyperboloid of two sheets. Figure 4.11 shows the interpretation of the ξ's in terms of a classical Ewald sphere of diameter κ. Here surfaces G_0 and G_H are the loci of points exactly k_0^* from the origin O and rel point H. For some arbitrary point A in reciprocal space ζ_0 is the difference between the line from A to O and the portion of the AO line going from the surface G_0 to O. Similarly ζ_H is the difference between the line AH and a line from G_H to O. But not all positions A give valid values of ζ_0 and ζ_H. Only those positions of A which lie on the *dispersion surface,* shown in Fig. 4.11, are allowed. The dispersion surface—a hyperboloid of two sheets satisfying the dispersion equation—is thus the locus of all resp points which give value to D_H/D_0. We note that for each direction of $\mathbf{k}_0$ and given polarization (via ξ) there are *two* solutions to D_H/D_0—that is, there exist two forward-directed waves, with wave vector magnitudes k_0^α and k_0^β and two diffracted rays, with wave vectors $\mathbf{k}_H^\alpha$ and $\mathbf{k}_H^\beta$.

What we have now is the ratio of H and O fields *within the crystal.* What

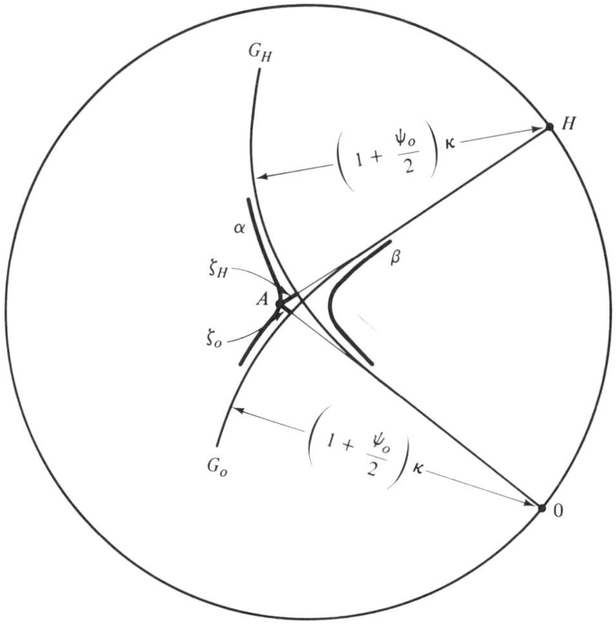

Figure 4.11. Geometrical relations in resp. G_O and G_H are spherical surfaces $(1+\frac{\psi_O}{2})\kappa$ from respectively O (the origin of resp) and H (a relnode). α and β are the two hyperboloid solutions to the dispersion equation; together they are known as the dispersion surface. The reciprocal distances ζ_O and ζ_H run from any point on α or β toward O and H, respectively, and terminate on G_O and G_H. In this figure the distances ζ_O/κ and ζ_H/κ are greatly exaggerated.

we would really need is the relation between O,H fields leaving the crystal and the O field entering. We proceed now to determine these quantities.

We must begin by examining the conditions at the entrance and exits surfaces. We note that the total electric displacement across any position $\mathbf{r}=\boldsymbol{\tau}$ on an external surface is given by

$$\mathbf{D}_0^e e^{i(\omega t-2\pi\boldsymbol{\kappa}\cdot\boldsymbol{\tau})}+\mathbf{D}_H^e e^{i[\omega t-2\pi(\boldsymbol{\kappa}+\mathbf{r}_H^*)\cdot\boldsymbol{\tau}]}=\sum_P\left[\mathbf{D}_0^p e^{i(\omega t+2\pi\mathbf{k}_0^p\cdot\boldsymbol{\tau})}+\mathbf{D}_H^p e^{i(\omega t+2\pi\mathbf{k}_H^p\cdot\boldsymbol{\tau})}\right] \tag{4.51}$$

where the superscript e denotes the external waves and p ($p=1,2$) indicates the internal waves associated with the two solutions to the dispersion equation. Noting that no discontinuity in wave phase may occur across the external surface, we see from (4.51) that

$$\begin{aligned}\boldsymbol{\kappa}\cdot\boldsymbol{\tau}&=\mathbf{k}_0^p\cdot\boldsymbol{\tau}\\(\boldsymbol{\kappa}+\mathbf{r}_H^*)\cdot\boldsymbol{\tau}&=\mathbf{k}_H^p\cdot\boldsymbol{\tau}\end{aligned} \tag{4.52}$$

Let us now relate internal to external wave vectors by definition of two deviation vectors $\boldsymbol{\delta}_0^p$ and $\boldsymbol{\delta}_H^p$

$$\mathbf{k}_0^p = \boldsymbol{\kappa} + \boldsymbol{\delta}_0^p \tag{4.53a}$$

$$\mathbf{k}_H^p = \boldsymbol{\kappa} + \mathbf{r}_H^* + \boldsymbol{\delta}_H^p \tag{4.53b}$$

Substituting (4.53a) into (4.52), we find

$$\boldsymbol{\delta}_0^p \cdot \boldsymbol{\tau} = \boldsymbol{\delta}_H^p \cdot \boldsymbol{\tau} = 0 \tag{4.54}$$

For the case of the external wave entering a plane surface, with the origin of coordinates located in that surface, both deviation vectors are perpendicular to the external surface.

Substituting (4.53a and b) and (4.54) into (4.51) and cancelling terms containing $e^{i(\omega t + 2\pi \boldsymbol{\kappa}\cdot\boldsymbol{\tau})}$, we have

$$\mathbf{D}_0^e + \mathbf{D}_H^e e^{2\pi i \mathbf{r}_H^* \cdot \boldsymbol{\tau}} = \sum_p \left[\mathbf{D}_0^p + \mathbf{D}_H^p e^{2\pi i \mathbf{r}_H^* \cdot \boldsymbol{\tau}}\right] \tag{4.55}$$

Rearranging terms, we now have

$$\left[\mathbf{D}_0^e - \sum_p \mathbf{D}_0^p\right] + e^{2\pi i \mathbf{r}_H^* \cdot \boldsymbol{\tau}} \left[\mathbf{D}_H^e - \sum \mathbf{D}_H^p\right] = 0 \tag{4.56}$$

Since the complex exponential term is in general nonzero, the two bracketed terms must independently equal zero. Hence the relation between fields across an incident boundary is

$$\begin{aligned} \mathbf{D}_0^e &= \sum \mathbf{D}_0 \\ \mathbf{D}_H^e &= \sum \mathbf{D}_H^p \end{aligned} \tag{4.57}$$

We shall see that the internal and external vectors differ little in either magnitude or direction. Hence

$$\begin{aligned} D_0^e &\simeq \sum_p D_0^p \\ D_H^e &\simeq \sum_p D_H^p \end{aligned} \tag{4.58}$$

4.5 INTENSITY RELATIONS IN THE DYNAMICAL THEORY

We shall first consider Laue geometry, again for the case of two internal beams. *Laue geometry* means only that the beam enters and leaves two different surfaces. The experiment is sketched in Fig. 4.12. At the entrant surface we have

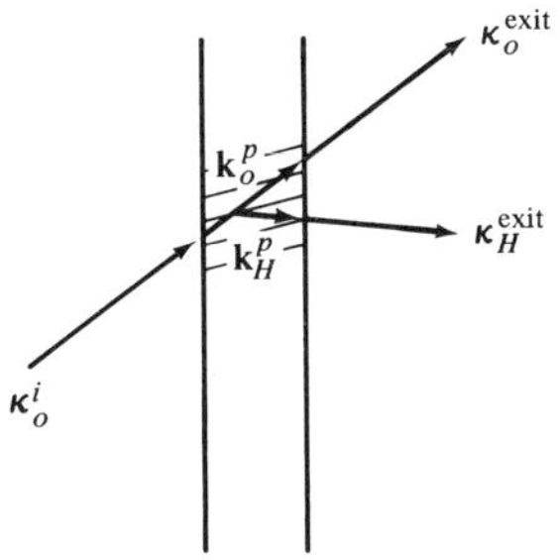

Figure 4.12. The Laue geometry.

$$\begin{aligned} D_0^i &= D_0^\alpha + D_0^\beta \\ D_H^i &= 0 = D_H^\alpha + D_H^\beta \end{aligned} \tag{4.59}$$

where the superscript i denotes the incident wave and where α and β denote the waves given by the two sheets of the dispersion surface. Now recall that

$$\begin{aligned} X^\alpha &\equiv \frac{D_H^\alpha}{D_0^\alpha} = -\frac{\kappa\psi_H \sin\xi}{2\zeta_H^\alpha} \\ X^\beta &\equiv \frac{D_H^\beta}{D_0^\beta} = -\frac{\kappa\psi_H \sin\xi}{2\zeta_H^\beta} \end{aligned} \tag{4.60}$$

Inserting (4.60) into (4.59), we have

$$\begin{aligned} D_0^i &= \frac{D_H^\alpha}{X^\alpha} + \frac{D_H^\beta}{X^\beta} \\ 0 &= D_H^\alpha + D_H^\beta \end{aligned} \tag{4.61}$$

Solving (4.61) for D_H^α and D_H^β, we have

$$\begin{aligned} D_H^\alpha &= \frac{\begin{vmatrix} D_0^i & \dfrac{1}{X^\beta} \\ 0 & 1 \end{vmatrix}}{\begin{vmatrix} \dfrac{1}{X^\alpha} & \dfrac{1}{X^\beta} \\ 0 & 1 \end{vmatrix}} = \frac{X^\alpha X^\beta D_0^i}{(X^\beta - X^\alpha)} \\ D_H^\beta &= \frac{\begin{vmatrix} \dfrac{1}{X^\alpha} & D_0^i \\ 1 & 0 \end{vmatrix}}{\begin{vmatrix} \dfrac{1}{X^\alpha} & \dfrac{1}{X^\beta} \\ 0 & 1 \end{vmatrix}} = -\frac{X^\alpha X^\beta D_0^i}{(X^\beta - X^\alpha)} \end{aligned} \tag{4.62}$$

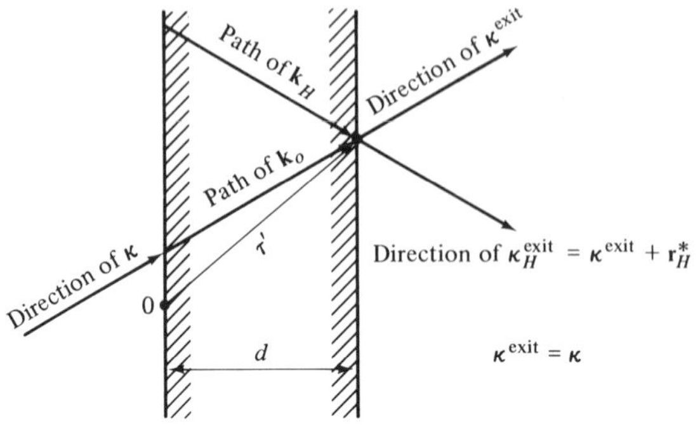

Figure 4.13. Geometry for computing exit wave fields.

Clearly the internal D_H^α and D_H^β fields are exactly 180° different in phase. (This is seen from their opposition in sign).

Using these values for the internal field we can now relate to the fields $\mathbf{D}_o^{exit}$ and $\mathbf{D}_H^{exit}$ beyond the exit surface. Consider the sketch of Fig. 4.13. For points τ' on the external surface, (4.51) is given by

$$\mathbf{D}_0^{exit}\, e^{i(\omega t+2\pi\kappa\cdot\tau')} + \mathbf{D}_H^{exit}\, e^{i[\omega t+2\pi(\kappa+r_H^*)\cdot\tau']}$$
$$= \sum_p \mathbf{D}_0^p\, e^{i(\omega t+2\pi\kappa\cdot\tau'+2\pi\delta_0^p d)} + \sum_p \mathbf{D}_H^p\, e^{i(\omega t+2\pi\kappa\cdot\tau'+2\pi r^*\cdot\tau'+2\pi\delta_H^p d)} \qquad (4.63)$$

where d is the crystal thickness and the $2\pi\delta^p d$ terms account for the change of phase, relative to $\kappa \cdot \tau$, as the internal waves traverse the crystal. Employing the same sequence of steps as in the preceding section and using the symbols

$$C_0^p = e^{2\pi i\delta_0^p d} \qquad (4.64)$$
$$C_H^p = e^{2\pi i\delta_H^p d}$$

we find

$$D_0^{exit} \cong C_0^\alpha D_0^\alpha + C_0^\beta D_0^\beta \qquad (4.65a)$$

$$D_H^{exit} \simeq C_H^\alpha D_H^\alpha + C_H^\beta D_H^\beta \qquad (4.65b)$$

Using our results relating D_H^α and D_H^β to the incident field D_0 (4.62), we have

$$\frac{D_H^{exit}}{D_0^i} = \frac{X^\alpha X^\beta C_H^\alpha}{X^\beta - X^\alpha} - \frac{X^\alpha X^\beta C_H^\beta}{X^\beta - X^\alpha} \qquad (4.66)$$

In general the relation between a field and an intensity is (intensity) = $|\text{field}|^2$. Thus

$$\frac{I_H^{exit}}{I_0^i} = \left|\frac{X^\alpha X^\beta (C_H^\alpha - C_H^\beta)}{X^\beta - X^\alpha}\right|^2 \qquad (4.67)$$

We shall return to the question of evaluation of the parameters in this intensity expression below.

One phenomenon is immediately approachable using (4.67). Figure 4.14*a* shows a magnified portion of a diffraction spot from a wedge-shaped very perfect silicon crystal. Figure 4.14*b* indicates the relation of the diffracting planes to the wedge and the incident beam. What we observe is a set of fringes—*pendellösung* fringes—lying parallel to the diffracting planes. The reason we should expect such fringes lies in the $(C_H^\beta - C_H^\alpha)$ term in (4.67). This quantity is

$$C_H^\beta - C_H^\alpha = e^{2\pi i \delta_H^\beta d} - e^{2\pi i \delta_H^\alpha d} \tag{4.68}$$

The complex exponentials on the right vary sinusoidally as d increases along the length of the wedge. Consequently, through the $(C_H^\beta - C_H^\alpha)$ term the intensity of the diffracted wave must vary periodically along the length of the wedge—hence the *pendellösung* fringes.

Consider a crystal of thickness d in the Bragg geometry, as in Fig. 4.15. Here $D_H^{\text{exit}} = 0$. Thus

$$D_0^0 = D_0^\alpha + D_0^\beta = \frac{D_H^\alpha}{X^\alpha} + \frac{D_H^\beta}{X^\beta} \tag{4.69a}$$

$$D_H^0 = D_H^\alpha + D_H^\beta \tag{4.69b}$$

$$D_0^{\text{exit}} = C_0^\alpha D_0^\alpha + C_0^\beta D_0^\beta \tag{4.69c}$$

$$D_H^{\text{exit}} = 0 = C_H^\alpha D_H^\alpha + C_H^\beta D_H^\beta \tag{4.69d}$$

Using (4.69a), (4.69b), and (4.69d), we have three linear algebraic equations in the three unknowns D_H^0, D_H^α, D_H^β. The solution for D_H^0 is then

$$D_H^0 = \frac{\begin{vmatrix} \dfrac{1}{X^\alpha} & \dfrac{1}{X^\beta} & D_0^0 \\ 1 & 1 & 0 \\ C^\alpha & C^\beta & 0 \end{vmatrix}}{\begin{vmatrix} \dfrac{1}{X^\alpha} & \dfrac{1}{X^\beta} & 0 \\ 1 & 1 & 1 \\ C_H^\alpha & C_H^\beta & 0 \end{vmatrix}} = \frac{X^\alpha X^\beta (C_H^\beta - C_H^\alpha)}{X^\alpha C_H^\alpha - X^\beta C_H^\beta} D_0^0 \tag{4.70}$$

Finally, for the Bragg geometry

$$\frac{I_H^0}{I_0^0} = \left| \frac{X^\alpha X^\beta (C_H^\beta - C_H^\alpha)}{X^\alpha C_H^\alpha - X^\beta C_H^\beta} \right|^2 \tag{4.71}$$

What is needed now is an evaluation of the terms involved in (4.67) and (4.71). Both X^{α} and X^{β} contain ψ_H and ψ_0 (the latter via ζ_H) and k_0 (also via ζ_H). Recall that the ψ_H are the Fourier coefficients of the total polarizability of the crystal

$$\psi(\mathbf{r}) = \sum_H \psi_H e^{2\pi i \mathbf{r}^* \cdot \mathbf{r}} \tag{4.35}$$

According to (1.30)

$$\psi_H = \frac{1}{V} \int \psi(\mathbf{r})\, e^{-2\pi i \mathbf{r}^* \cdot \mathbf{r}}\, dv_{\mathbf{r}} \tag{4.72}$$

where V is the volume of the crystal. The electron field which interacts with the crystal is

$$\mathbf{E} = \mathbf{E}_0 e^{i(\omega_0 t + 2\pi \mathbf{k}_0 \cdot \mathbf{r})} \tag{4.73}$$

Now the force F on an electron is given by

$$\mathbf{F} = -e\mathbf{E} = -e\mathbf{E}_0 e^{i(\omega_0 t + 2\pi \mathbf{k}_0 \cdot \mathbf{r})} = m_e \ddot{x} \tag{4.74}$$

where m_e is the mass of the electron and $\ddot{\mathbf{x}}$ is its acceleration. The solution to (4.74) is

$$\mathbf{x} = \frac{e}{m_e \omega_0^2} \mathbf{E}_0 e^{i(\omega_0 t + 2\pi \mathbf{k}_0 \cdot \mathbf{r})} \tag{4.75}$$

and the electric moment is

$$-ex = \frac{e^2}{m_e \omega_0^2} \mathbf{E}_0 e^{i(\omega_0 t + 2\pi \mathbf{k}_0 \cdot \mathbf{r})} \tag{4.76}$$

Then the electric moment α_e induced in the electron by unit field is

$$\alpha_e = \frac{ex}{E} = -\frac{e^2}{m_e \omega_0^2} \tag{4.77}$$

By definition, the polarizability $P(\mathbf{r})$ of the material at position $\mathbf{r}$ is given by

$$P(\mathbf{r}) = E \alpha_e \rho_e(\mathbf{r}) \tag{4.78}$$

By definition, the dielectric constant $\epsilon(\mathbf{r})$ is

$$\epsilon(\mathbf{r}) = \frac{D}{E} = 1 + 4\pi \left[\frac{P(\mathbf{r})}{E}\right] = 1 + \psi(\mathbf{r}) \tag{4.79}$$

Combining (4.77), (4.78), and (4.79), we have

$$\psi(\mathbf{r}) = -4\pi \left(\frac{e^2}{m_e \omega_0^2}\right) \rho_e(\mathbf{r}) \tag{4.80}$$

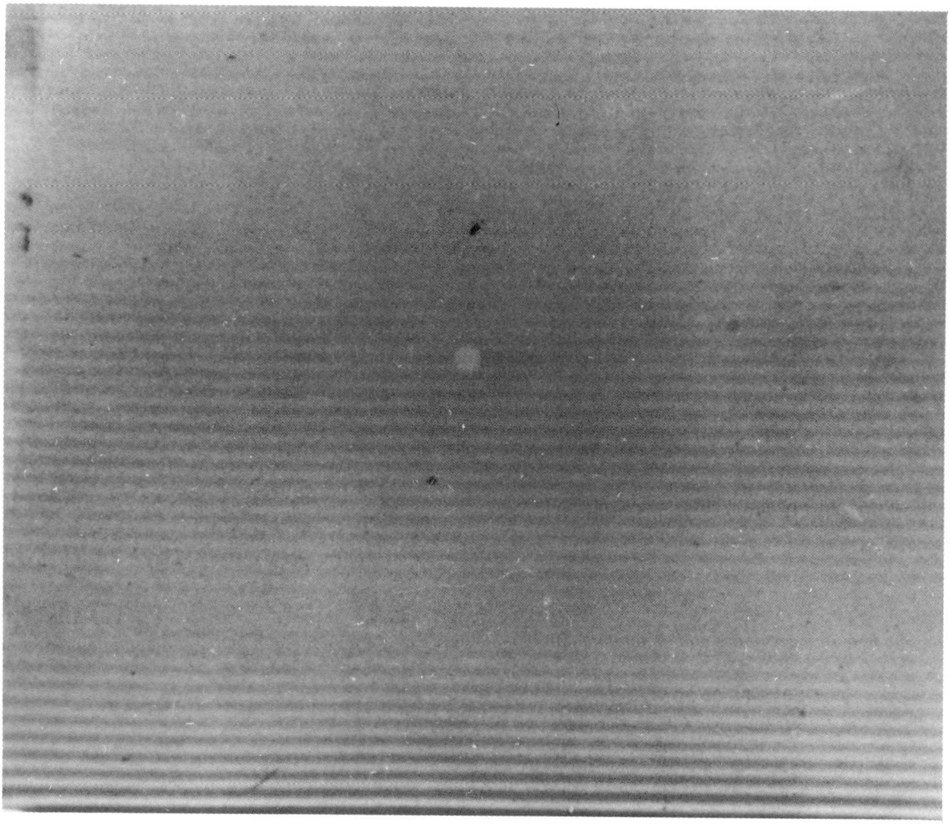

(*a*)

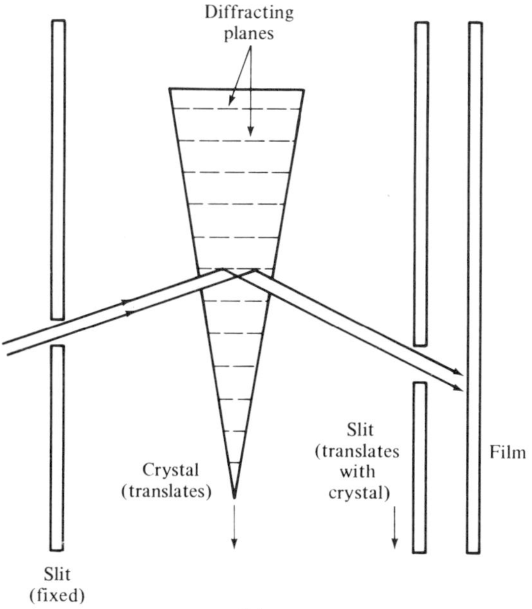

(*b*)

Figure 4.14. *Pendellösung* fringes from a silicon crystal wedge: *(a)* Laue geometry x-ray diffraction topograph [N. Kato, *Acta Cryst.*, **A25**, 119 (1969)]; *(b)* schematic representation of the experiment.

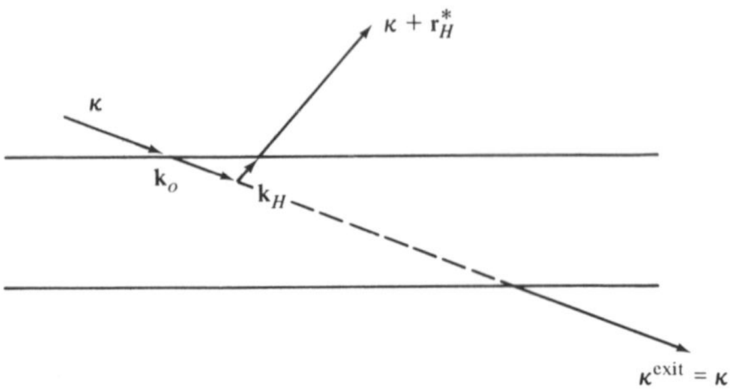

Figure 4.15. The Bragg geometry.

From its definition, the structure factor is written, using (4.80)

$$F_H = \int_{\text{cell}} \rho_e(\mathbf{r}) e^{2\pi i \mathbf{r}_H^* \cdot \mathbf{r}} dv_{\mathbf{r}} = -\frac{1}{4\pi}\left(\frac{m_e\omega_0^2}{e^2}\right)\int_{\text{cell}} \psi(\mathbf{r}) e^{2\pi i \mathbf{r}_H^* \cdot \mathbf{r}} dv_{\mathbf{r}} \quad (4.81)$$

Substituting (4.74) into (4.81), we have

$$\begin{aligned} F_H &= -\frac{1}{4\pi}\left(\frac{m_e\omega_0^2}{e^2}\right)\int_{\text{cell}} \sum_L \psi_L e^{2\pi i \mathbf{r}_L^* \cdot \mathbf{r}} e^{2\pi i \mathbf{r}_H^* \cdot \mathbf{r}}\, dv_{\mathbf{r}} \\ &= -\frac{1}{4\pi}\left(\frac{m_e\omega_0^2}{e^2}\right)\sum_L \psi_L \int_{\text{cell}} e^{2\pi i (\mathbf{r}_H^* + \mathbf{r}_L^*)\cdot \mathbf{r}}\, dv_{\mathbf{r}} \end{aligned} \quad (4.82)$$

The integral on the right was evaluated in Chap. 1 and has value $V_c\delta(\mathbf{r}_H^* + \mathbf{r}_L^*)$, where V_c is the volume of the unit cell. Thus

$$F_H = -\frac{1}{4\pi}\left(\frac{m_e\omega_0^2}{e^2}\right) V_c\psi_H \quad (4.83)$$

Rewriting ω_0 as $2\pi c\kappa$, we have

$$\psi_H = -\frac{1}{\pi}\left(\frac{e^2}{m_e c^2}\right)\frac{1}{V_C\kappa^2} F_H \quad (4.84)$$

The Fourier coefficient ψ_H is just a constant times the structure factor F_H.

One result easily obtained from this formulation is the magnitude of the index of refraction of x-rays. By definition, the index of refraction n is precisely the square root of the dielectric constant ϵ. Hence, by (4.79), ignoring squares of small terms, $n = 1 + \psi_o/2$. By (4.84)

$$\psi_0 = -\frac{1}{\pi}\left(\frac{e^2}{m_e c^2}\right)\frac{Z}{V_C\kappa^2} = -\frac{1}{\pi}\left(\frac{e^2}{m_e c^2}\right)\frac{N}{\kappa^2} \quad (4.85)$$

where Z is the number of electrons per unit cell and N is the number of electrons per unit volume. Using the appropriate values of the constants and using CuK$_\alpha$ radiation ($\lambda = 1.54$ Å), we find that

$$\psi_0\,(\mathrm{CuK}_\alpha) = -2.13 \times 10^{-29}\,N$$

For sodium chloride, $N = 6.0 \times 10^{23}\mathrm{cm}^{-3}$. Hence, for NaCl

$$\psi_0^{\mathrm{NaCl}}\,(\mathrm{CuK}_\alpha) = -1.28 \times 10^{-5}$$

Thus the index of refraction is less than unity and is of the order of 10^{-5} for a typical material.

Using the result that the deviation vectors $\boldsymbol{\delta}_0^\alpha$, $\boldsymbol{\delta}_0^\beta$, $\boldsymbol{\delta}_H^\alpha$, and $\boldsymbol{\delta}_H^\beta$ are all normal to the incident surface, we can now return to the dispersion surface to obtain a nice graphical model of the relationships involved here. Recall that $\mathbf{k}_0^p = \boldsymbol{\kappa} + \boldsymbol{\delta}_0^p$. In Fig. 4.16*a* we draw, from the resp origin, the loci of all external incident vectors $\boldsymbol{\kappa}$. This is a sphere about the origin. The allowable $\mathbf{k}_0^\alpha$ must, for a given incident $\boldsymbol{\kappa}$, be found by drawing from the point $\boldsymbol{\kappa}$ a vector normal to the external surface. The intersection point A of this surface normal vector with the $\mathbf{k}_0^\alpha$ hyperboloid defines then the appropriate value of $\mathbf{k}_0^\alpha$, as indicated in Fig. 4.16*b*. Thus $\mathbf{k}_0^\alpha$ is defined uniquely by $\boldsymbol{\kappa}$, the dispersion equation, and the incident surface normal. Continuing the surface normal vector from the κ position to its intersection point B with the β hyperboloid, we obtain the appropriate value of $\mathbf{k}_0^\beta$. Clearly the vectors from A and B to the rel point H define the diffracted wave vectors $\mathbf{k}_H^\alpha$ and $\mathbf{k}_H^\beta$. The deviations ζ_0^p, ζ_H^p are also shown.

We notice in the construction of Fig. 4.16*b* that ζ_0^α and ζ_0^β are of opposite sign. The implication here is that the diffracted intensity decreases as the sum of $|\zeta_H^\alpha|$ and $|\zeta_H^\beta|$ increases, as we shall now see. For the Laue geometry our intensity function is

$$\frac{I_H^{\mathrm{exit}}}{I_0^i} = \left|\frac{X^\alpha X^\beta\,(C_H^\alpha - C_H^\beta)}{X^\beta - X^\alpha}\right|^2 \tag{4.67}$$

Substituting from (4.60) for X^α and X^β, we have

$$\frac{I_H^{\mathrm{exit}}}{I_0^i} = \frac{\kappa \sin \xi}{2}\left|\frac{\psi_H(C_H^\alpha - C_H^\beta)}{\zeta_H^\alpha - \zeta_H^\beta}\right|^2 \tag{4.86}$$

In this, ψ_H varies only very slowly with incident angle relative to ζ_H^α, ζ_H^β and $C_H^\alpha - C_H^\beta$ is a periodic function which depends, as we have seen, on the slab thickness. Thus, aside from the thickness effect, intensity is controlled by $\zeta_H^\alpha - \zeta_H^\beta$. Since ζ_H^α and ζ_H^β are of opposite sign

$$\frac{I_H^{\mathrm{exit}}}{I_0^i} \propto \frac{1}{(|\zeta_H^\alpha| + |\zeta_H^\beta|)^2} \tag{4.87}$$

It is apparent then that the smaller the *sum* of the absolute deviation of points A and B in Fig. 4.16*b* from the H line, the greater the diffracted intensity. Clearly the greatest intensity will occur for $\boldsymbol{\kappa}$ originating at the Laue point.

Recall again that our intensity formulae require evaluation of the quantities X^α and X^β, where X^α and X^β themselves depend on ψ_H (now known) and ζ_H^α or ζ_H^β. We shall now see how the ζ_H are determined. From (4.44a), we have

$$\zeta_0^\alpha = \frac{1}{2\kappa}\,[(1-\psi_0)k_0^2 - \kappa^2] \simeq \frac{1}{2\kappa}\,[k_0^2 - (1+\psi_0)\kappa^2) \tag{4.88}$$

We now define Δ_0^α, by

$$k_0^\alpha \equiv (1+\Delta_0^\alpha)\kappa \tag{4.89}$$

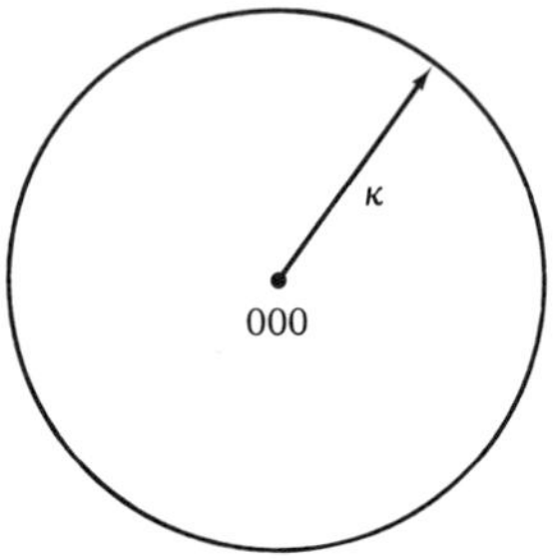

(a)

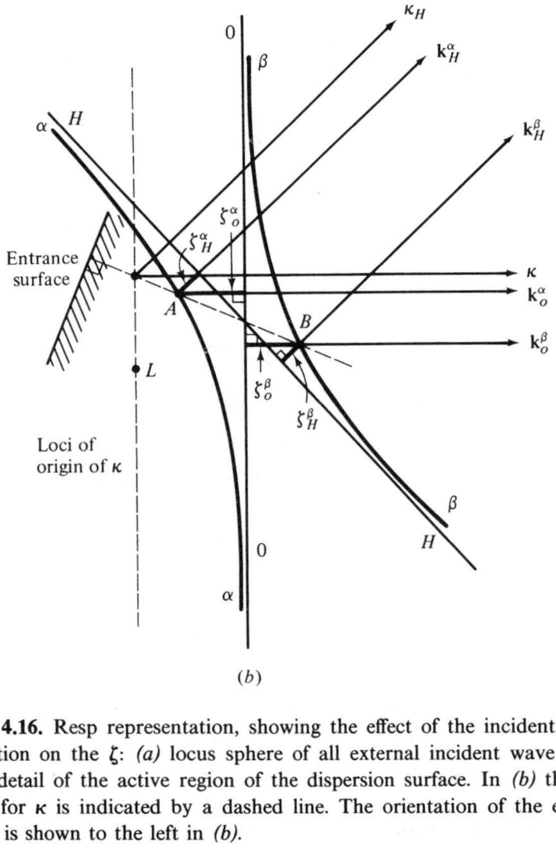

(b)

Figure 4.16. Resp representation, showing the effect of the incident surface orientation on the ζ: *(a)* locus sphere of all external incident wave vectors $\boldsymbol{\kappa}$; *(b)* detail of the active region of the dispersion surface. In *(b)* the locus sphere for $\boldsymbol{\kappa}$ is indicated by a dashed line. The orientation of the entrance surface is shown to the left in *(b)*.

Inserting (4.89) into (4.88), we find

$$\zeta_0^\alpha = \frac{1}{2\kappa}\,[(1+\Delta_0^\alpha)^2\kappa^2 - (1-\psi_0)\kappa^2] \simeq \frac{\kappa}{2}\,(2\,\Delta_0^\alpha - \psi_0) \tag{4.90}$$

Recall also that $\mathbf{k}_o^\alpha$ is written

$$\mathbf{k}_0^\alpha = \boldsymbol{\kappa} + \boldsymbol{\delta}_0^\alpha \tag{4.53a}$$

where $\boldsymbol{\delta}_0^\alpha$ is normal to the entrant surface. Therefore, combining (4.88) and (4.53a)

$$(k_0^\alpha)^2 = \kappa^2 + 2\kappa\gamma_0\delta_0^\alpha = \kappa^2 + 2\Delta_0^\alpha\kappa^2 \tag{4.91}$$

where γ_0 is the cosine of the angle between the incident beam and the entrant surface normal. In these terms

$$\Delta_0^\alpha = \frac{\gamma_0\delta_0^\alpha}{\kappa} \tag{4.92}$$

For the internal H wave we have a similar set of relations

$$\mathbf{k}_H^\alpha = \mathbf{k}_0^\alpha + \mathbf{r}_H^* = \boldsymbol{\kappa} + \boldsymbol{\delta}_0^\alpha + \mathbf{r}_H^* \tag{4.93}$$

and

$$k_H^\alpha \equiv (1+\Delta_H^\alpha)\kappa \tag{4.94}$$

Combining (4.91) and (4.94) and neglecting a term in $(\delta_0^\alpha)^2$, we find

$$\Delta_H^\alpha = \left(1 + \frac{\mathrm{r}_H^*}{\kappa}\frac{\gamma_n^H}{\gamma_0}\right)\Delta_0^\alpha + \frac{1}{\kappa^2}\,[(\mathrm{r}_H^*)^2 - 2\kappa r_H^*\gamma_H] \tag{4.95}$$

where γ_n^H is the cosine of the angle between the entrance surface normal and $\mathbf{r}_H^*$ (the normal to the diffracting plane) and γ_H is the cosine of the angle between the incident wave vector and $\mathbf{r}_H^*$ ($\gamma_H = \sin\theta_H$). Setting

$$\begin{aligned} \frac{1}{b} &\equiv 1 + \frac{\hat{\mathbf{n}}\cdot\mathbf{r}_H^*}{\hat{\mathbf{n}}\cdot\boldsymbol{\kappa}} = 1 + \frac{r_H^*}{\kappa}\frac{\gamma_n^H}{\gamma_0} \\ a &\equiv \frac{1}{k_0^2}\,[(\mathrm{r}_H^*)^2 + 2\boldsymbol{\kappa}\cdot\mathbf{r}_H^*] \simeq \frac{2}{\kappa^2}\,[(\mathrm{r}_H^*)^2 - 2\kappa r_H^*\gamma_H] \end{aligned} \tag{4.96}$$

we rewrite (4.95)

$$\Delta_H^\alpha = \frac{\Delta_0^\alpha}{b} + \frac{a}{2} \tag{4.97}$$

In the same manner as for (4.88)

$$\zeta_H^\alpha = \frac{1}{2\kappa}\,[k_H^2 - (1+\psi_0)\kappa^2] = \frac{\kappa}{2}\,(2\Delta_H^\alpha - \psi_0) = \frac{\kappa}{2}\left(2\,\frac{\Delta_0^\alpha}{\mathrm{b}} + \mathrm{a} - \psi_0\right) \tag{4.98}$$

We observe that the dispersion equation

$$\zeta_0 \zeta_H = \frac{\kappa^2}{4} \psi_H \psi_{-H} \sin^2 \xi \tag{4.46}$$

is now quadratic in Δ_0^α

$$(2\Delta_0^\alpha - \psi_0)\left(\frac{2}{b}\Delta_0^\alpha + a - \psi_0\right) = \psi_H \psi_{-H} \sin^2 \xi \tag{4.99}$$

Equation 4.99 can be solved directly for Δ_0^α. The result is

$$\left.\begin{matrix}\Delta_0^\alpha \\ \Delta_0^\beta\end{matrix}\right\} = \frac{1}{2}\left\{\psi_0 - \left[\frac{(1-\mathrm{b})\psi_0}{2} + \frac{ab}{2}\right] \pm \sqrt{b\psi_H\psi_{-H} + \left[\frac{(1-b)\psi_0}{2} + \frac{ab}{2}\right]^2}\right\} \tag{4.100}$$

Thus we now have all the information needed in order to compute the diffracted intensities for a given experimental geometry. It is convenient to call attention at this point to the parameter b. This parameter carries in it the dependence on the explicit scattering angle, via the γ_0 and γ_n^H terms.

We can explicitly now find the values of X^α and X^β. According to (4.47)

$$X^\alpha = \frac{D_H^\alpha}{D_0^\alpha} = -\frac{2\zeta_0}{\kappa\psi_{-H}\sin\xi} \tag{4.47}$$

Using (4.90)

$$X^\alpha = -\frac{2}{\kappa\psi_{-H}\sin\xi} \cdot \frac{\kappa(2\,\Delta_0^\alpha - \psi_0)}{2} = \frac{2\,\Delta_0^\alpha - \psi_0}{\psi_{-H}\sin\xi} \tag{4.101}$$

We recognize now, from (4.90) and (4.98) that

$$\begin{aligned}\zeta_0 &= \frac{\kappa}{2}(2\,\Delta_0^p - \psi_0) = \frac{\kappa\psi_{-H}\sin\xi}{2} X^p \\ \zeta_H &= \frac{\kappa}{2}\left[\frac{1}{b}(2\,\Delta_0^p - \psi_0) + a - \left(1 - \frac{\psi_0}{b}\right)\right] \\ &= \frac{k\psi_{-H}\sin\xi}{2b} X^p + \left[a - \left(1 - \frac{\psi_0}{b}\right)\right]\end{aligned} \tag{4.102}$$

Substituting (4.102) into the dispersion equation, we find

$$\left.\begin{matrix}X^\alpha \\ X^\beta\end{matrix}\right\} = \frac{-\dfrac{(a-b)}{2}\psi_0 + \dfrac{ab}{2} \pm \sqrt{b\psi_H\psi_{-H} + \left[\dfrac{(1-b)}{2}\psi_0 + \dfrac{ab}{2}\right]^2}}{\psi_{-H}} \tag{4.103}$$

Equation (4.103) is not a very simple formulation for relating D_H^p/D_0^0 to the explicit value of the scattering vector **s**. Nevertheless this form for X^α, X^β has been input to intensity formulas such as (4.67) and (4.71) to compute the theoretical intensities for various materials.

Before proceeding to a description of solutions to intensity formulae, a very useful anciliary result can be obtained, almost by inspection, from (4.100). The term b in (4.100) can be negative under some circumstances, which we shall inspect. Considering (4.96) and looking, e.g., at the case where $r_H^* > \kappa$ (always true for back reflection geometry), we see that b will be negative whenever γ_n^H is negative, since γ_0 is always the cosine of an acute angle. Thus b is negative whenever the angle between the surface normal and the H-plane normal is obtuse—that is, when we are in the Bragg, or back-reflection, geometry. Consequently, in *all* cases in which the beam enters and leaves the same surface the value of b is less than zero. For these cases the possibility exists that Δ_0^α or Δ_0^β will be complex. It is apparent that the critical value of b, call it b^*, to create a complex Δ_0 is

$$|b^*| = \left| \frac{1}{\psi_H \psi_{-H}} \left[\frac{(1 - b^*)\psi_0}{2} + \frac{ab^*}{2} \right]^2 \right| \tag{4.104}$$

Solution to (4.104) yields

$$|b^*| = 2\, \psi_H \psi_{-H} \sin \xi \frac{|\gamma_H|}{\gamma_0} \tag{4.105}$$

In obtaining (4.104) we have used the approximations to (4.96)—good when H is on or near the Ewald sphere

$$\begin{aligned} b &\approx \frac{\gamma_0}{|\gamma_H|} \\ a &\approx 2(\theta_B - \theta) \sin 2\theta_B \end{aligned} \tag{4.106}$$

Here θ_B is the exact Bragg angle and θ is any incident angle. The complex Δ_0^α in turn calls for a complex $\mathbf{k}_0$. Now the electric strength of the forward-directed ray is $\mathbf{D}_0^\alpha \exp[i(\omega t - \mathbf{k}_0 \cdot \mathbf{r})]$. A complex $\mathbf{k}_0$ therefore leads to a forward-directed ray whose magnitude must either increase or decrease as it proceeds through the crystal. If we consider the infinitely thick crystal, the field strength would increase without bound—clearly an inconsistency. Hence, the energy contained in the forward wave must die out as it proceeds into the crystal. Since we have not allowed a true absorption to occur in this model, the energy which entered with the forward wave must now be found in the diffracted wave. What we then have is a condition of total reflection.

In terms of the dispersion surface the condition of total reflection is demonstrated in Fig. 4.17. This is the case when the surface normal lies between the two sheets of the dispersion surface and no intersection with either the α or β sheet can occur. For this set of conditions there is no real solution. Now consider Fig. 4.18. From this figure we see that the breadth $\Delta^*(2\theta)$ of the region of total reflection is given by

$$\kappa\, \Delta^*(2\theta) = \frac{2\chi^*}{\sin \theta} \tag{4.107}$$

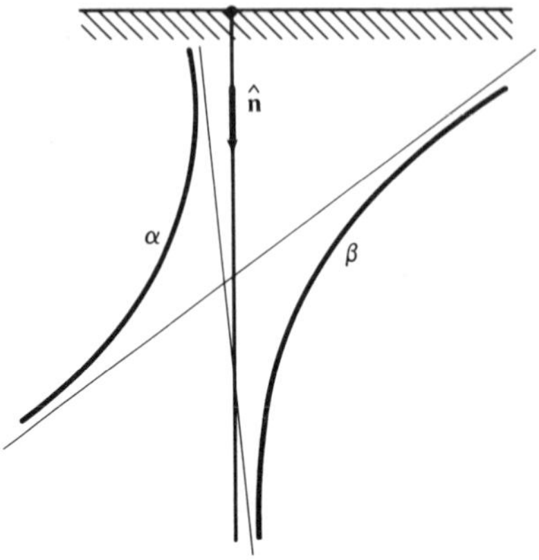

Figure 4.17. Geometrical condition for total reflection.

where $2\chi^*$ is the closest approach of the two sheets of the dispersion surface. By the geometry of Fig. 4.19 we see that

$$\zeta_0^* = \chi^* \cos\theta \tag{4.108}$$

Thus, in terms of ζ_0^*, ζ_H^*, the values of ζ_0, ζ_H at the exact Bragg angle, we have

$$\Delta^*(2\theta) = \frac{2\zeta_0^*}{\kappa \sin\theta \cos\theta} = \frac{4\zeta_0^*}{\kappa \sin 2\theta} \tag{4.109}$$

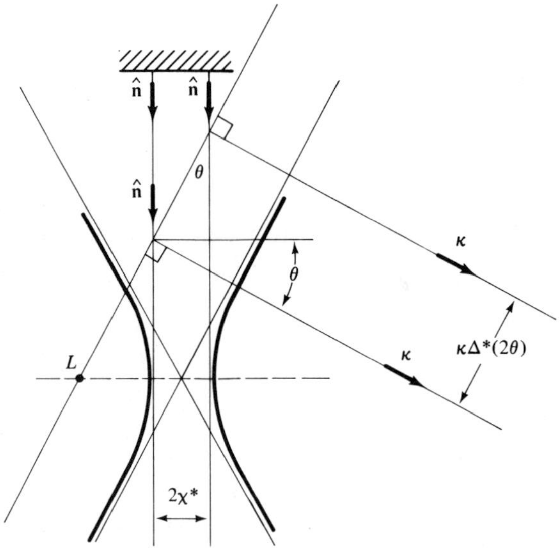

Figure 4.18. Relationship between $\Delta^*(2\theta)$ and χ^*, the resp distance between α and β.

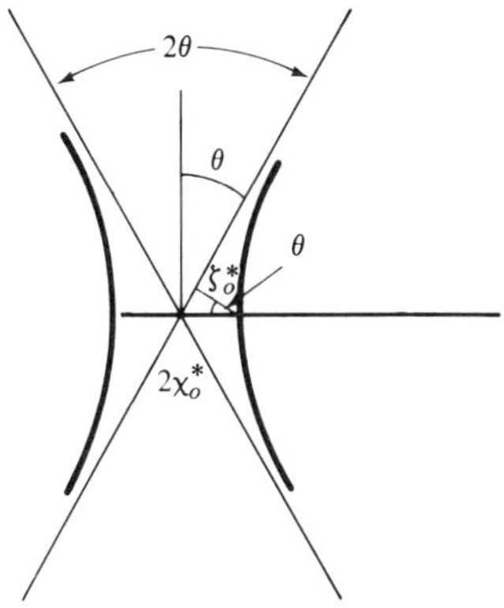

Figure 4.19. Relationship between ζ_0^* and χ^*.

For the exact Bragg condition, the dispersion equation, (4.46) yields

$$\zeta_0^* = \frac{\kappa}{2} |\psi_H| \sin \xi \tag{4.110}$$

Thus

$$\Delta^*(2\theta) = \frac{2|\psi_H| \sin \xi}{\sin 2\theta} \tag{4.111}$$

Inserting (4.84) for ψ_H, we have

$$\Delta^*(2\theta) = \frac{2}{\pi}\left(\frac{e^2}{m_e c^2}\right)\frac{\sin \xi}{V_c \kappa^2}\frac{|F_H|}{\sin 2\theta} \tag{4.112}$$

Using the appropriate values for the universal constants, we have, for perpendicular polarization ($\sin \xi = 1$)

$$\Delta^*(2\theta) = 1.79 \times 10^{-13}\ \text{cm}\,\frac{|F_H|}{V_c \kappa^2 \sin 2\theta} \tag{4.113}$$

As an example, we take the rock salt (200) reflection ($F = 84$, $a = 5.628 \times 10^{-8}$, $V_c = 3.17 \times 10^{-22}$ cm³) and CuK_α radiation ($\lambda = 1/\kappa = 1.54 \times 10^{-8}$ cm). In this case the region of total reflection is $\Delta^*(2\theta) = 1.67 \times 10^{-5}$ rad = *3.44 seconds of arc!*

The integrated reflecting power from a very perfect crystal can be estimated from the foregoing. When the full intensity expression is derived from (4.71) and (4.103), the result is as shown in Fig. 4.20. Here the intensity is plotted versus $\Delta\theta$ for a symmetrical Bragg reflection. The units of the ordinate are normalized for all material; i.e., the units of θ are in $|\psi_H|/\sin 2\theta$. We observe that the intensity rises very abruptly to unity (complete reflection) as we approach the Laue condition. Therefore to a first approximation the integrated intensity would be

$$I_{hkl}^{\text{INT}} = \int I d\theta \simeq 1 \cdot \Delta(2\theta) = \frac{2}{\pi}\left(\frac{e^2}{m_e c^2}\right)\frac{\sin \xi}{V_c \kappa^2}\frac{|F_H|}{\sin 2\theta} \tag{4.114}$$

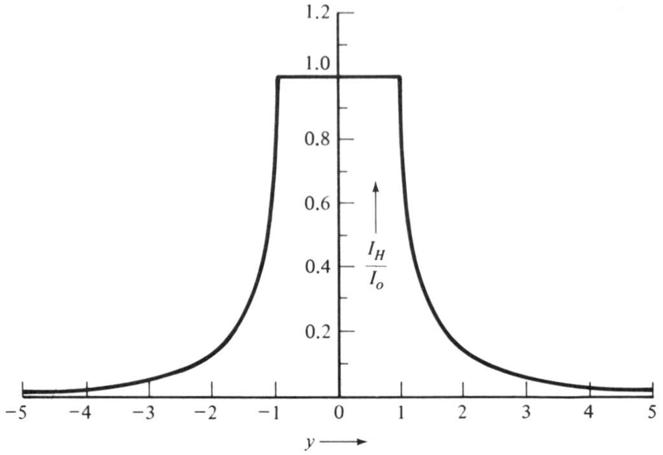

Figure 4.20. The diffracted intensity I_H/I_O, plotted against $\Delta(2\theta)$. The units of $\Delta(2\theta)$ are $|\psi_H|/\sin 2\theta$.

The correct calculation, which takes the wings of the curve into account yields

$$I_{hkl}^{\text{INT}} = \left(\frac{e^2}{m_e c^2}\right) \frac{\sin \xi}{V_c \kappa^2} \frac{|F_H|}{\sin 2\theta} \tag{4.115}$$

When the incident beam is not polarized, we average the intensities for $\xi = 90°$ and $\xi = 90° - 2\theta$. Thus

$$I_{hkl}^{\text{INT}} = \left(\frac{e^2}{m_e c^2}\right) \frac{\lambda^2}{\sin 2\theta} \left(\frac{1 + |\cos 2\theta|}{2}\right) \frac{|F_H|}{V_c} \tag{4.116}$$

Let us now compare this result with that of the kinematic (single-scattering) theory. The single-scattering result was

$$(I_{hkl}^{\text{INT}})_{\text{kin}} = \left(\frac{e^2}{m_e c^2}\right)^2 \frac{\lambda^3}{\sin 2\theta} \left(\frac{1 + \cos^2 2\theta}{2}\right) \frac{|F_H|^2}{2\mu\rho V_c} \tag{4.117}$$

The ratio of kinematic and dynamic results is then

$$\frac{(I_{hkl}^{\text{INT}})_{kin}}{(I_{hkl}^{\text{INT}})_{dyn}} = \left(\frac{e^2}{m_e c^2}\right) \frac{\lambda}{V_c} \frac{1}{2\mu} \frac{1 + \cos^2 2\theta}{1 + |\cos 2\theta|} |F_H| \tag{4.118}$$

For rock salt the value of this ratio is 5.3. As the number of electrons per unit cell increases, $|F_H|$ increases and the degradation of intensity due to primary extinction decreases.

It is useful to point out here that two new quantities—an extinction coefficient and an extinction length—may be defined for perfect crystals. As we have just seen, in the region of total reflection the wave vector $\mathbf{k}_0$ includes an imaginary portion δ_{0i}, where δ_{0i} is equal, via (4.92), to $\kappa\Delta_{0i}^{\alpha}/\gamma_0$. By (4.100)

$$\delta_{0i} = \frac{\kappa}{\gamma_0}\left\{b|\psi_H|^2 + \left[\frac{(1-b)\psi_0}{2} + \frac{ab}{2}\right]^2\right\}^{1/2} \quad (4.119)$$

At the center of the region of total reflection, $b \simeq \gamma_0/\gamma_H = -1$ (4.104) and

$$\left[\frac{(1-b)\psi_0}{2} + \frac{ab}{2}\right] = 0 \quad (4.120)$$

Thus at the perfect Bragg condition

$$\delta_{0i} = i\kappa|\psi_H|\hat{n}/\gamma_0 \quad (4.121)$$

where $\hat{n}$ is the surface normal. Inserting into the wave equation $\mathbf{D}_0^{\max}$ $\exp(2\pi i\mathbf{k}_0 \cdot \mathbf{r})$, we obtain the beam attenuation factor $A = |D_0|^2/|D_0^{\max}|^2$ as a function of depth d as

$$A = e^{-4\pi\kappa|\psi_H|d} \quad (4.122)$$

We thus define an extinction coefficient μ_{ext}—similar to a linear absorption coefficient μ_a', as

$$\mu_{\text{ext}} = 2\pi\kappa|\psi_H| = 5.64 \times 10^{-13}\frac{\lambda}{V_c}|F_H| \quad (4.123)$$

We compare in Table 4.1 the values of μ_{ext} and μ_a (for the exact Laue condition) for the (200) reflections of several FCC metals, using CuK_α radiation. The extinction coefficient is seen to be several times greater than the linear absorption coefficient. The ratio of the two decreases at higher atomic numbers. This is expected since the extinction coefficient goes as Z (through $|F_H|$), whereas the absorption coefficient goes as Z^3.

TABLE 4.1
EXTINCTION AND LINEAR ABSORPTION COEFFICIENTS FOR THE CuK_α (200) REFLECTION OF SOME FCC METALS

Material	*Atomic number*	μ_{ext}, cm^{-1}	μ_a', cm^{-1}	$\frac{\mu_{\text{ext}}}{\mu_a'}$	Extinction distance, h, in microns
Al	13	4,680	131	35.8	2.14
Cu	29	15,320	472	32.4	0.65
Ag	47	17,900	2290	7.8	0.56
Pt	78	35,400	4290	8.2	0.48

The extinction distance h may be defined as the penetration at which the forward-directed intensity has been attenuated to $1/e$ of its initial value. Thus $h = 1/\mu_{\text{ext}}$. The extinction distances for Al, Cu, Ag, and Pt are shown in Table 4.1. These values are generally of the order of one micron. What we learn from this is that ordinary absorption of the beam does not become very

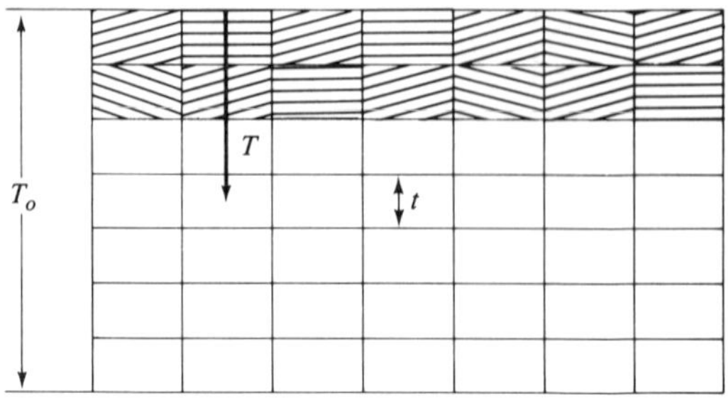

Figure 4.21. The mosaic crystal of average block thickness t_0.

important for perfect crystals, because the beam has totally reflected within a depth so small that absorption can have little effect.

We have now seen how primary extinction affects the intensity of scattering by a perfect crystal. Suppose now that the crystal were slightly imperfect. As a model, let us suppose that the crystal is built up of blocks, each of which are perfect, but each of which is misoriented sufficiently far with respect to its neighbors that the depth of dynamical interaction will be limited to the dimensions of each block. A model for such a *mosaic crystal* is shown in Fig. 4.21. If we consider the block-limiting entities as either subgrain boundaries about perfect crystal or as the effect of gentle continuous lattice curvature by individual dislocations the model appears to be a good representation of reality. Let us obtain now an expression for the integrated reflecting power of a mosaic crystal.

To obtain the reflecting power for the Bragg geometry we take a crystal of thickness T_0 and assume an average block thickness t_0, as indicated in Fig. 4.21. Each block will totally reflect an integrated intensity $P(t)$ of the incident beam. At some depth T the amount of power taken from the incident-directed wave is given by

$$dP_0 = -\mu_a' P_0 \frac{dT}{\gamma_0} - \frac{I_{hkl}^{\text{INT}} P_0}{t_0} dT + \frac{I_{hkl}^{\text{INT}} P_H}{t_0} dT \tag{4.124}$$

Here the first term on the right accounts for normal absorption, the second term accounts for total reflection into the diffracted direction and the third term accounts for redirection of the diffracted beam entering from the bottom. The I_{hkl}^{INT} terms in (4.124) are (4.74) (Laue geometry) or (4.71) (Bragg geometry) integrated over 2θ.[4] Similarly the power directed into reflection in this layer is

$$dP_H = \mu_a' P_H \frac{dT}{|\gamma_n^H|} + \frac{I_{hkl}^{\text{INT}} P_H}{t_0} dT - \frac{I_{hkl}^{\text{INT}} P_0}{t_0} dT \tag{4.125}$$

[4] Only if the mosaic block size is larger than the extinction distance can the I_{hkl}^{INT} approach our thick crystal solution (4.116). Otherwise the entire (4.67) or (4.71) must be used to obtain I_{hkl}^{INT}. The case of a very small mosaic block thickness is, however, also easily obtained and will be considered in the following paragraphs.

These two simultaneous equations can be solved for the symmetrical Bragg case using the boundary conditions

$$\begin{aligned} P_0(T) &= P_0(0) \text{ at } T=0 \\ P_H(T) &= 0 \text{ at } T=T_0 \end{aligned} \tag{4.126}$$

The solution yields for the integrated reflection

$$\frac{P_H(0)}{P_0(0)} = \frac{\sigma \dfrac{\mu_a'}{\gamma_0} - U}{\sigma} - \frac{U\left[\sigma + \dfrac{\mu_a'}{\gamma_0} - U\right] e^{-UT_0}}{\sigma\left[\left(\sigma + \dfrac{\mu_a'}{\gamma_0}\right)\sinh UT_0 + \mathrm{U}\cosh UT_0\right]} \tag{4.127}$$

where

$$\begin{aligned} \sigma &\equiv \frac{I_{hkl}^{\mathrm{INT}}}{t_0} \\ U &\equiv \left(\sigma + \frac{\mu_a'}{\gamma_0}\right)^2 - \sigma^2 \end{aligned} \tag{4.128}$$

For a very thick crystal, the exponent on the right of (4.127) vanishes and we have

$$\frac{P_H(0)}{P_0(0)} = \frac{\sigma + \dfrac{\mu_a'}{\gamma_0} - \left(\sigma + \dfrac{\mu_a'}{\gamma_0}\right)^2 - \sigma^2}{\sigma} \tag{4.129}$$

The corresponding result for the Laue geometry is

$$\frac{P_H(T_0)}{P_0(0)} = \sinh\left[\sigma T_0 e^{-\left(\frac{\mu'_a}{\gamma_0} + \sigma\right) T_0}\right] \tag{4.130}$$

Equations (4.129) and (4.130) are simple in form, but require prior determination of the σ terms. These terms call for integration of the intensity functions (4.67) or (4.71) over 2θ and then an averaging of the integration result over some reasonable range of mosaic block thicknesses. Recall that the intensity functions both contain *pendellösung* terms—oscillations of intensity through the thickness. Thus, the thickness t_0 is important in determining the magnitude of I_{hkl}^{INT} and, for a real crystal, with a range of t_0, the average I_{hkl}^{INT} requires close attention to the *pendellösung* term. A good discussion of the semiempirical treatment of this problem can be found in W. H. Zachariasen, *Theory of X-ray Diffraction in Crystals* (Dover, N.Y., 1945). This work is beyond the scope of the present text. It is sufficient to note that in the case of an adequately small mosaic block thickness the kinematic result is obtained. For mosaic block thicknesses in excess of about

$$(t_0)_{\mathrm{crit}} \simeq \frac{2\sin^2\theta}{1+|\cos 2\theta|}\left(\frac{m_e c^2}{e^2}\right)\frac{V_c}{\lambda}\frac{1}{|F_H|} \tag{4.131}$$

one can no longer use the kinematic result. To look at a number, $(t_0)_{\text{crit}}$ for the 100 reflection of a simple cubic crystal of parameter $a = 3.00$ Å and using $\lambda = 1.54$ Å yields $(t_0)_{\text{crit}} \simeq 0.00082/|F_H|$. Taking $z = 25$ as a middle-range value, $|F_H|_{z=25} \simeq 20$. In this case, then, $(t_0)_{\text{crit}} \simeq 4 \times 10^{-5}$cm. If we now identify a mosaic block as the domain between neighboring dislocations, we see that the critical dislocation density for this case is about 10^8 lines/cm^2. Metal crystals typically have grown-in dislocation densities of this magnitude (or greater), whereas ionic salts and covalently bonded crystals, if produced with any care can fall well below this defect level.

4.6 THE BORRMANN EFFECTS

In 1950 (after previous experiments dating from 1941), Borrmann published a remarkable x-ray photograph.[5] A very perfect, 0.27-cm thick calcite ($CaCO_3$) crystal was set in the Laue geometry, at the precise Laue condition, using CuK$_\alpha$ radiation. A film was placed beyond the exit surface, to record any possible forward or diffracted x-radiation. Using tabulated mass absorption coefficients, the beam should have been attenuated to about 10^{-12} its initial intensity. Nevertheless, after exposure of one hour not only one, but three distinct spots were found on the film. The positions of the x-ray spots, relative to the incident geometry were as sketched in Fig. 4.22. The spot P appeared as a continuation

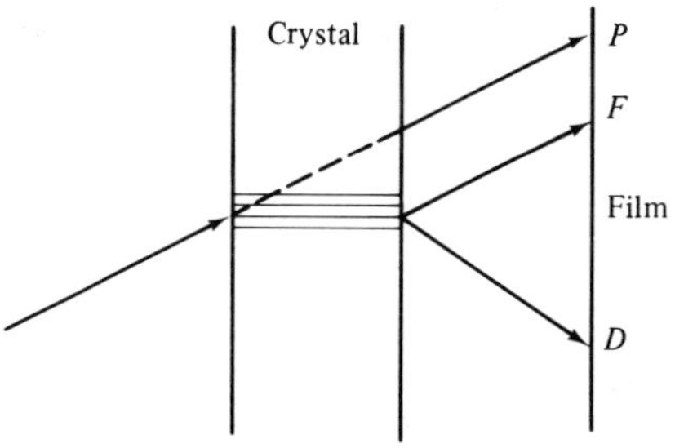

Figure 4.22. The geometry of the Borrmann effect.

of the primary beam. The spots D and F were found at positions such that they must have arisen from energy transported across the crystal along the diffraction plane (!!) and then separated into forward and diffracted directions at the exit surface. In this section we shall look at the nature of these two effects of Borrmann: the transport of energy along the diffraction plane and the anomalously low absorption at or near the exact Laue condition.

As we shall see, it is convenient to first examine the energy transport. Formally this problem can be handled by a Poynting vector approach.[6] However, we can more easily treat the concepts involved in another way. The total electrical

[5] G. Borrmann, *Z. Physik*, **127**, 297 (1950).

[6] B. W. Batterman and H. Cole, *Rev. Mod. Phys.*, **36**, 681 (1964).

displacement $\mathbf{D(r)}$ contained in the portion of the wave energy which is interacting with the crystal is the sum of the forward and diffracted amplitudes $\mathbf{D}_0$ and $\mathbf{D}_H$

$$\mathbf{D(r)} = \mathbf{D}_0 e^{2\pi i(\nu t + \mathbf{k}_0 \cdot \mathbf{r})} + \mathbf{D}_H e^{2\pi i(\nu t + \mathbf{k}_H \cdot \mathbf{r})} \tag{4.132}$$

Using Fig. 4.23, we write

$$\begin{aligned} \mathbf{k}_0 &= \mathbf{k}_\parallel - \mathbf{h} \\ \mathbf{k}_H &= \mathbf{k}_\parallel + \mathbf{h} \end{aligned} \tag{4.133}$$

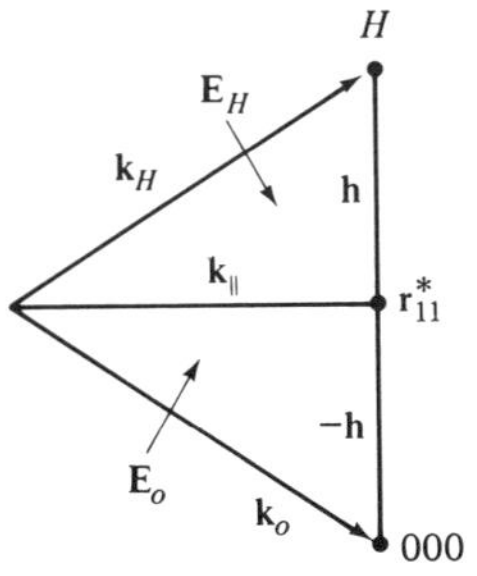

Figure 4.23. Vector relations required in the analysis of the Borrmann effect.

Inserting (4.132) into (4.131), we have

$$\begin{aligned} \mathbf{D(r)} &= \mathbf{D}_0 e^{2\pi i(\nu t + \mathbf{k}_\parallel \cdot \mathbf{r})} e^{-2\pi i \mathbf{h}\cdot\mathbf{r}} + \mathbf{D}_H e^{2\pi i(\nu t + \mathbf{k}_\parallel \cdot \mathbf{r})} e^{2\pi i \mathbf{h}\cdot\mathbf{r}} \\ &= e^{2\pi i(\nu t + \mathbf{k}_\parallel \cdot \mathbf{r})} (\mathbf{D}_0 e^{-2\pi i \mathbf{h}\cdot\mathbf{r}} + \mathbf{D}_H e^{2\pi i \mathbf{h}\cdot\mathbf{r}}) \end{aligned} \tag{4.134}$$

For the polarization of $\mathbf{D}_0$ and $\mathbf{D}_H$ normal to the plane of incident and diffracted rays we have

$$D^\perp(\mathbf{r}) = e^{2\pi i(\nu t + \mathbf{k}_\parallel \cdot \mathbf{r})} \left(e^{-2\pi i \mathbf{h}\cdot\mathbf{r}} + \frac{D_H^\perp}{D_0^\perp} e^{2\pi i \mathbf{h}\cdot\mathbf{r}} \right) \mathbf{D}_0^\perp \tag{4.135}$$

The ratio $D_H^\perp / D_0^\perp$ at the exact Bragg condition is, according to (4.46), (4.47), and (4.91)

$$\left(\frac{D_H^\perp}{D_0^\perp} \right)_{\text{exact}} = \frac{\kappa \psi_H}{2\zeta_H} = \pm \frac{\psi_H}{|\psi_H|} = \pm \frac{F_H}{|F_H|} \tag{4.136}$$

Examination of Fig. 4.16*b* shows that ζ_H is positive in the α-sheet and negative in β; consequently the sign in (4.136) is $+$ for α and $-$ for β.

Now consider the structure factor of some simple unit cells. For simple cubic, with an origin chosen at some arbitrary position x,y,z

$$F_H^{\text{sc}} = f e^{2\pi i (hx + ky + lz)} \tag{4.137}$$

Only if the origin is chosen at one of the atomic sites is $F_H^{\text{sc}} = f$. In any case $|F_H^{\text{sc}}|^2 = f^2$. But if and only if the unit cell origin is chosen at an atomic site will $F_H^{\text{sc}}/|F_H^{\text{sc}}| = 1$. Similarly, the structure factor for BCC is, in general

$$F_H^{\mathrm{BCC}} = \begin{cases} 2f_e e^{2\pi i(hx_1+ky_1+lz_1)} \text{ for } h+k+l \text{ even} \\ 0 \text{ for } h+k+l \text{ odd} \end{cases} \tag{4.138}$$

Again, only when the origin is chosen at an atomic site and the axes chosen along cube edges will be $F_H^{\mathrm{BCC}}/|F_H^{\mathrm{BCC}}| = 1$. This argument can be extended to all centrosymmetric unit cells. Using this argument

$$\left(\frac{D_H^{\perp}}{D_0^{\perp}}\right)^{\alpha}_{\text{exact}} = +1$$
$$\left(\frac{D_H^{\perp}}{D_0^{\perp}}\right)^{\beta}_{\text{exact}} = -1 \tag{4.139}$$

if the origin of coordinates is chosen at an atomic position in the diffracting plane. Substituting into (4.135), we now find

$$[\mathbf{D}^{\perp}(\mathbf{r})]^{\alpha} = e^{2\pi i(\nu t+\mathbf{k}_{\|}\cdot\mathbf{r})}(\cos 2\pi\mathbf{h}\cdot\mathbf{r})\mathbf{D}_0^{\perp})^{\alpha}$$
$$[\mathbf{D}^{\perp}(\mathbf{r})]^{\beta} = ie^{2\pi i(\nu t+\mathbf{k}_{\|}\cdot\mathbf{r})}(\sin 2\pi\mathbf{h}\cdot\mathbf{r})(\mathbf{D}_0^{\perp})^{\beta} \tag{4.140}$$

On the right of either expression (4.140) the first term is a wave propagating along the diffracting plane. The second term is a standing wave normal to the diffracting plane. Thus while the field propagates along the diffracting plane—explaining one of the Borrmann effects—its magnitude varies periodically along the normals to the diffracting planes.

The last finding allows us qualitatively to understand Borrmann's anomalous absorption. First let us find the intensities associated with $\mathbf{D}^{\perp}(\mathbf{r})$. Here, using $\mathbf{h} = \mathbf{r}_H^*/2$

$$|[\mathbf{D}^{\perp}(\mathbf{r})]^{\alpha}|^2 = [(D_0^{\perp})^{\alpha}]^2 \sin^2(\pi\mathbf{r}_H^*\cdot\mathbf{r})$$
$$|[\mathbf{D}^{\perp}(\mathbf{r})]^{\beta}|^2 = [(D_0^{\perp})^{\beta}]^2 \cos^2(\pi\mathbf{r}_H^*\cdot\mathbf{r}) \tag{4.141}$$

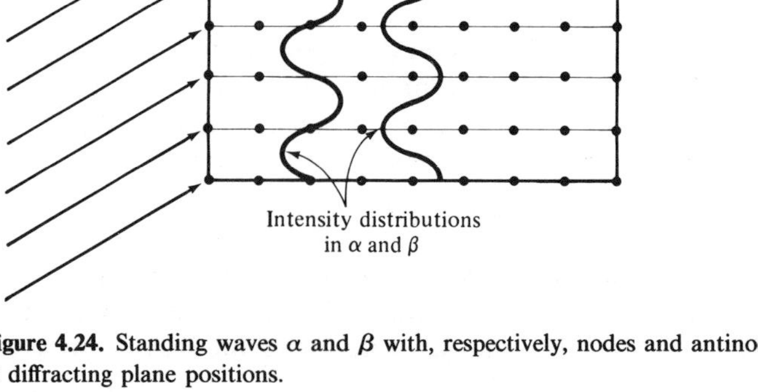

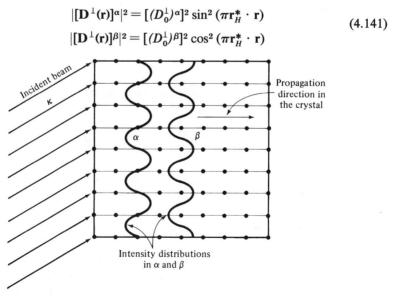

Figure 4.24. Standing waves α and β with, respectively, nodes and antinodes at diffracting plane positions.

Recalling that our origin of $\mathbf{r}$ is in the H-plane and that $\mathbf{r}_H^*$ is normal to the H plane, then $\mathbf{r}_H^* \cdot \mathbf{r} = d_H r_H$, where d_H is the distance perpendicular to an H-plane. Thus, as indicated in Fig. (4.24) the intensity of the α wave has its nodes directly on the diffracting planes, whereas the β wave has its antinodes at the diffracting planes. Recalling that, in general, absorption deals with the transfer of energy to the atomic masses, it is evident that the β wave will efficiently transfer such energy. On the other hand, the interaction of the α wave with the atom cores is very slight, the wave having its nodes at the mean core positions. Consequently the α wave will be little attenuated as it proceeds through the crystal.

The preceding argument was set forth using the perpendicular polarization of the wave. The same approach yields the same result for the parallel polarization, although one must carry through a larger number of geometric terms.

4.7 STRAIN FIELD CONTRAST

The prime useful manifestation of dynamical interactions relates to x-ray topography. As mentioned in Sect. 4.1, the difference in scattering intensity from perfect and defective regions within a crystal permits the direct observation of locallized strain regions in crystals. In fact, one can determine the direction and magnitude of such strain fields by these methods. In the following, we shall first describe the methods of x-ray topography. Then we shall look at what can be inferred from the topographic images. Reviews by Lang[7] and by Authier[8] can be consulted for further treatment.

Topographs can be taken in either Bragg (back-reflection) or Laue (transmission) geometry. In Bragg geometry the method is termed "Berg-Barrett"; in the Laue geometry, the instrumentation is often associated with A. R. Lang. Figures 4.25 and 4.26 show the two geometrical arrangements. (We shall look briefly at some variants of these later.)

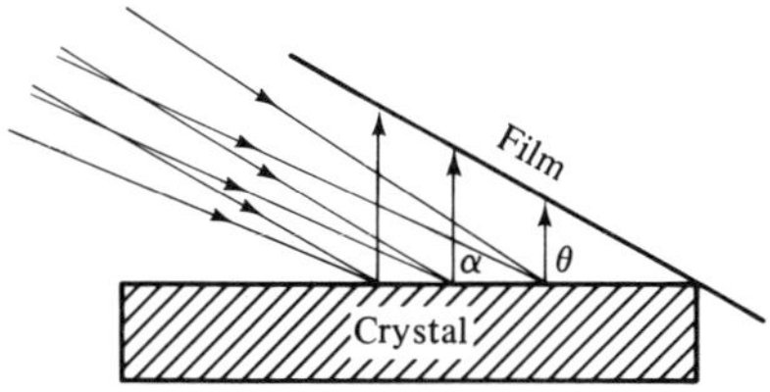

Figure 4.25. The Berg-Barrett geometry.

In the Berg-Barrett method a slightly divergent beam is incident at the Bragg angle onto a relatively perfect crystal (e.g., a dislocation density less than some 10^6 lines/cm^2). The beam is divergent so that the full kinematic

[7] A. R. Lang, in *Modern Diffraction and Imaging Techniques in Materials Science,* S. Amelinckx, R. Gevers, G. Remaut, J. van Landuyt, editors (North-Holland Publishing Co., Amsterdam, 1970), pp. 407–79.

[8] A. Authier, *ibid,* pp. 481–520.

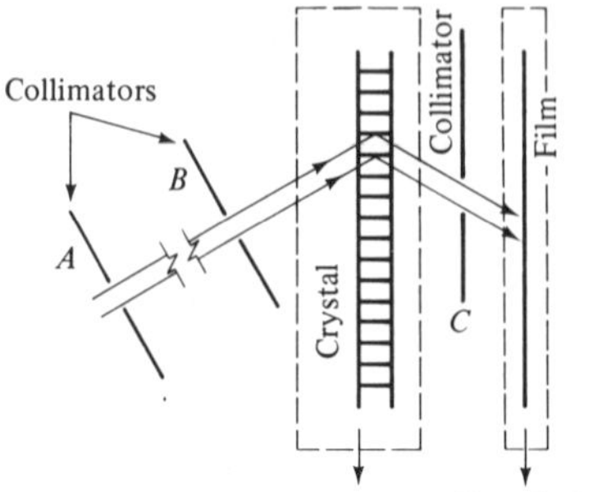

Figure 4.26. The Laue (Lang) geometry.

integrated intensity will be obtained from the defective regions. The integrated intensities of the perfect crystal regions will be quite low and the regions of lattice distortion stand out in vivid contrast (see Fig. 4.8).

Figure 4.27 shows four Berg-Barrett topographs of the same region of a zinc single crystal. In this case the exposed face was the basal plane, (0001). The reflections are from the pyramidal planes, of the form $(11\bar{2}3)$. Notice that no dislocation image is present in all topographs. This is the behavior which we should intuitively expect. Consider two extreme cases of diffraction from an edge-dislocated region, as in Fig. 4.28. For case (a) there is no distortion of the diffracting plane, whereas in (b) there is. In fact, the cases (a) and (b) are the extreme cases of diffracting plane distortion. Any other plane, such, for example, as the dashed plane in Fig. 4.28*a*, would show an intermediate degree of distortion. In (a), $\mathbf{s} \cdot \mathbf{b} = 0$, whereas in (b), $\mathbf{s} \cdot \mathbf{b} = 2b \sin \theta/\lambda$. The normalized function $(\mathbf{s}/|s|) \cdot (\mathbf{b}/|b|)$ runs through a range of zero to one between the extremes. The value of this normalized quantity is conventionally taken as a measure of the ability of a dislocation to enhance the local diffracted intensity.

This concept is frequently used to determine Burgers' vectors in topography. If no dislocation image is seen in diffraction from a given *hkl* plane, whereas it had been seen in other reflections, the dislocation must have its Burgers' vector in that *hkl* plane. If no disappearance of dislocation image is noted, then application of the $\mathbf{s} \cdot \mathbf{b}$ rule to the observed intensities can allow one to *estimate* the direction of the Burgers' vector.

While this $\mathbf{s} \cdot \mathbf{b}$ concept has been discussed in the context of the Bragg geometry, it is equally valid for the transmission case.

The appearance of transmission topographs depends on the crystal thickness *s*. For example, compare Figs. 4.29*a* and *b*. These represent topographs taken through silicon wafers 500 and 4000 microns thick, respectively. In (a) the dislocation images are dark, and doubled. In (b) the images are white and very diffuse. The thin crystal case is similar to the imaging in Berg-Barrett topography; the locally strained regions effectively direct more intensity into the diffracted beam.

To understand the contrast effect for thick crystals, we need to return first briefly to our discussion of the direction of energy flow in a crystal. Recall

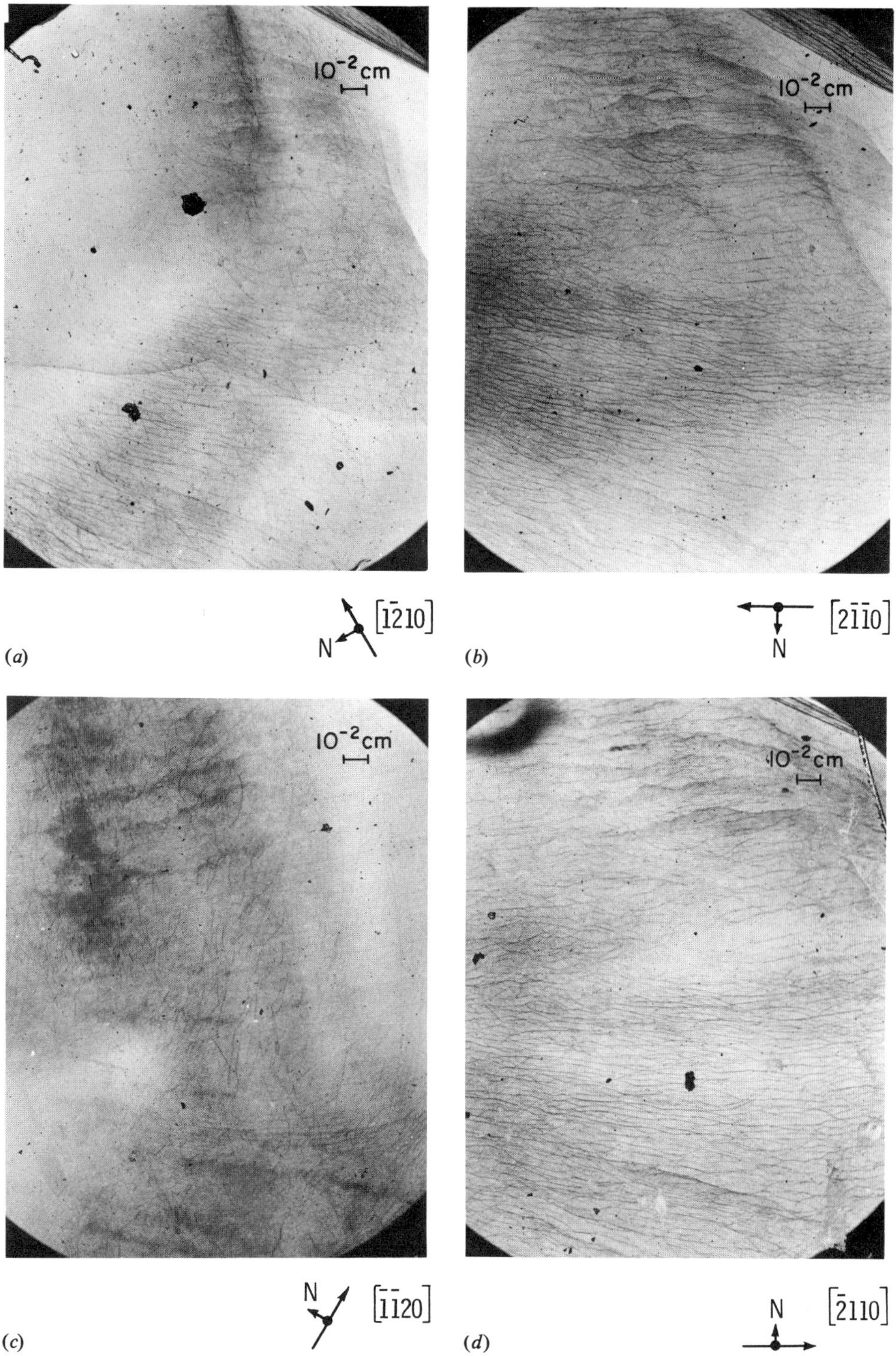

Figure 4.27. Berg-Barrett topographs of the same region of a zinc crystal. Each topograph was made using a different $(11\bar{2}3)$ reflection. [J. M. Schultz and R. W. Armstrong, *Phil. Mag.*, **10,** 497 (1964)]

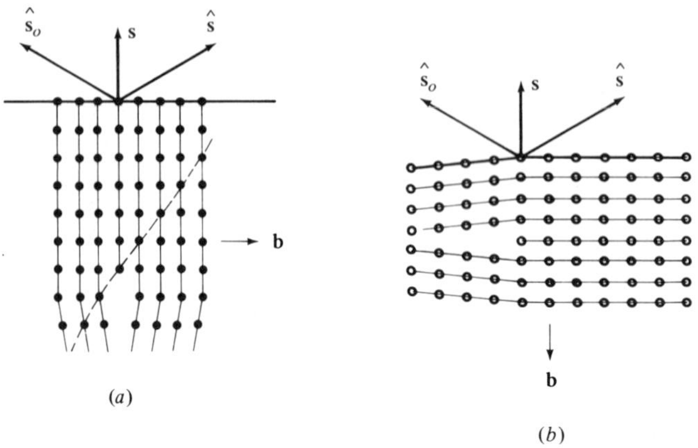

Figure 4.28. Schematic illustration of diffraction from the elastically distorted regions near edge dislocations whose Burgers' vectors lie *(a)* parallel and *(b)* perpendicular to the diffracting plane. In *(a)* no topographic image should form, because the diffracting planes themselves are not distorted.

that we have seen that, for a crystal set at the perfect center of the diffraction condition, the x-ray energy flowed along the diffracting planes. The case of an incident beam *not* passing through the Laue point (i.e., off the perfect condition) is less easy to analyze. In fact, one must then consider that the beam is (usually) not perfectly plane either and that one excites some *area* of the dispersion surfaces, as indicated in Fig. 4.30. It turns out that, in such a case, the energy of the packet of waves propagates normal to the excited portion of the dispersion surface.[9] This result is consistent with our previous one, in which the wave energy travelled perpendicular to the dispersion surface for the perfect diffraction condition. Recall now that the bend of the dispersion surface is really very tight; if the dispersion surface and the Ewald sphere were drawn to the same scale, the dispersion surface bend would appear almost like a sharp kink. Any reasonable divergence in the incident beam should result, then, in excitation of the entire bend. For this case, the propagation of energy through the crystal will follow the path sketched in Fig. 4.31; paths will fill the triangle *ABC*. Paths will be most dense near *AB* and *AC*, since these represent the approach of the dispersion surfaces to their asymptotes. For a very thick crystal, only the path *AD* will be active, as only here do the nodes of the wave from one of the dispersion surfaces fall on the atomic sites. For such a case, an imperfection in the path *AD* will cause the energy there to be normally absorbed, and a white region should appear on the film. Dislocation images should appear as narrow white images on a dark background. Figure 4.32 is an example, a topograph from a relatively very thick sheet (0.4 mm) of silicon iron. The relative diffuseness of the dislocation images in Fig. 4.29*b* can be explained using the

[9] See, for example, the following for an analysis of the physics involved: B. W. Batterman and H. Cole, *Rev. Mod. Phys.*, **36**, 681–717 (1964).

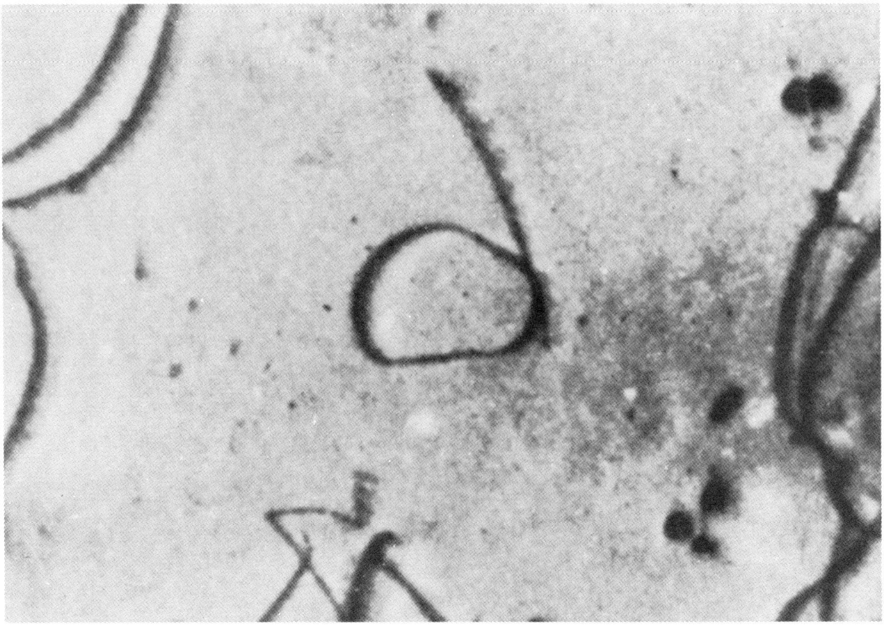

(*a*)

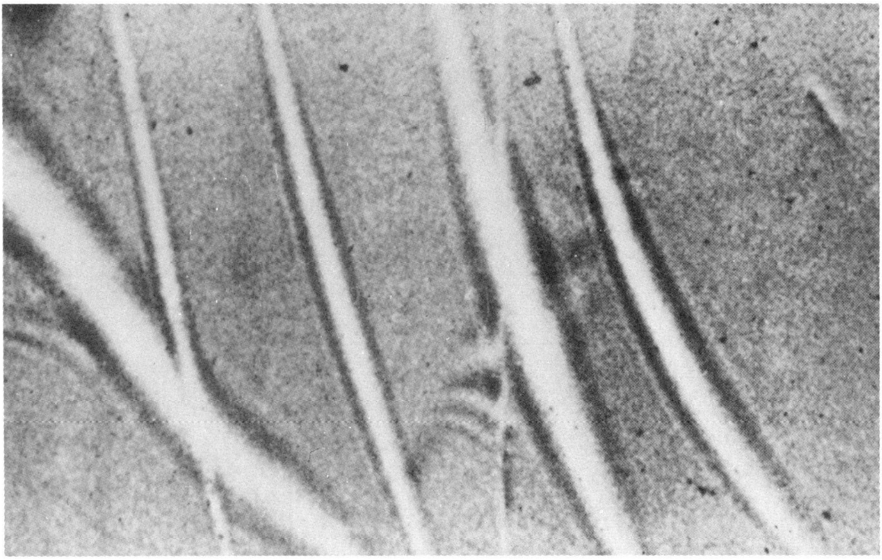

(*b*)

Figure 4.29. Crystal thickness effect in Lang topographs. In *(a)* the crystal is 500 microns thick, in *(b)* 4000 microns thick. [A. Authier, in J. B. Newkirk and G. R. Mallett, eds., *Advances in X-ray Analysis,* **10,** 9 (Plenum Press, New York, 1967)]

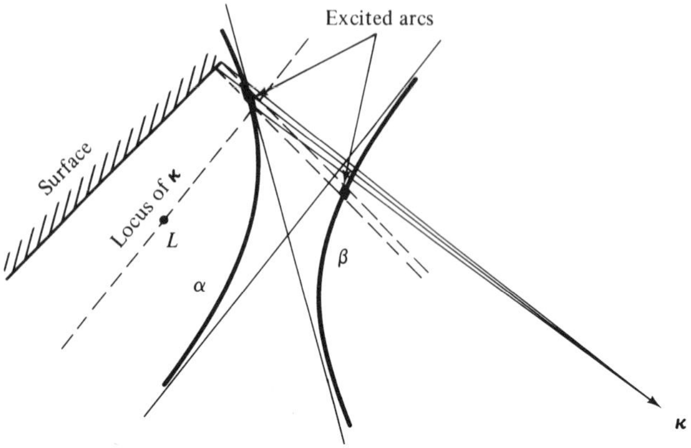

Figure 4.30. Excitation of a portion of the dispersion surface by a slightly divergent beam.

same set of concepts. Here the thickness t of the crystal is not so large, relative to $1/\mu_a'$, as is the case for Fig. 4.30. Energy paths other than AD are still active. Hence, any point in the crystal will be intersected by many paths, since the crystal surface is bathed uniformly by the x-ray beam and the triangle ABC is thereby shifted along the entrance surface. The effect of the many intersections of a defect region will be to smear its image.

Mention should be made of the translation indicated in Fig. 4.26. For good resolution, the primary beam should be highly collimated and the collimator C should also be very narrow. But in this case only a small section of the specimen will be imaged. What is done, then, is to translate specimen and film together past the collimators. The large topograph thus formed is termed a "projection topograph," contrasted to a "section topograph" (taken without translation).

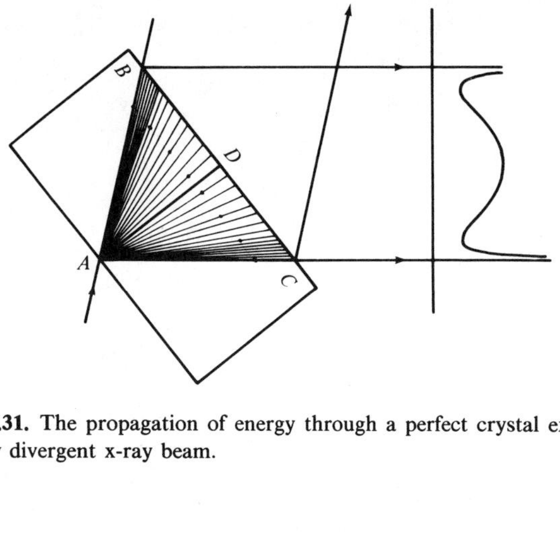

Figure 4.31. The propagation of energy through a perfect crystal excited by a slightly divergent x-ray beam.

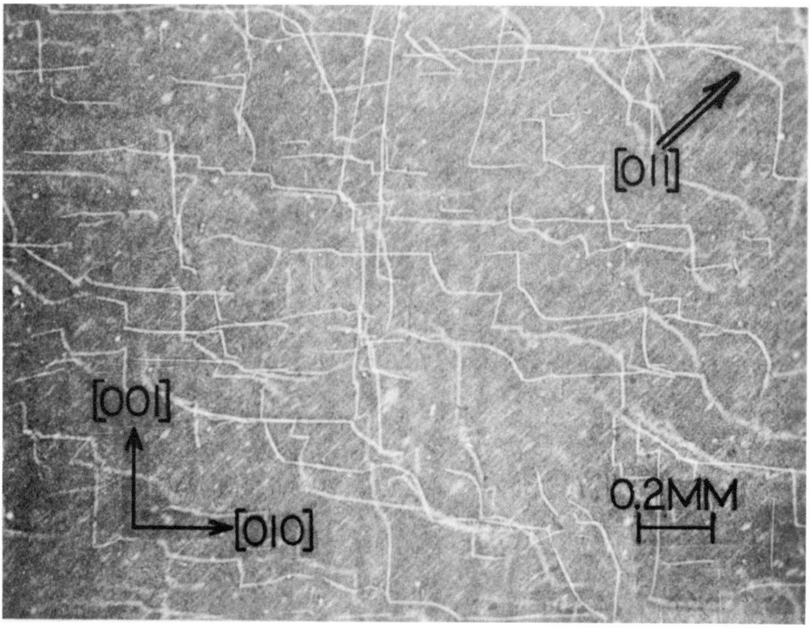

Figure 4.32. Dislocation lines within a relatively thick (100) sheet of an iron = 3% silicon alloy crystal, observed using (011) reflection in transmission (Co radiation). [C. Cm. Wu and B. Roessler, *Phys. Stat. Sol.*, **A8**, 571 (1971)]

In order to measure details of the strain fields about dislocation lines, Bonse[10] devised a modification of the Berg-Barrett method. The modification is sketched in Fig. 4.33. Here crystal II is the specimen. Crystal I, a rather perfect crystal, is set at the Bragg angle to monochromatize and make parallel the beam incident on crystal II. Thus a beam with the few seconds divergence characteristic of a dynamical diffraction is played on crystal II. Because of this small divergence, defective regions reduce rather than enhance the diffracted intensity. It is possible to nearly directly photograph the strain fields about defects. Very delicate rotations of the specimen in the neighborhood of the Bragg angle can place different parts of the strain field into reflection. X-ray topography has also been useful in observing twin boundaries and magnetic domains.

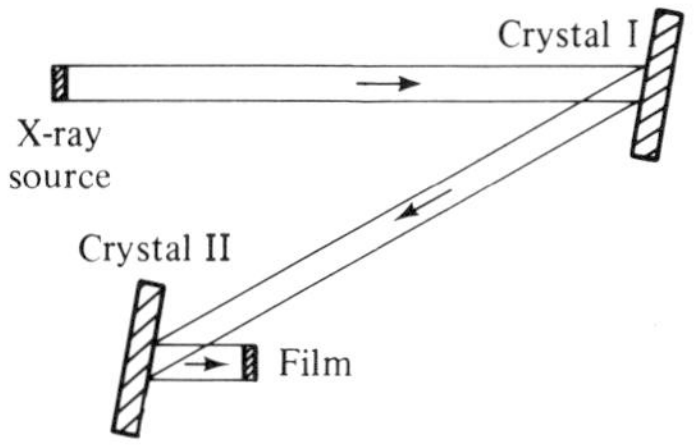

Figure 4.33. The Bonse method of back-reflection x-ray topography. Here the beam is made very plane, by diffraction from crystal I, before impinging on the specimen (crystal II).

[10] U. Bonse, *Z. Phys.*, **153**, 278–96 (1958).

4.1. We have derived expressions for the scalar deviations Δ_0^P, Δ_H^P of the internal wave vectors k_0^P and k_H^P from the external incident wave vector κ:

$$k_0^P = \kappa\left(1 + \Delta_0^P\right)$$

$$k_H^P = \kappa\left(1 + \Delta_H^P\right) = \kappa\left(1 + \frac{\Delta_0^P}{b} + \frac{a}{2}\right)$$

(a) At the exact Laue conditions, with the crystal set in symmetrical Laue geometry, $k_0^P = k_H^P$. Prove this.

(b) Given that at the exact Laue conditions these scalar wave differences are equal, show that the spacing Λ of *pendellösung* fringes in a symmetric wedge of angle α is (approximately)

$$\frac{\pi V_c \kappa \cos\theta}{4\left(\dfrac{e^2}{m_e c^2}\right)|F_H| \tan\dfrac{\alpha}{2}}$$

4.2. Derive an approximate expression for the natural line breadth, $\Delta^*(2\Delta)$, of a perfect crystal in the symmetric geometry. Is this value much different in magnitude from the Bragg geometry case?

4.3. LiF, NaCl, and KBr all have the same crystal structure. The effect of multiple scattering on the symmetric Bragg (111) reflection intensities of these three decreases in the order shown. Likewise, diamond, Si, and Ge all have the same type unit cell and have primary extinction effects decreasing in the order given. Lattice parameters are given below.

(a) Present an analytical argument which explains this ordering. Cite then a general rule regarding the effect of multiple scattering on materials of similar structure.

(b) Describe in simple physical terms why you would expect the result you obtained. (Hint: What terms "play off" against each other in your analysis?)

(c) Compute and compare extinction and absorption distances for all three materials in each series.

Material	*Lattice Parameter*
LiF	4.02
NaCl	5.63
KBr	6.59
C(diamond)	3.57
Si	5.43
Ge	5.66

4.4. Transmission x-ray topographs are taken for a symmetrical silicon wedge. One topograph, *A*, is taken near the tip of the wedge, where the crystal is much thinner than the extinction distance. Another topograph, *B*, is taken toward the base of the wedge, where the thickness is several times the normal absorption distance.

(a) Describe the differences, if any, in the appearance of dislocation images at *A* and at *B*.

(b) Describe the differences, if any, in the appearance of *pendellösung* fringes at *A* and at *B*.

4.5. A single crystal of copper is prepared as a thin film for electron microscopy. The incident electron beam enters along (010). A slip dislocation, whose Burgers' vector must be in the family $\frac{1}{2}a$ <101> is observed in (100) and (110) dark-field micrographs, but not in a (001) micrograph. What precisely is the Burgers' vector of this dislocation?

4.6. Should the contrast due to strain fields in Berg-Barrett x-ray topography improve or get worse with increasing atomic number? Why?

4.7. How broad should the image of a dislocation in an x-ray topograph be? More explicitly, consider the case of a Berg-Barrett topograph of an iron crystal, assuming a symmetric geometry in order to estimate the line breadth of total reflection. Consider a screw dislocation intersecting the surface at a right angle. You may assume isotropic elasticity and ignore surface effects. Based on your answer, what should be an order of magnitude for the limiting density of dislocations (lying parallel to the surface) such that dislocation images overlap and resolution of individual dislocations is destroyed?

BIBLIOGRAPHY

1. R. W. ARMSTRONG and C. CM. WU, "X-ray Diffraction Microscopy," in J. L. McCall and W. M. Mueller, eds., *Tools and Techniques for Microstructural Analysis,* Plenum Publishing Corporation, N.Y. (1973).
2. A. AUTHIER, in S. AMELINCK, ET AL., eds., *Techniques in Materials Science,* North-Holland, Amsterdam (1970).
3. L. V. AZAROFF, R. KAPLOW, N. KATO, R. J. WEISS, A. J. C. WILSON, and R. A. YOUNG, *X-ray Diffraction,* Chs. 3–5, McGraw-Hill Book Company, N.Y. (1974).
4. B. W. BATTERMAN and H. COLE, "Dynamical Diffraction of X-rays by Perfect Crystals," *Reviews of Modern Physics, Vol. 36,* p. 681 (1964).
5. A. H. COMPTON and S. K. ALLISON, *X-rays in Theory and Experiment, 2nd Ed.,* Ch. 6, Van Nostrand Reinhold Company, N.Y. (1935).
6. J. M. COWLEY, *Diffraction Physics,* Chs. 8–11, 13, and 15, North-Holland, Amsterdam (1973).
7. C. G. DARWIN, "The Reflection of X-rays from Imperfect Crystals," *The Philosophical Magazine, Vol. 43,* p. 800 (1922).
8. P. A. EGELSTAFF, *Thermal Neutron Scattering,* Ch. 3, Academic Press, Inc., N.Y. (1965).

9. P. P. EWALD, "Group Velocity and Phase Velocity in X-ray Crystal Optics," *Acta Crystallographica, Vol. 11,* p. 888 (1958).

10. P. P. EWALD, "Crystal Optics for Visible Light and X-rays," *Reviews of Modern Physics, Vol. 37,* p. 46 (1965).

11. P. B. HIRSCH, A. HOWIE, R. B. NICHOLSON, D. W. PASHLEY, and M. J. WHELAN, *Electron Microscopy of Thin Crystals,* Chs. 8–12, Plenum Publishing Corporation, N.Y. (1967).

12. R. W. JAMES, *The Optical Principles of the Diffraction of X-rays,* Chs. 1, 6, and 8, G. Bell & Sons, Ltd., London (1954).

13. LAUE, M. T. F. VON, *Röntgenstrahlinterferenzen, 3. Ausg.,* Ch. 5, Akademische Verlagsgesellschaft, Frankfurt (1960).

14. T. RYMER, *Electron Diffraction,* Ch. 6, Methuen, London (1970).

15. L. H. SCHWARTZ and J. B. COHEN, *Diffraction from Materials,* Ch. 8, Academic Press, Inc., N.Y. (1977).

16. J. C. SLATER, "Interaction of Waves in Crystals," *Reviews of Modern Physics, Vol. 30,* p. 197 (1958).

17. B. E. WARREN, *X-ray Diffraction,* Ch. 14, Addison-Wesley Publishing Co., Inc., Reading, MA (1969).

18. W. H. ZACHARIASEN, *Theory of X-ray Diffraction in Crystals,* Dover Publications, Inc., N.Y. (1967).

Crystal Structure Analysis 5

5.1 EFFECT OF SYMMETRY IN CRYSTAL STRUCTURE ANALYSIS

We saw in Sect. 3.4 that the intensity (including systematic absences) of diffraction peaks depends upon the placement of atoms within the unit cell. One of the principal uses of diffraction is to deduce these placements from the intensities. This process is termed *crystal structure analysis.*

In principle, the procedure works in the following way. Recall our expression for the reflecting power of a rotating crystal diffraction peak

$$P_{hkl} = \frac{r_e^2}{R_o^2}\frac{V}{V_c^2}\frac{\lambda^3}{2\sin 2\theta}\left(\frac{1+\cos^2 2\theta}{2}\right) m_{hkl}|F_{hkl}|^2 \tag{5.1}$$

(For Weissenberg or precession photographs, or if a diffractometer is used, overlapping peaks are avoided and the multiplicity factor m_{hkl} is equal to unity.) All the structural information is stored in the structure factor, F_{hkl}. Relative values of the $|F_{hkl}|$ are then given by

$$|F_{hkl}|_{\text{rel}} = (I_{hkl})_{\text{meas}} \Big/ \left[\frac{1+\cos^2 2\theta_{hkl}}{\sin 2\theta_{hkl}}\right] \cdot m_{hkl} \tag{5.2}$$

The right side of (5.2) is determined experimentally. A set of $|F_{hkl}|_{\text{rel}}$ is thus found, each hkl reflection providing one value of $|F_{hkl}|_{\text{rel}}$. The relation

$$F_{hkl} = \sum_{n=1}^{N'} f_n e^{2\pi i(hX_n + ky_n + lZ_n)} \tag{5.3}$$

relates the structure factors F_{hkl} to the atomic positions X_n, Y_n, and Z_n of the N' atoms of the unit cell. A possible set of placements X_n, Y_n, Z_n are assumed and the $|F_{hkl}|$ of the peaks computed. If the computed relative $|F_{hkl}|$ values coincide with the measured set, within a desired accuracy, the crystal structure is said to be determined. In practice the initial placements do not predict the measured $|F_{hkl}|_{\text{rel}}$ and several iterations on the atomic positions are required to achieve a fit.

In order to get a fit between computed and measured values, one has to correctly fit all N' atoms of the unit cell. For a complex material (e.g., a mineral or a large organic system), this can be a prodigious feat. The structure analyst is, however, assisted by symmetry relations in the unit cell. That is, not all atomic positions are independent of the others; some are interrelated by symmetry operations.

In order to see this more clearly, we consider the structure of barium titanate, $BaTiO_3$. A unit cell of $BaTiO_3$ is shown in Fig. 5.1. The unit cell contains

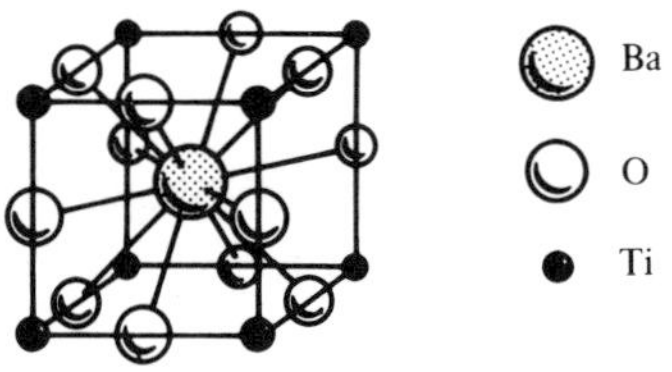

Figure 5.1. Crystal structure of $BaTiO_3$ above 393° K.

one stoichiometry unit: 1 Ba, 1 Ti, and 3 O. The positions of the Ba and Ti are unique (here at 0,0,0 and ½,½,½). The oxygen atoms are all related by symmetry operations. The *x*-, *y*-, and *z*-axes all are axes of four-fold symmetry; the crystal can be rotated 90°, 180°, 270°, and 360° about any of these axes with no change. This means that only one oxygen position determines the entire set; the other oxygens are found by applying the symmetry operations to the original oxygen. Thus one needs to have found only three independent positions in this unit cell. This example is illustrative of the general situation that a prior knowledge of the symmetry of the crystal reduces the number of independent position parameters. Fortunately, the symmetry within the unit cell is one of the most readily available pieces of information.

5.2 SYMMETRY IN CRYSTALS AND SPACE GROUP DETERMINATION

The symmetry of a crystal is a combination of its *macroscopic* and its *microscopic symmetry*. Consider Fig. 5.2. This figure shows a two-dimensional "crystal." The unit cell contains two large white circles and two small dark ones. The unit cell outlines show a twofold symmetry; we can rotate the unit cell 180° and 360° about its center and not see a change. The atoms within the cell, however, do not show this symmetry; upon rotation of 180° about the center,

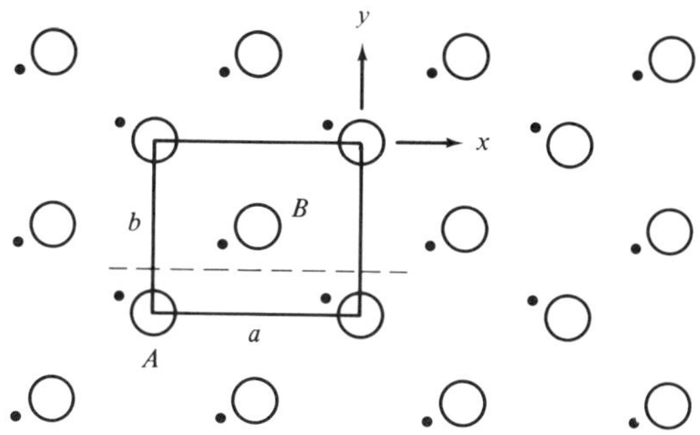

Figure 5.2. Two-dimensional "crystal" exhibiting twofold symmetry.

the small, dark circle associated with the group B comes to a new position and the old placements are *not* replicated.

In this case, macroscopic properties measured along $+x$ are the same as those along $-x$, and properties along $+y$ are the same as those measured along $-y$. Suppose the circles represent atoms in a two-dimensional array. The alternate rows of atomic groups along x differ only in the placement of the dark circles (atoms) above or below the line connecting the light circles, as seen in Fig. 5.3. The properties of the two types of rows may differ, but there are the same number of rows of one type as of the other. It makes then no difference whether properties are measured along $+x$ or $-x$. The same argument holds for $+y$ and $-y$. The macroscopic properties of this structure are said to exhibit twofold symmetry.

On the atomic level, a different symmetry is observed. Consider in Fig. 5.2 the groups at A and at B. The group at B can be generated from that at A by first rotating the crystal about the dashed line located at $\frac{1}{4}b$ and then translating that group by $\frac{1}{2}a$ along the x-axis. This sequence of operations is shown in Fig. 5.4. This operation is called a twofold screw axis and is one example of a microscopic symmetry operator.

It is convenient to describe macroscopic symmetry operations in terms of a *point lattice.* The point lattice is constructed by replacing each unit cell of the crystal by one point, located similarly within each unit cell. This is illustrated in Fig. 5.5. A unit cell of the point lattice is shown by dashed lines. A necessary characteristic of a point lattice is that *all points have identical environments.*

Macroscopic symmetry formally relates to the symmetry of the point lattice. The macroscopic symmetry operators transform a point lattice into itself. There are three distinct types of macroscopic symmetry operators: *axes of rotation, inversion through a point,* and *mirror through a plane.*

The only rotation symmetries which are capable of maintaining long-range two-dimensional periodicity are two-, three-, four-, and sixfold rotations. Two-dimensional point lattices consistent with these symmetries are shown in Fig. 5.6. The unit cells of the lattices are indicated by dashed lines. The positions of the symmetry axes are given on unit cells alongside the point lattices. (It

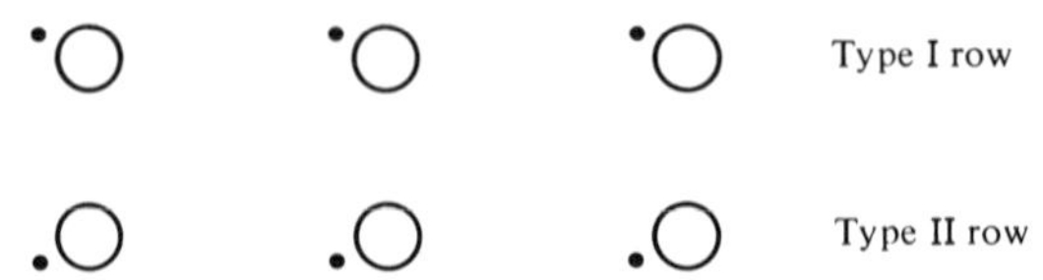

Figure 5.3. Type I and Type II rows in the crystal of Fig. 5.2.

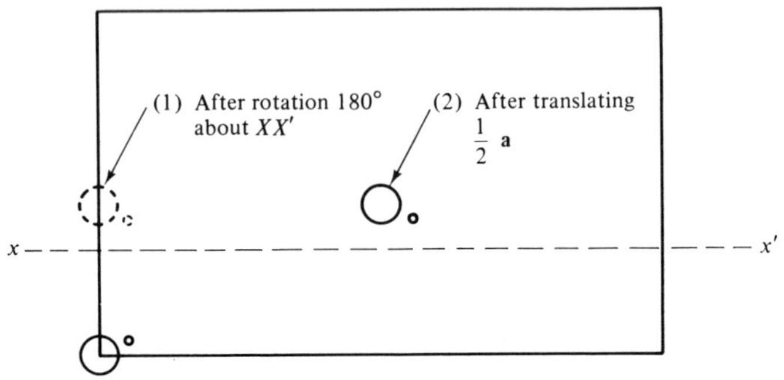

Figure 5.4. Twofold screw axis operation.

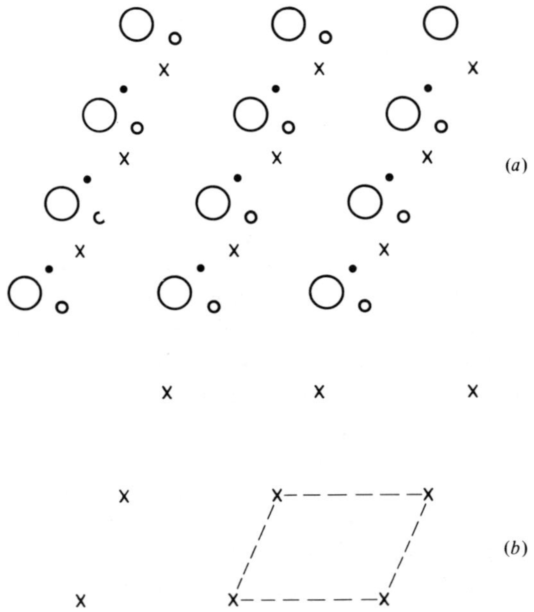

Figure 5.5. Relationship between a crystal and its point lattice: *(a)* crystal with three types of atoms (○○●) and x's marking points of the point lattice; *(b)* the point lattice for *(a)*.

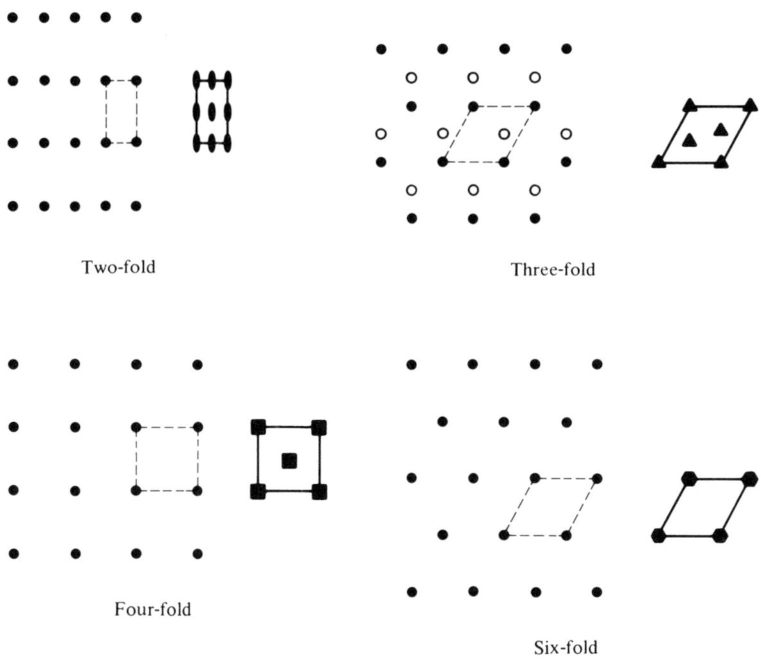

Figure 5.6. Two-dimensional point lattices showing two-, three-, four-, and sixfold rotation symmetries. In the threefold case, two superposed layers are shown, one (○) in the plane of the page and one (●) above the page.

should be mentioned that no two-dimensional lattices show threefold symmetry. The threefold system shown is composed of two layers, denoted by the closed and opened circles). The other distinct macroscopic symmetry operation is the *center of inversion.* The center of inversion is any point within a lattice (but not necessarily a lattice point), such that for every lattice point located at *x, y, z* from the inversion center, there exists another point at $\bar{x}, \bar{y}$ and $\bar{z}$. (Overbars denote *negative.*) The operation is shown in Fig. 5.7.

The center of inversion can be combined with any of the axes of rotation, to give rotation-inversion symmetries. The symbols for these symmetry operators are $\bar{1}, \bar{2}, \bar{3}, \bar{4}$, and $\bar{6}$. ($\bar{1}$ is identically the center of inversion itself.) Point distributions consistent with these operators are shown in Fig. 5.8.

A *mirror plane* replicates a point on one side of the plane at an equivalent position on the other side of the plane. This operation is indicated in Fig. 5.9. The symbol for a mirror plane is **m.** Note also that $\bar{2}$ is identical to *m.*

When sketching symmetrical relationships the symbols used for $2,3,4,6,\bar{1},\bar{3},\bar{4}$, and $\bar{6}$ are ⬮, ▲, ■, •, ◬, ⊡ and ⬡. The pictorial symbol for a mirror plane is a heavy line (the mirror seen edge on).

Compound symmetry operators can be formed by placing a mirror plane perpendicular to a rotation axis. The new symmetries are shown in Fig. 5.10.

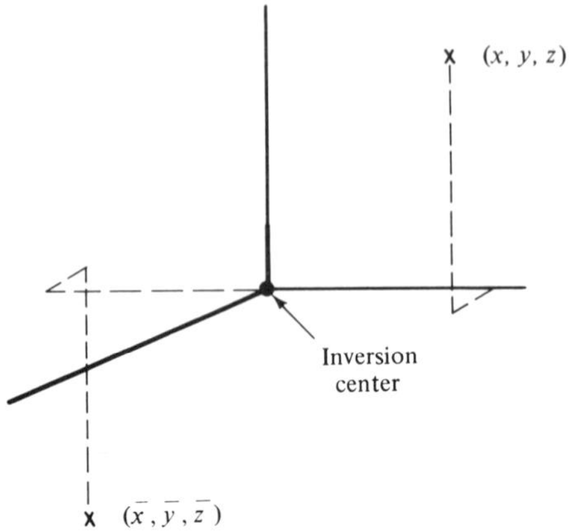

Figure 5.7. Center of inversion relating points at *(x, y, z)* and $(\bar{x}, \bar{y}, \bar{z})$.

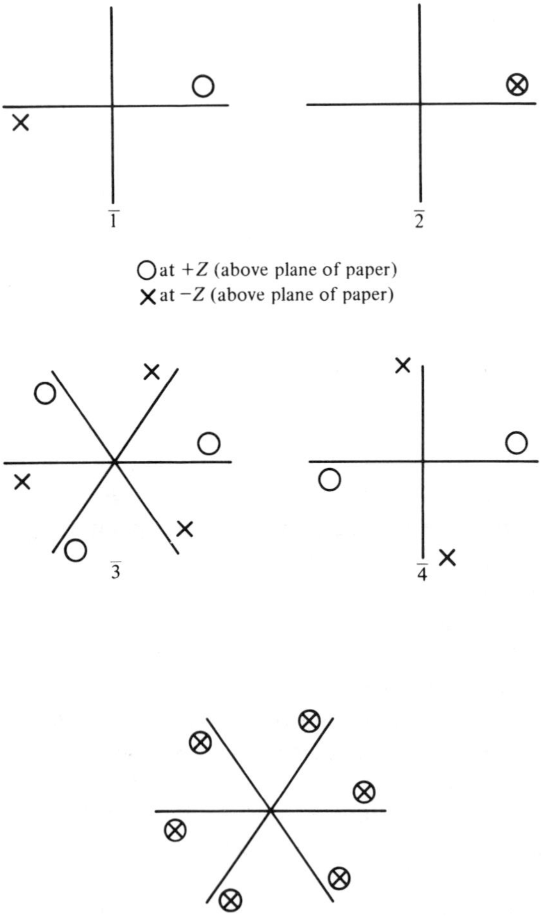

Figure 5.8. Rotation-inversion operations $\bar{1}$, $\bar{2}$, $\bar{3}$, $\bar{4}$, and $\bar{6}$.

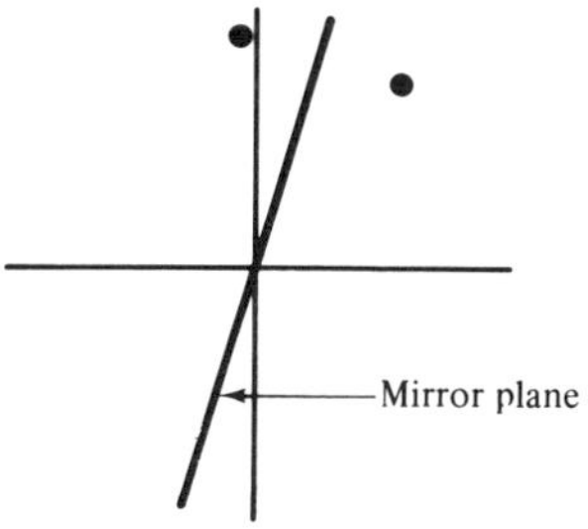

Figure 5.9. Mirror plane.

(In this figure, the mirror plane is, of course, the plane of the paper.) The symbols for such systems are p/m, where p is the rotation symmetry.

We are usually concerned with the complete symmetry about a point. For example, there may be not one, but three twofold axes, all perpendicular to each other, operating simultaneously at a point. This situation is depicted in Fig. 5.11. Here all three twofold symmetries are simultaneously met. Any such allowable collection of symmetry operations at a point is termed a *point group*. All of the symmetry operations described above are allowable, nonredundant point groups. The symbol for the point group of Fig. 5.11 is 222. In general, the symbol is either a triad describing the symmetries of three rotation axes or a single symbol (e.g., $\bar{3}$,$\bar{4}$,$2/m$). There are precisely 32 nonredundant point groups.

The 32 point groups are conveniently grouped according to minimum symmetry. These sets of point groups constitute the seven crystal systems. The seven crystal systems, their minimum symmetries, and the relationships among the unit cell lengths and angles are shown in Table 5.1.

The symmetry elements of the 32 point groups are listed and drawn stereographically in Fig. 5.12. It should be pointed out that the axes to which the point groups are referred have conventional settings, depending on the crystal class. The conventional axis relations are shown in Fig. 5.13.

For each point group, one may associate a set of *equivalent points*, related mutually by the symmetry of the point group. For instance, the equivalent

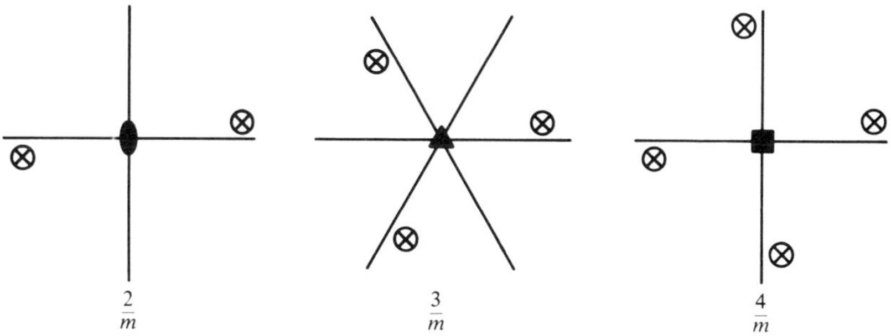

Figure 5.10. Examples of compound symmetry operators: $2/m$, $3/m$, and $4/m$.

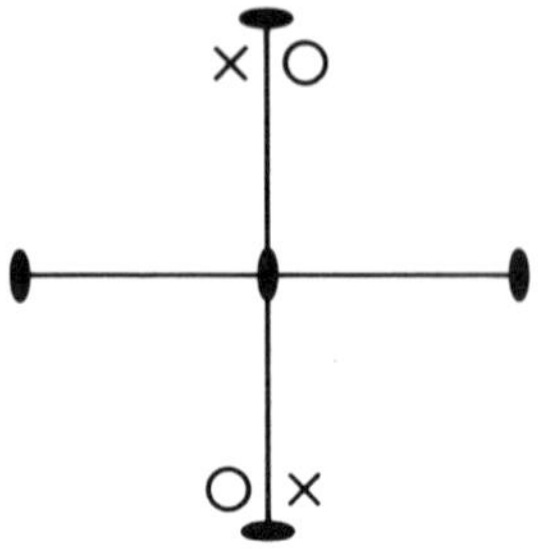

Figure 5.11. Point group 222.

points for point group 2 are (x,y,z) and $(\bar{x},\bar{y},z)$, as shown in Fig. 5.14*a*. The equivalent points for 222 are (x,y,z), $(x,\bar{y},\bar{z})$, $(\bar{x},\bar{y},z)$, and $(\bar{x},y,\bar{z})$.

Space lattices can be conveniently classified in terms of their crystal system and *Bravais lattice.* Any space group can be described in terms of a large number of possible unit cells. Conventionally, however, one choses the unit cell which exhibits the largest symmetry. The outlines of that cell obey the symmetry requirements of one of the crystal systems. Thus, any point lattice can be described in terms of its crystal system. Further, the lattice points within that most symmetric cell may exhibit one of only four possible modes of arrangement, consistent with the requirement that all points have equivalent environment. Consider Fig. 5.15, which shows four different space lattices, all of which have orthorhombic symmetry (three mutually perpendicular axes of twofold symmetry). In each case, the environment of all lattice points is identical, but the basic arrangements and numbers of points in each cell differ. The positions of the equivalent points for the four different arrangements are given in Table 5.2.

Not all crystal systems can exhibit all four lattice types; in most cases there are redundancies. Figure 5.16 shows, for instance, that face-centered tetragonal can be resolved into body-centered tetragonal. In this case the lattice is conventionally termed body-centered. Table 5.3 lists the 14 nonredundant arrangements: the Bravais lattices. For each Bravais lattice type, only certain Bragg reflections

TABLE 5.1
THE SEVEN CRYSTAL SYSTEMS

Crystal system name	*Minimum symmetry*	*Unit cell dimension relationships*
Triclinic	None	$a \neq b \neq c; \alpha \neq \beta \neq \gamma$
Monoclinic	One twofold axis	$a \neq b \neq c; \alpha = \gamma = 90°, \beta \neq 90°$
Orthorhombic	Three twofold axes	$a \neq b \neq c; \alpha = \beta = \gamma = 90°$
Rhombohedral	One threefold axis	$a = b = c; \alpha = \beta = \gamma \neq 90°$
Tetragonal	One fourfold axis	$a = b \neq c; \alpha = \beta = \gamma = 90°$
Hexagonal	One sixfold axis	$a = b \neq c; \alpha = \beta = 90°, \gamma = 120°$
Cubic	Four threefold axes	$a = b = c; \alpha = \beta = \gamma = 90°$

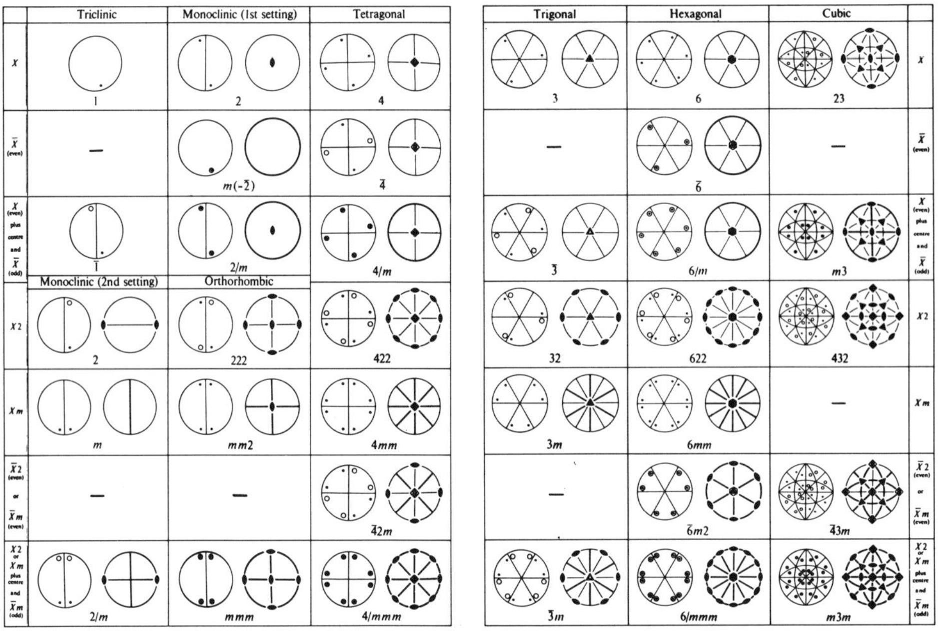

Figure 5.12. Stereographic representation of the 32 point groups. Each pair of stereograms shows, on the left, the poles of a general form and, on the right, the symmetry elements of the point group. Planes of symmetry are indicated by bold lines. (Reprinted from International Tables for X-ray Crystallography, Vol. I, with the permission of the International Union of Crystallography.)

exhibit nonzero intensity. For instance, within a body-centered crystal, each atom of type j at x,y,z has a counterpart at $x + ½$, $y + ½$, $z + ½$. The structure factor can then be written

$$F_{hkl} = \sum_{j=1}^{N_c/2} f_j \left\{ e^{2\pi i[hx_j + ky_j + lz_j]} + e^{2\pi i[h(x_j+1/2) + k(y_j+1/2) + l(z_j+1/2)]} \right\}$$
$$= \sum_{j=1}^{N_c/2} f_j e^{2\pi i(hx_j + ky_j + lz_j)} [1 + e^{\pi i(h+k+l)}] \qquad (5.4)$$

This structure factor is zero when $h + k + l$ is odd, but finite otherwise. Similarly, one can show that the only reflections which can appear for face-centered systems are those with $h, k,$ and l all even or all odd; mixed indices produce $F_{hkl} = 0$. The diffraction conditions associated with the four lattice types are given in the last column of Table 5.2.

The microscopic symmetry operations are combinations of rotation axes or a mirror plane with a lattice translation; the microscopic symmetries are defined only for atoms arranged on periodic lattices. Screw axes combine rotation axes with translations along those axes. Glide planes combine mirror planes with translations parallel to those planes.

The symbol for a screw axis is n_m, where n is the rotation symmetry and m/n is the fractional lattice translation. For instance, with the 4_1 operation

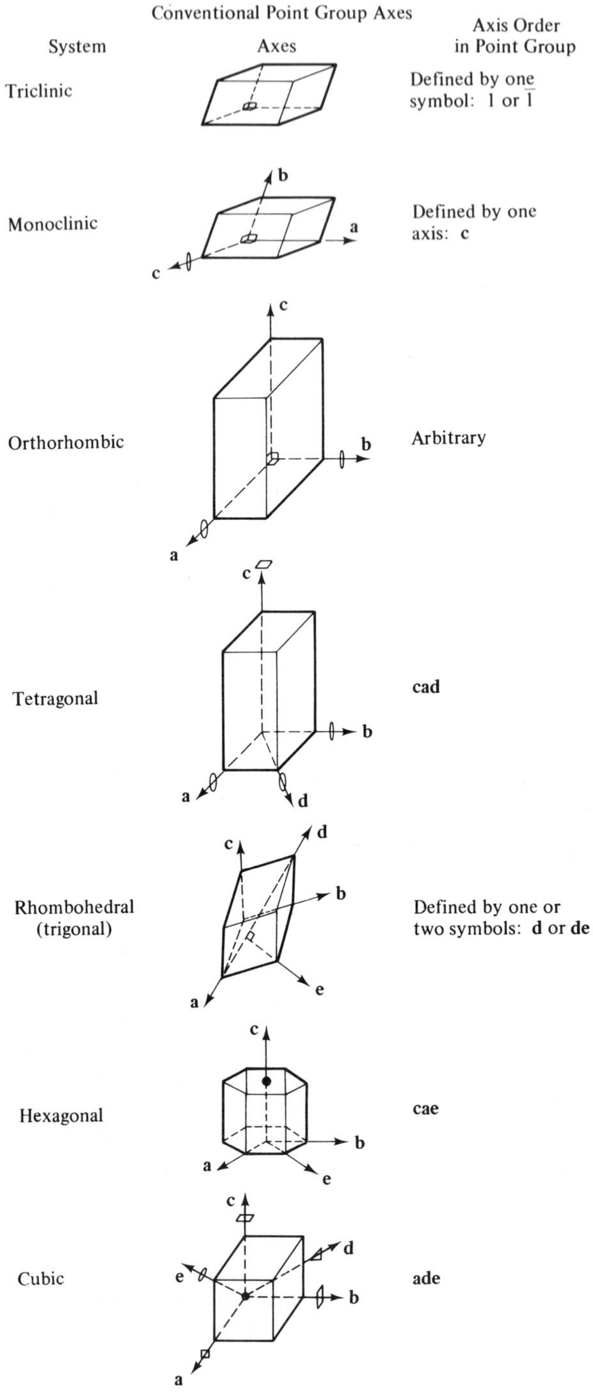

Figure 5.13. Conventional point group axes and symmetry elements.

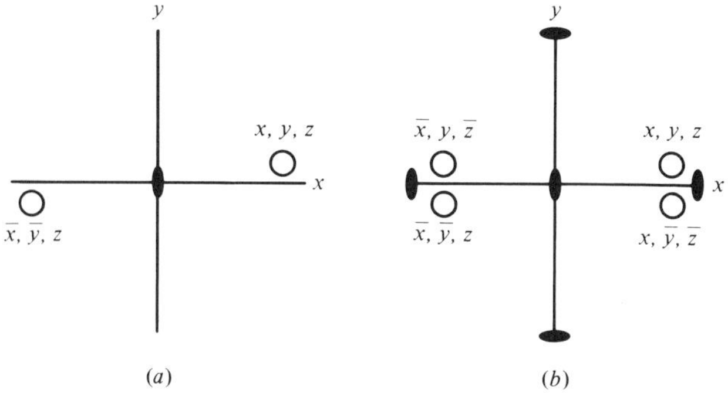

Figure 5.14. Equivalent points for space groups 2 and 222.

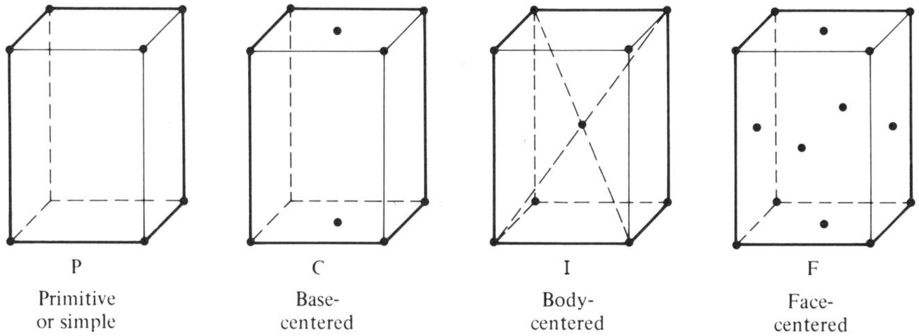

Figure 5.15. The four lattice types.

TABLE 5.2
EQUIVALENT POINTS AND DIFFRACTION CONDITIONS FOR THE FOUR LATTICE TYPES OF FIG. 5.15

Symbol	*Name*	*Equivalent points*	*Diffractions conditions*
P	Primitive or simple	0,0,0	All *hkl*
C	Base-centered (three possibilities)	(a) 0,0,0; ½,½,0, (b) 0,0,0; ½,0,½ (c) 0,0,0; 0,½,½	(a) $h+k=$ even (b) $h+l=$ even (c) $k+l=$ even
I	Body-centered	0,0,0; ½,½,½	$h+k+l=$ even
F	Face-centered	0,0,0; ½,½,0; ½,0, ½; 0,½,½	*h,k,l* all even or all odd

about **c**, every atom has one like atom rotated 90° about **c** and translated **c**/4, one rotated 180° about **c** and translated **c**/2, and one rotated 270° about **c** and translated 3**c**/4. This is shown in Fig. 5.17*a*. The four positions are generated, starting at ①, in the sequence ①, ②, ③, ④, using 90° rotations, coupled with translations of **c**/4. Figure 5.17 shows the three possible fourfold screw axes, 4_1, 4_2, and 4_3. Note that 4_3 is generated, beginning at ① in the

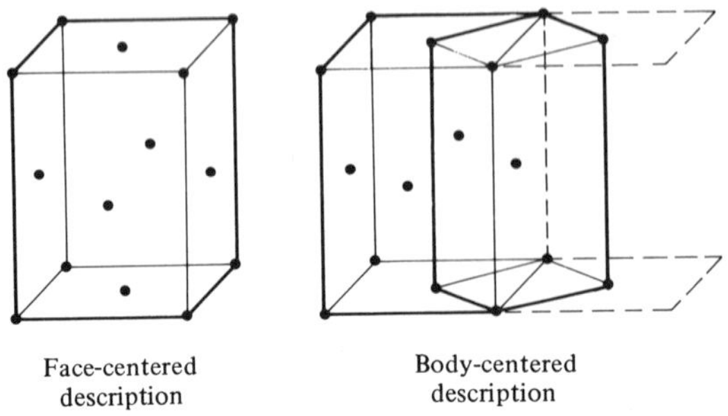

Figure 5.16. Identity of face-centered tetragonal and body-centered tetragonal.

TABLE 5.3
THE 14 BRAVAIS LATTICES

Crystal class	*Lattice types*
Cubic	P,I,F
Tetragonal	P,I
Hexagonal	P
Rhombohedral	P
Orthorhombic	P,C,I,F
Monoclinic	P,I
Triclinic	P

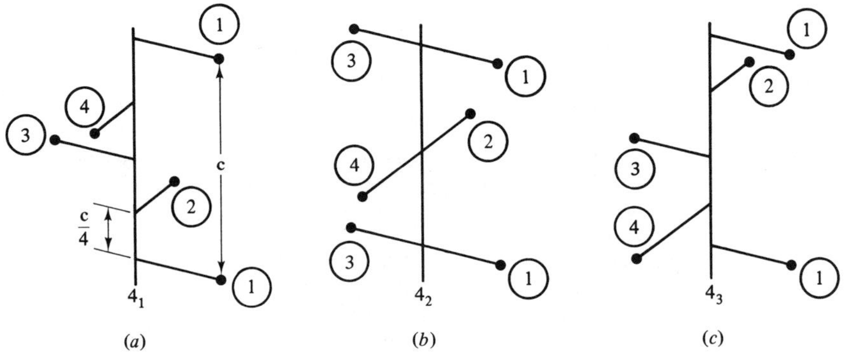

Figure 5.17. The fourfold screw axes.

sequence ①, ②, ③, ④ shown. The second 90° rotation places a point at 3**c**/2, but since all cells must be equivalent, there is a similar point at **c**/2. Similarly, the next rotation places a point at level 9**c**/4—consequently also at 5**c**/4 and **c**/4. Observe that 4_3 is the opposite-handed equivalent of 4_1.

Figure 5.18 illustrates all possible screw axis operators. Again, observe that several pairs of these are merely of opposite hand.

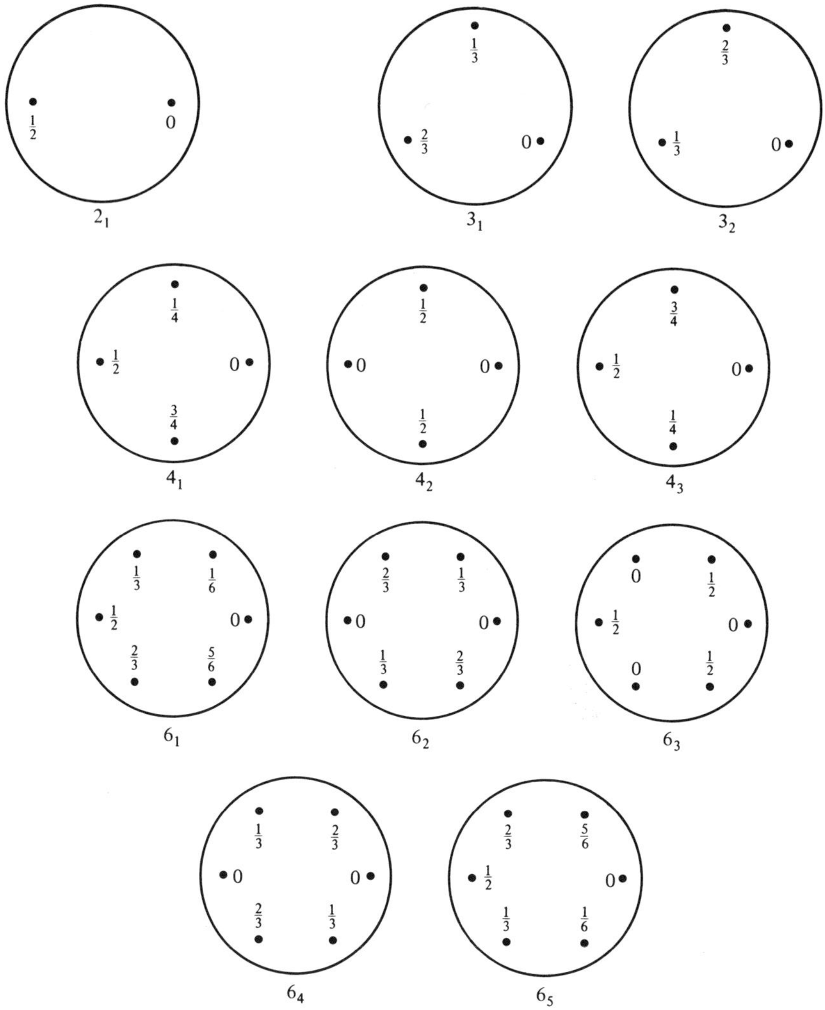

Figure 5.18. The screw axes. Fractional numbers indicate the level, normal to the paper, of the point. This number is a fraction of the lattice vector in that direction.

The effect of screw axes is to extinguish certain of the $00l$ reflections (here we have taken **c** along the screw axis). This is seen as follows. For each atom of type j at x_j, y_j, and z_j there are others related to it by the screw operations. For instance, for 4_2 there are, in addition to the atom at x_j, y_j, z_j, others at $\bar{x}_j, \bar{y}_j, z_j$; $y_j, x_j, z_j + \frac{1}{2}$; and $\bar{y}_j, \bar{x}_j, z_j + \frac{1}{2}$. Since atoms of *all* types must exhibit this symmetry, the structure factor can be written

$$F_{hkl} = \sum_{j=1}^{N_c/4} f_j [e^{2\pi i(hy_j + kx_j + lz_j)} + e^{2\pi i(-hx_j - ky_j + lz_j)} + e^{2\pi i(hy_j + kx_j + lz_j + l/2)} + e^{2\pi i(-hy_j - kx_j + lz_j + l/2)}] \tag{5.5}$$

For $00l$ reflections

$$F_{00l} \sum_{j=1}^{N_c/4} f_j[2e^{2\pi i l z_j}(1 + e^{\pi i l})] = \begin{cases} \sum_{j=1}^{N_c} f_j \text{ if } l = \text{even} \\ 0 \quad \text{if } l = \text{odd} \end{cases} \tag{5.6}$$

Thus, in this case 002,004,006, . . . will appear, whereas 001,003,005, . . . will not. The reason for this behavior is that we have interposed a plane of atoms between those separated by **c**-translation. First-order scattering from the interposed plane is exactly out of phase with the first-order scattering from its neighbors.

Similar, specific $00l$ extinctions are found for the other screw axes. These microscopic symmetry operators on the three crystallographic axes can thus be determined experimentally by observation of systematic extinctions in $h00$, $0k0$, and $00l$.

Glide planes combine a mirror operator with a translation. Such a combined operation is illustrated in Fig. 5.19 for the case of a mirror normal to **c** followed by translation of **a**/2. The symbol for this microscopic symmetry element is *a*. Had the translation been **b**/2, the symbol *b* would be used. A list of all allowed glide plane translations and their symbols is found in Table 5.4.

Again we find that each glide plane has its own specific set of extinctions. For instance, the glide plane *a* must exhibit only even *h* reflections for $h00$. Geometrically this is seen from the planes of atoms inserted at $a/2$ between other planes of similar type. The conditions for systematic extinctions are actually more general than this. Analytically, for each atom of type *j* at x,y,z, there is another at $x + ½, y, \bar{z}$. The structure factor is then

$$F_{hkl} = \sum_{j}^{N_c/2} f_j \left[e^{2\pi i (hx_j + ky_j + lz_j)} + e^{2\pi i (hx_j + \frac{h}{2} + ky_j - lz_j)} \right] \tag{5.7}$$

For $hk0$ reflections

$$\begin{aligned} F_{hk0} &= \sum_{j}^{N_c/2} f_j e^{2\pi i (hx_j + ky_j)} (1 + e^{\pi i h}) \\ &= \begin{cases} 2 \sum_{j}^{N_c/2} f_j e^{2\pi i (hx_j + ky_j)} & \text{for } h = \text{even} \\ 0 \quad \text{for } h = \text{odd} \end{cases} \end{aligned} \tag{5.8}$$

Thus, the **a** glide plane symmetry shows systematic extinction for all $hk0$ with h odd.

In a crystal, more than one microscopic symmetry element can exist. *Space groups*—the collection of simultaneously allowable symmetry elements on a space lattice—are defined. There are precisely 230 nonredundant space groups. (This can be proved by geometrical argument or by group theory). A few examples of space groups are shown in Fig. 5.20.

You may notice that the positions of the symmetry axes are different for

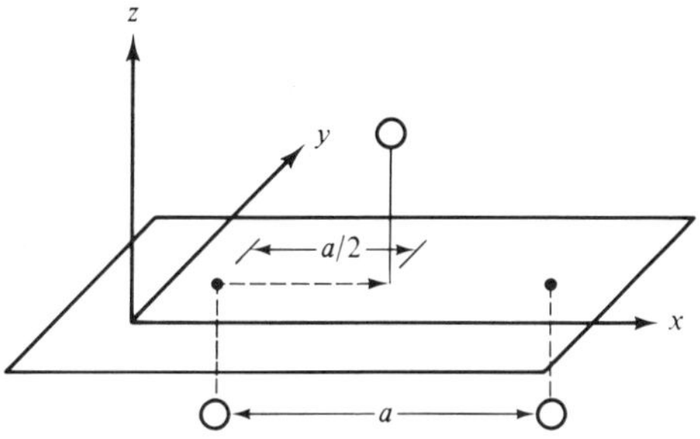

Figure 5.19. Glide plane *a*, normal to lattice vector **c**.

different space groups. For example, in $P222$ the twofold rotation axes intersect, where in $P2_12_12_1$ they do not. The displacements of the axes control the symmetry. The correct placements can be predicted by matrix analysis of the equivalent points.[1] This analysis is beyond the scope of the present text. The symbols used for space groups are related to those used for point groups. A space group symbol is of the form *Apqr*. Here *A* represents the lattice type; the symbol *A* can be *P, C, I,* or *F*. The symbols *p,q,* and *r* represent the symmetry corresponding to the three point group axes. The crystallographic axes to which p,q,r correspond are similar to those of the point groups. Table 5.5 specifies this ordering.

Table 5.4
Glide Planes

Translation	*Symbol*
$\mathbf{a}/2$	*a*
$\mathbf{b}/2$	*b*
$\mathbf{c}/2$	*c*
$\frac{\mathbf{b}}{2}+\frac{\mathbf{c}}{2}$	
$\frac{\mathbf{a}}{2}+\frac{\mathbf{c}}{2}$	*n*
$\frac{\mathbf{a}}{2}+\frac{\mathbf{b}}{2}$	
$\frac{\mathbf{a}}{4}+\frac{\mathbf{b}}{4}$	
$\frac{\mathbf{b}}{4}+\frac{\mathbf{c}}{4}$	
	d
$\frac{\mathbf{a}}{4}+\frac{\mathbf{c}}{4}$	
$\frac{\mathbf{a}}{4}+\frac{\mathbf{b}}{4}+\frac{\mathbf{c}}{4}$	

[1] See, for example, G. Burns and A. M. Glazer, *Space Groups for Solid State Scientists* (Academic Press, New York, 1979).

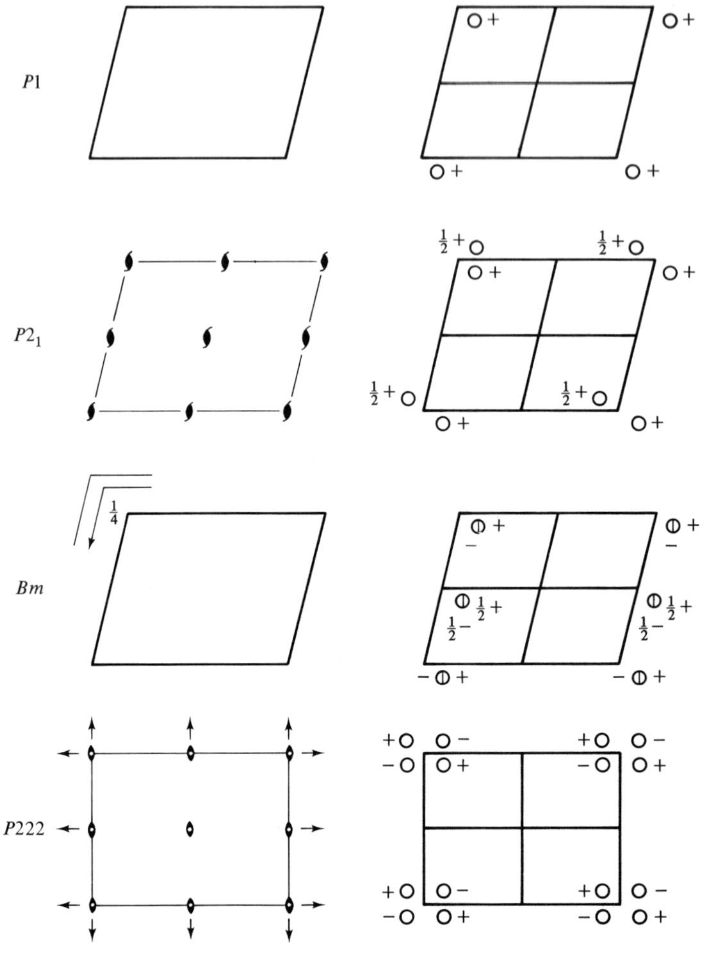

Figure 5.20. Representative space groups.

TABLE 5.5
DIRECTIONS ASSOCIATED WITH *Apqr* FOR SPACE GROUPS

System	*1st symbol*	*2nd symbol*	*3rd symbol*
Triclinic			
Monoclinic	[010]		
Orthorhombic	[100]	[010]	[001]
Tetragonal	[001]	[100]	[110]
Cubic	[100]	[111]	[110]
Hexagonal	[001]	[100]	[210]
Rhombohedral	[001]	[100]	[210]

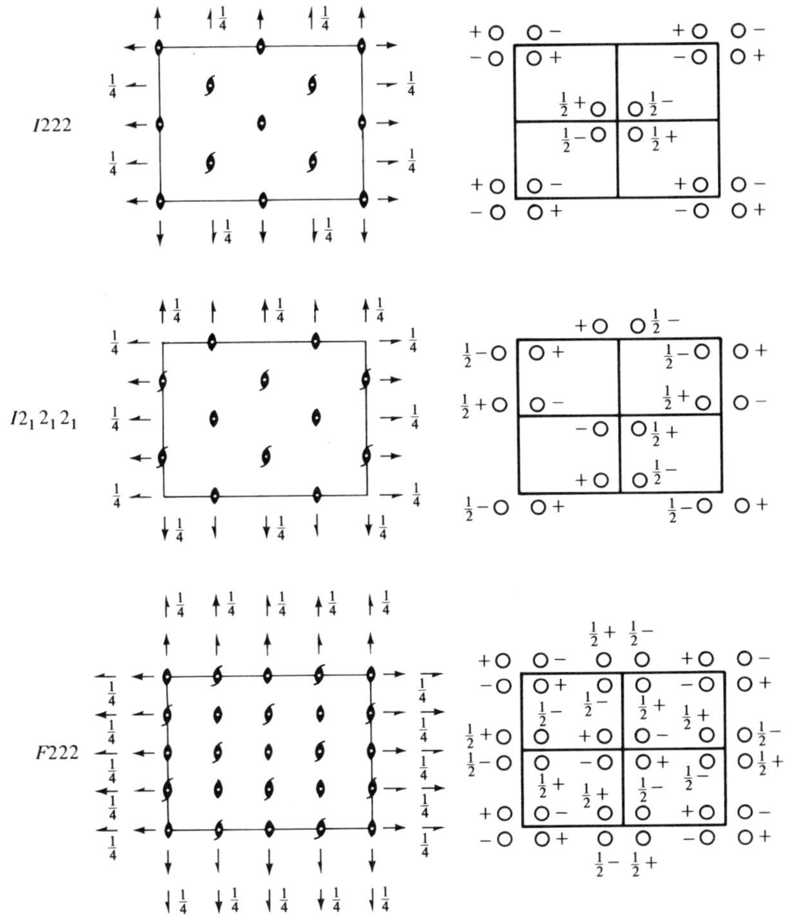

Figure 5.20. (Continued).

Corresponding to each space group is a specific macroscopic symmetry and a specific set of nonzero reflections. Thus measurement of the macroscopic symmetry and systematic extinctions defines the lattice type and the complete symmetry within the crystal. The problem of determining the complete crystal structure can thereby be reduced to managable proportions. That is, if there are eight equivalent positions corresponding to the full symmetry, the number of independent atomic positions to be determined is reduced by a factor of eight.

To assist in this procedure, there exists a full compilation of space groups and their allowable reflections. This is Vol. 1 of the *International Tables for X-ray Crystallography.* Table 5.6 is one page of space group information from these tables. For each space group more than one set of allowable reflections is listed. These relate to *special positions* of the atoms; i.e., some atoms lie along a specific symmetry axis, plane, or point, thereby reducing the total number of symmetrically related points.

TABLE 5.6[a]

Pnnm D_{2h}^{12}

No. 58 $P\ 2_1/n\ 2_1/n\ 2/m$ $m\ m\ m$ Orthorhombic

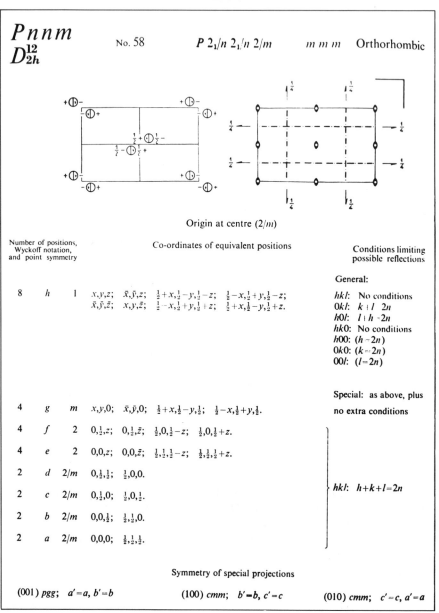

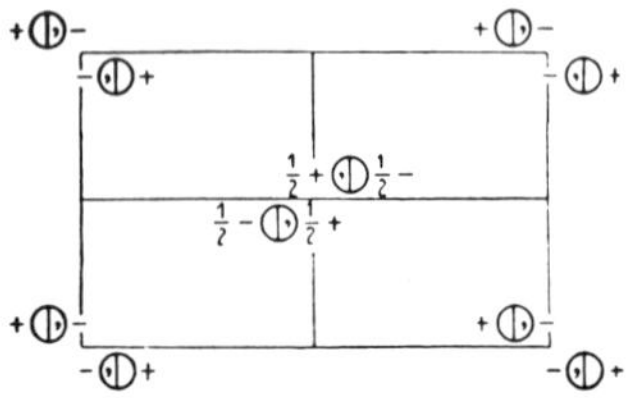
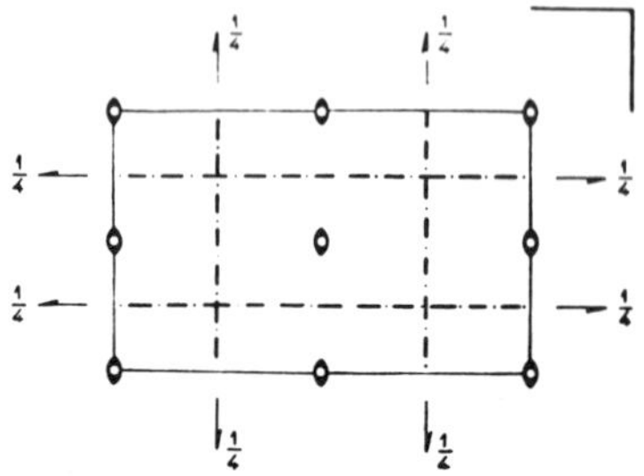

Origin at centre (2/*m*)

Number of positions, Wyckoff notation, and point symmetry			Co-ordinates of equivalent positions	Conditions limiting possible reflections
				General:
8	*h*	1	$x,y,z;\ \bar{x},\bar{y},z;\ \tfrac{1}{2}+x,\tfrac{1}{2}-y,\tfrac{1}{2}-z;\ \tfrac{1}{2}-x,\tfrac{1}{2}+y,\tfrac{1}{2}-z;$ $\bar{x},\bar{y},\bar{z};\ x,y,\bar{z};\ \tfrac{1}{2}-x,\tfrac{1}{2}+y,\tfrac{1}{2}+z;\ \tfrac{1}{2}+x,\tfrac{1}{2}-y,\tfrac{1}{2}+z.$	*hkl*: No conditions 0*kl*: $k+l=2n$ *h*0*l*: $l+h=2n$ *hk*0: No conditions *h*00: ($h=2n$) 0*k*0: ($k=2n$) 00*l*: ($l=2n$)
				Special: as above, plus
4	*g*	*m*	$x,y,0;\ \bar{x},\bar{y},0;\ \tfrac{1}{2}+x,\tfrac{1}{2}-y,\tfrac{1}{2};\ \tfrac{1}{2}-x,\tfrac{1}{2}+y,\tfrac{1}{2}.$	no extra conditions
4	*f*	2	$0,\tfrac{1}{2},z;\ 0,\tfrac{1}{2},\bar{z};\ \tfrac{1}{2},0,\tfrac{1}{2}-z;\ \tfrac{1}{2},0,\tfrac{1}{2}+z.$	} *hkl*: $h+k+l=2n$
4	*e*	2	$0,0,z;\ 0,0,\bar{z};\ \tfrac{1}{2},\tfrac{1}{2},\tfrac{1}{2}-z;\ \tfrac{1}{2},\tfrac{1}{2},\tfrac{1}{2}+z.$	
2	*d*	2/*m*	$0,\tfrac{1}{2},\tfrac{1}{2};\ \tfrac{1}{2},0,0.$	
2	*c*	2/*m*	$0,\tfrac{1}{2},0;\ \tfrac{1}{2},0,\tfrac{1}{2}.$	
2	*b*	2/*m*	$0,0,\tfrac{1}{2};\ \tfrac{1}{2},\tfrac{1}{2},0.$	
2	*a*	2/*m*	$0,0,0;\ \tfrac{1}{2},\tfrac{1}{2},\tfrac{1}{2}.$	

Symmetry of special projections

(001) *pgg*; $a'=a,\ b'=b$ (100) *cmm*; $b'=b,\ c'=c$ (010) *cmm*; $c'=c,\ a'=a$

[a] Reprinted from International Tables for X-ray Crystallography, Vol. I, with the permission of the International Union of Crystallography.

5.3 THE FIRST STEPS IN STRUCTURE ANALYSIS

The first steps in structure analysis are done almost routinely. These steps are: (1) identification of the macroscopic symmetry and crystal class, (2) determination of the lattice constants, and (3) computation of the number of atoms per unit cell.

Any intensive property of the crystal will manifest its macroscopic symmetry. If the crystal has been slowly grown, it will exhibit well-defined facets. The symmetry of the facets will then define the macroscopic symmetry of the crystal. Furthermore, the axes of specific symmetry are located by inspection. Most books on mineralogy give numerous examples of crystals and of their symmetry. If the facets are not well formed, the investigator resorts to other physical properties. Some examples would be optical birefringence, optical absorption, or electrical conductivity. In these cases, one must observe the symmetry in a three-dimensional property map.

Very often, crystals with ill-formed or no facets are crystals of near isotropy—as is the case for most metals. Such crystals are usually cubic, hexagonal, or tetragonal. It is possible to determine if a crystal possesses one of these symmetries by x-ray diffractometry. The principle here is that for such high symmetries, the d-spacings must follow simple rules, with only one or two unknown parameters

Cubic $$d_{hkl} = \frac{a}{\sqrt{h^2 + k^2 + l^2}} = \frac{a}{\text{Integer}}$$

Tetragonal $$d_{hkl}^{-2} = \frac{h^2 + k^2}{a^2} + \frac{l^2}{c^2} = \frac{\text{Integer}}{a^2} + \frac{\text{Integer}}{c^2}$$

Hexagonal $$d_{hkl}^{-2} = \frac{4}{3}\,(h^2 + hk + k^2)\frac{1}{a^2} + \frac{l^2}{c^2}$$
$$= \frac{4}{3a^2} \times \text{Integer} + \frac{\text{Integer}}{c^2}$$

If a quantity a or the quantities a and c can be determined such that one of the above formulas hold for all reflections, then one has found both the crystal class and the lattice parameters.

Once one has identified the crystal symmetry, one obtains the lattice constants through x-ray rotation patterns. Here the rotation axes are **a, b,** and **c.** The layer line spacings give **a*, b*,** and **c*.** Knowing **a*, b*,** and **c*** and the angles between them (from the property measurement), one obtains the volume of the unit cell via the relation

$$V_c = \mathbf{a} \cdot (\mathbf{b} \times \mathbf{c}) \tag{5.9}$$

The macroscopic density, ρ, of the crystal must also be found, using any appropriate technique. The stoichiometry must also be determined, using any chemical method. Using V_c and ρ, one then computes the number N_c of molecules in the unit cell

$$N_c = \frac{N_A \rho V_c}{M} \tag{5.10}$$

where N_A is Avogadro's number and M is the molecular weight of the molecule (or stoichiometric unit). One now knows how many atomic positions must be determined (exactly N_c). By determination of the systematic extinctions of the crystal, one reduces the number of *independent* atomic positions to a number less than N_c.

Experimentally, one uses some variant of the simple rotating crystal method. A variant must be used to avoid overlapping reflections. Modern crystal structure laboratories use automated four-circle x-ray goniometers which rotate the crystal through the full 4π solid angle and automatically record intensity as a function of resp position. Moving-film methods represent an older technology which is still appropriate for the analysis of the structure of simpler crystals. The experimental methods are described in books on x-ray crystallography or structure analysis.

5.4 THE PATTERSON FUNCTION

Analysis of diffraction data yields in a fairly direct manner the complete set of interatomic distances within the unit cell. The descriptor of the interatomic distances is termed the *Patterson function.* The visual representation of the interatomic distances is termed a *Patterson map.* Such information represents the maximum information that can be directly drawn from diffraction data; higher levels of information require further information about the material.

The Patterson function is nothing more than the autocorrelation function $P(\mathbf{u})$

$$P(\mathbf{u}) = \int \rho(\mathbf{r})\rho(\mathbf{r} + \mathbf{u})dv \tag{5.11}$$

where $\mathbf{u}$ is the vector distance between two volume elements. It was recognized by A. L. Patterson in 1934 that this function can be useful in structure analysis and was introduced by him for that purpose. The Patterson function is derived by Fourier transformation of the observed intensities via

$$P(\mathbf{u}) = \int \frac{I(\mathbf{s})}{I_e(s)}\, e^{-2\pi i \mathbf{s}\cdot\mathbf{u}}\, dv_{\mathbf{u}} \tag{5.12}$$

The Patterson function peaks at vector positions $\mathbf{u}$ for which both $\rho(\mathbf{r})$ and $\rho(\mathbf{r} + \mathbf{u})$ have finite magnitude. That is, *peaks in the Patterson function denote*

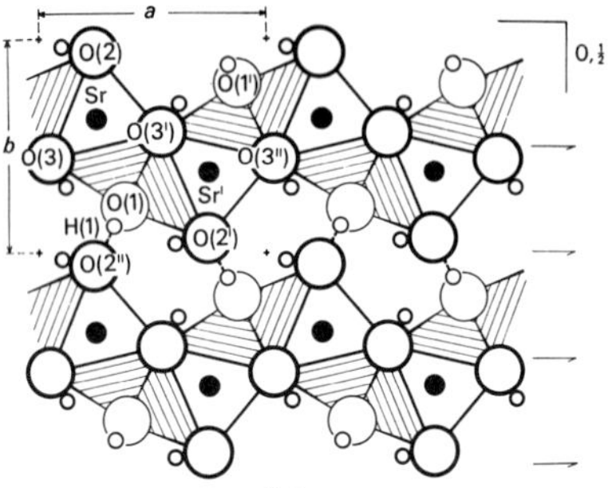

(*a*)

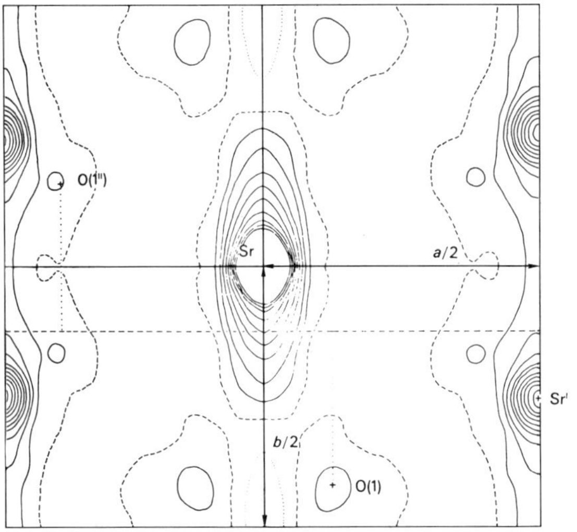

(*b*)

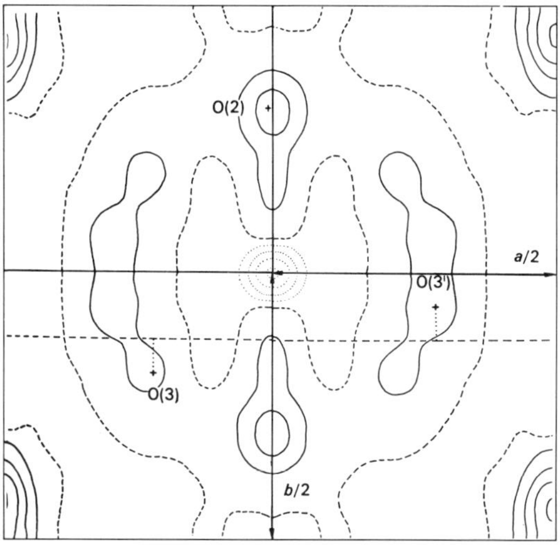

(*c*)

Figure 5.21. *(a)* Sketch of the crystal structure of $Sr(OH)_2 \cdot H_2O$, viewed along the *c*-axis. The heights of the several atoms along *z* are: Sr and O(1) in $z = 0$; O(2) and O(3) and the associated *H*'s in $z = \frac{1}{2}$. Four unit cells are shown in this drawing; *(b)* Patterson map, $P(U'_x, U'_y, 0)$ at the level $z = 0$. The origin is taken on a Sr atom. Crosses in *(b)* and *(c)* show interatomic distances derived from the final structure analysis; *(c)* Patterson map, $P(U'_x, U'_y, \frac{1}{2})$, at the level $z = \frac{1}{2}$. The origin is here again the Sr atom at the center. [H. Bärnigshausen and J. Weidlein, *Acta Cryst.*, **22**, 252 (1967)]

interatomic separations. When one maps the Patterson function, one obtains a contour map in which the distance of each peak from the origin is an interatomic distance. One has from this map a "jigsaw puzzle" containing all interatomic positions. It remains then to piece this information together to extract real crystal structure information.

Computations of the Patterson function are structured as follows. For a crystalline solid, the observable **s** are the rel vectors $\mathbf{r}_{hkl}$. Thus

$$P(\mathbf{u}) = \sum_h \sum_k \sum_l \frac{I_{hkl}}{(I_e)_{hkl}} e^{-2\pi i(hU'_x + kU'_y + lU'_z)} \tag{5.13}$$

where $U'_x = u_x/a$, $U'_y = u_y/b$, and $U'_z = u_z/c$ are the normalized vector components of **u**. In order to determine $P(U'_x, U'_y, U'_z)$, a program must be written to determine the Fourier sum over h,k, and l for each real space position U'_x, U'_y, U'_z.

Since three-dimensional contour maps are difficult to construct, one may obtain two-dimensional Patterson maps at several depths through the unit cell. This is done by restricting $\mathbf{U}'$ to a plane. For instance, suppose a cut were made through the $z = 0$ plane. In this case

$$P(U'_x, U'_y, 0) = \sum_h \sum_k \sum_l \frac{I_{hkl}}{(I_e)_{hkl}} e^{-2\pi i(hU'_x + kU'_y)} \tag{5.14}$$

where U'_z has been set to zero everywhere. A set of such contour maps taken at different z levels can be useful in determining the full crystal structure of the material. Figure 5.21 shows the crystal structure of $Sr(OH)_2 \cdot H_2O$ and Patterson maps taken at levels $U'_z = 0$ and $U'_z = ½$.

More often, however, the projection of the three-dimensional Patterson function onto a plane is reported. The projection onto the xz-plane is

$$P(U'_x, U'_z) = \sum_h \sum_l \left[\int_0^1 \sum_k \frac{I_{hkl}}{(I_e)_{hkl}} e^{-2\pi i(hU'_x + kU'_y + lU'_z)} dU'_y \right] \tag{5.15}$$

Figure 5.22*a* shows the xz-plane Patterson projection of $W_4Nb_{26}O_{77}$. For comparison, a projection of the fully determined crystal structure is shown in Fig. 5.22*b*.

5.5 FOURIER SYNTHESIS; THE PHASE PROBLEM

5.5.1 General Attack

Recall the general expression for the structure factor

$$F_{hkl} = \int_{\text{cell}} \rho(\mathbf{r})\, e^{2\pi i(h\mathbf{a}^* + k\mathbf{b}^* + l\mathbf{c}^*)\cdot\mathbf{r}}\, dv_{\mathbf{r}} \tag{5.16}$$

Taking the inverse Fourier transform, we have

$$\begin{aligned} \rho(\mathbf{r}) &= \int F_{hkl}\, e^{2\pi i(h\mathbf{a}^* + k\mathbf{b}^* + l\mathbf{c}^*)\cdot\mathbf{r}}\, dv_{\mathbf{s}} \\ &= \frac{1}{V} \sum_h \sum_k \sum_l F_{hkl}\, e^{2\pi i(h\mathbf{a}^* + k\mathbf{b}^* + l\mathbf{c}^*)\cdot\mathbf{r}} \qquad (5.17) \\ &= \frac{1}{V} \sum_h \sum_k \sum_l F_{hkl}\, e^{2\pi i(hX + kY + lZ)} \end{aligned}$$

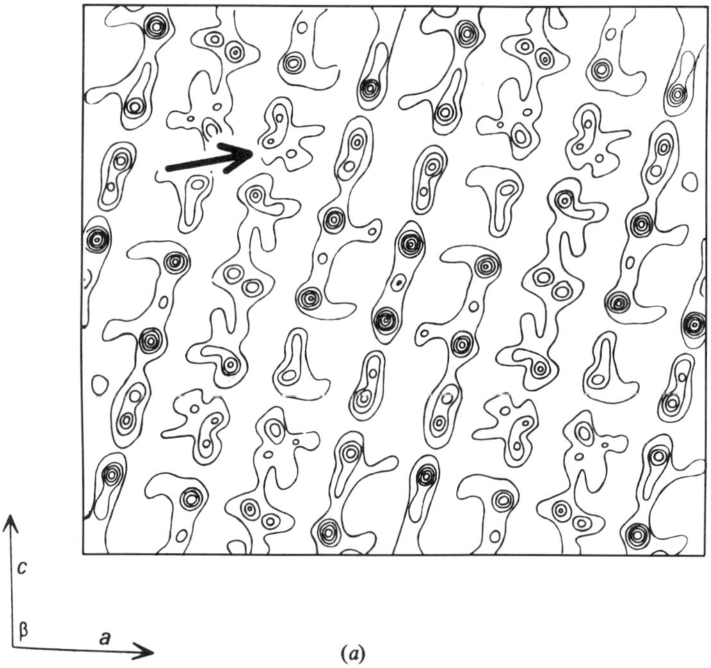

(*a*)

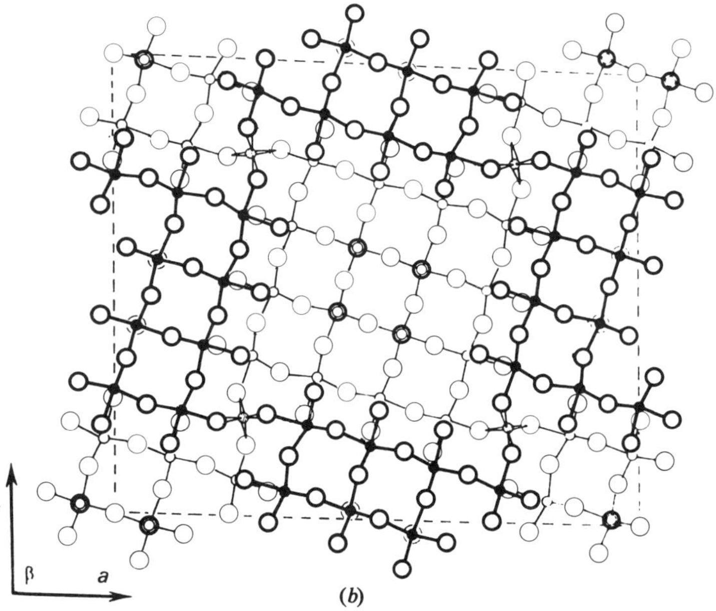

(*b*)

Figure 5.22. *(a) xz*-plane Patterson projection of $W_4Nb_{26}O_{77}$; *(b)* the final crystal structure. In *(b),* the small hatched circles are tetrahedral W, small open and black circles are octahedral metals, and the largest circles are oxygen. The heavier (or blacker) circles denote $y = 0$ and the lighter $y = \frac{1}{2}$. [S. Andersson, W. G. Mumme, and A. D. Wadsley, *Acta Cryst.,* **21,** 802 (1966)]

in which the position **r** within the unit cell is given by

$$\mathbf{r} = X\mathbf{a} + Y\mathbf{b} + Z\mathbf{c}$$

Thus, were the set of F_{hkl} known, the electron density distribution $\rho(\mathbf{r})$ within the unit cell could be found. This is the final step in crystal structure determination.

A problem exists, however; F_{hkl} cannot be completely determined from a diffraction experiment. F_{hkl} has real and imaginary parts (see Eqs. (5.3) or (5.16))

$$F_{hkl} \equiv F'_{hkl} + iF''_{hkl} \tag{5.18}$$

Rewritten

$$F_{hkl} = |F_{hkl}|e^{i\epsilon_{hkl}} \tag{5.19}$$

in which

$$\begin{aligned} |F_{hkl}|^2 &= (F'_{hkl})^2 + (F''_{hkl})^2 \\ \tan\epsilon &= F''_{hkl}/F'_{hkl} \end{aligned} \tag{5.20}$$

Only the magnitude, $|F_{hkl}|$, of the structure factor can be measured experimentally (see Eq. 3.60)

$$\frac{I_{hkl}}{(I_e)_{hkl}} = B_{hkl}|F_{hkl}|^2 \tag{5.21}$$

where B_{hkl} contains the Thomson radius of the electron, the specimen-to-detector distance, and the various corrections (Lorentz-polarization, absorption, Debye-Waller, and multiplicity). *The phase factor $e^{i\epsilon_{hkl}}$ cannot be directly determined.*

For crystals possessing a center of symmetry, the problem of determining the phase factor, exp $(i\epsilon_{hkl})$ is greatly simplified. Note that in this case for each atom located at X, Y, Z there exists a similar type atom at $-X, -Y, -Z$. Thus the structure factor can be written as the sum of pairs of terms

$$\begin{aligned} F_{hkl} &= \sum_j f_j\left[e^{2\pi i(hX_j+kY_j+lZ_j)} + e^{-2\pi i(hX_j+kY_j+lZ_j)}\right] \\ &= 2\sum_j f_j \cos\left[2\pi(hX_j + kY_j + lZ_j)\right] \end{aligned} \tag{5.22}$$

This is a *real* number. Thus $|F_{hkl}|$ can be either $+F_{hkl}$ or $-F_{hkl}$, depending on whether F_{hkl} is greater or less than zero. The phase factor, then, is either $+1$ or -1, corresponding to $\epsilon_{hkl} = 0$, or π. One need now only determine the proper *sign* for F_{hkl}, since $|F_{hkl}|$ is experimentally determined.

The skill of the professional crystal structure analyst lies in the determination of the phase factor. Several general methods are available. These will now be described briefly.

5.5.2 Heavy Atom Method

Suppose that the unit cell is composed of a large number of atoms. Suppose also that a certain atomic type, present only a small number of times in the cell, is much heavier than any of the others. A good criterion is

$$Z_H/\Sigma\, Z_i \geq 0.3 \tag{5.23}$$

where Z_H is the atomic number of the heavy atom and the summation is over all atoms in the repeating unit. In this case, the phase angle for most reflections turns out to be nearly that for the heavy atom alone.

This is seen in the following way. The structure factor, in general, is written

$$F_{hkl} = \sum_j f_j\, e^{2\pi i(hX_j + kY_j + lZ_j)} \tag{5.24}$$

Suppose now that there is only one heavy atom per cell and that we choose to place the origin of **r** at its center. Then

$$F_{hkl} = f_H + \sum_{j \neq H} f_j\, e^{2\pi i(hX_j + kY_j + lZ_j)} \tag{5.25}$$

where f_H is the atomic scattering factor for the heavy atom. If there are many other atoms, distributed throughout the cell, then the sum over $j \neq H$ contains a number of small terms, some positive and some negative. The sum, then, will be of a small absolute magnitude, compared with f_H. And the phase angle of most reflections, in this case, is approximately 0. The condition for cells containing more than one heavy atom is similar, except that for *some* reflections (for which the heavy atoms are out of phase with each other) the approximation will not apply at all.

What is done is the following. The phase angle is determined as if the heavy atoms alone controlled. Using these phase angles (usually 0 or π), a trial crystal structure is generated, via (5.17). The trial atomic positions generated in this step are then used to determine a more precise set of phase angles, and the crystal structure redetermined. This process may be repeated until no change in the crystal structure is produced by further iteration. Figure 5.23 is an example of a first trial and a final crystal structure so obtained.

5.5.3 The Replaceable Atom Method

In this case one deals with a multicomponent system in which one type of atom *(M)* can be replaced by another type *(N)*. If it may also be assumed that no appreciable structural rearrangement occurs thereby (i.e., the atoms N simply replace the M in a rigid framework), then one can utilize this method. We write the structure factor of the original, M-bearing, material as

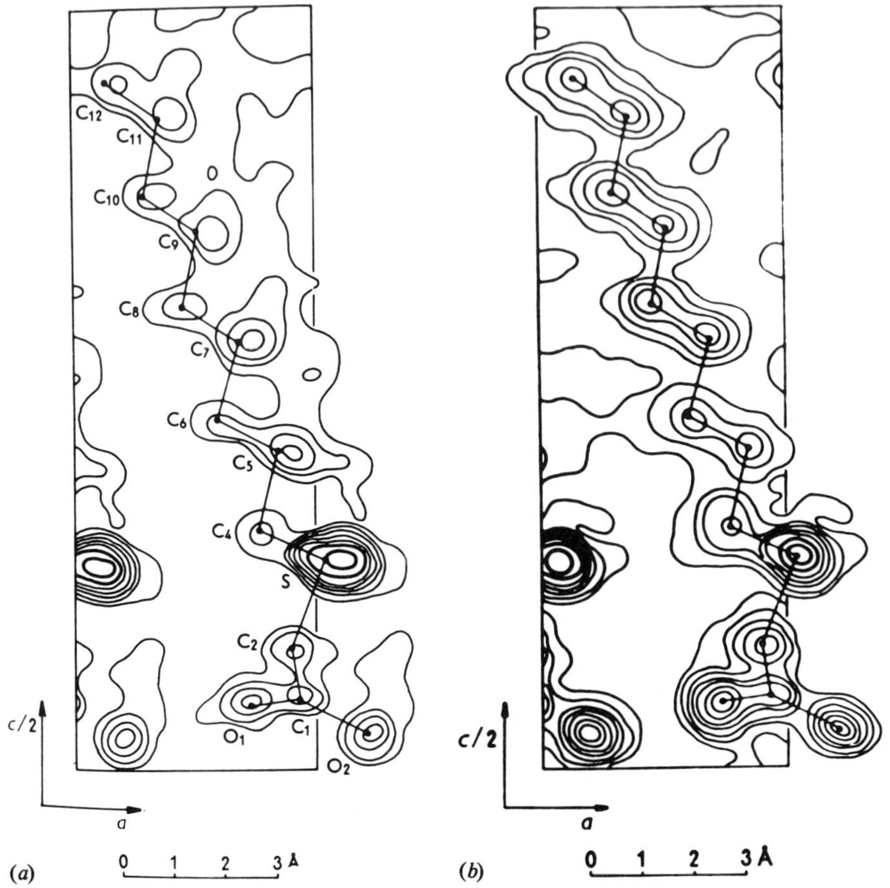

Figure 5.23. Initial try *(a)* and final *(b)* crystal structures of 3-thiadodecanoic acid. The projection is along the *b*-axis. Carbon, oxygen, and sulfur atoms are marked in *(a)*. In *(b)*, the contours are given at intervals of 1 electron/Å² intervals. [S. Abrahamson and A. Westerdahl, *Acta Cryst.*, **16,** 404 (1963)]

$$F_{hkl}^{CM} = F_{hkl}^{C} + F_{hkl}^{M} \tag{5.26}$$

where the superscript C denotes the components common to both M- and N-bearing materials. Similarly, the structure factor of the N-bearing material is

$$F_{hkl}^{CN} = F_{hkl}^{C} + F_{hkl}^{N} \tag{5.27}$$

and

$$\Delta F_{hkl} = F_{hkl}^{CN} - F_{hkl}^{CM} = F_{hkl}^{N} - F_{hkl}^{M} \tag{5.28}$$

Suppose now that the crystal is centrosymmetric. Then, as we have seen, the phase angles of the F_{hkl} can be only 0 or π and the values of the F_{hkl} can be only $+|F_{hkl}|$ or $-|F_{hkl}|$. The correct signs can be determined as follows:

(1) $F^N_{hkl} - F^M_{hkl}$ is computed, using a Patterson map, to determine the position of M or N. The difference

$$F^N_{hkl} - F^M_{hkl} \tag{5.29}$$

is exactly

$$F^N_{hkl} - F^M_{hkl} = 2(f_N - f_M) \sum_j \cos(2\pi hX_j + kY_j + lZ_j)$$

where the X_j, Y_j, Z_j are the noncentrosymmetrically related positions of the M (or N) atoms.

(2) $|F^{CM}_{hkl}|$ and $|F^{CN}_{hkl}|$ are determined from diffraction intensities.

(3) The four possible values of $F^{CN}_{hkl} - F^{CM}_{hkl}$ are computed. These are $+|F^{CN}_{hkl}| + |F^{CM}_{hkl}|$, $+|F^{CN}_{hkl}| - |F^{CM}_{hkl}|$, $-|F^{CN}_{hkl}| + |F^{CM}_{hkl}|$, and $-|F^{CN}_{hkl}| - |F^{CM}_{hkl}|$. The correct signs are then found by comparison with the experimentally determined $F^{CN}_{hkl} - F^{CM}_{hkl}$.

(4) Once the correct signs for the F_{hkl} are found, the crystal structure is computed via (5.17).

The procedure is considerably more difficult if the unit cell has no center of symmetry.

5.5.4 Direct Methods

The so-called "direct" methods are probabilistic schemes for assigning tentative phase factors to the several structure factors F_{hkl}. These methods require no heavy or special atoms and provide a first approximation to the crystal structure. The atom positions indicated in this first attempt can then be used in an iterative way to refine the phase factors and, consequently, the crystal structure determination. These methods, summarized by Buerger,[2] are based on nonsimple mathematical methods and are beyond the scope of this book.

5.5.5 Thermal Motion

The parameters $|F_{hkl}|$ are used in the determination of crystal structure. Experimentally, these $|F_{hkl}|$ are derived from intensity data. In properly extracting the $|F_{hkl}|$ from intensity data, one analytically corrects for polarization, multiplicity, the Lorentz factor, absorption, and thermal motion. All these corrections are straightforward, except for thermal motion. The parameters involved in the Debye-Waller (thermal) term are not known *a priori.* The thermal motions, as well as the mean atomic positions, are unknowns, to be determined by analysis of the data.

In principle, the parameters in the Debye-Waller term are left as free variables

[2] M. J. Buerger, *Contemporary Crystallography* (McGraw-Hill, New York, 1970).

and fit at the same time that the atomic positions are fit, using the methods outlined above. Modern computing software does the thermal parameter fitting routinely, using anisotropic Debye-Waller factors of the form

$$D^j_{hkl} = \exp[-(\beta^j_{11} h^2 + \beta^j_{22} k^2 + \beta^j_{33} l^2 + \beta^j_{12} hk + \beta^j_{23} kl + \beta^j_{31} lh)] \quad (5.30)$$

where the superscript j denotes the jth atom and the β^j_{mn} relate to the root mean square atomic displacements u_{mn} via

$$\begin{aligned} \beta_{11} &= 2\pi^2 u_{11} a^{*2} & \beta_{12} &= 4\pi^2 u_{12} a^* b^* \cos\gamma^* \\ \beta_{22} &= 2\pi^2 u_{22} b^{*2} & \beta_{23} &= 4\pi^2 u_{23} b^* c^* \cos\alpha^* \\ \beta_{33} &= 2\pi^2 u_{33} c^{*2} & \beta_{31} &= 4\pi^2 u_{31} c^* a^* \cos\beta^* \end{aligned} \quad (5.31)$$

$$\begin{aligned} \alpha^* &= \text{angle between } b^* \text{ and } c^* \\ \beta^* &= \text{angle between } c^* \text{ and } a^* \\ \gamma^* &= \text{angle between } a^* \text{ and } b^* \end{aligned}$$

The standard depiction of crystal structures places an ellipse at each atomic position. The distance from the center of each ellipse to a point on its surface is proportional to the root mean square atomic displacement in that direction. An example of a structure depicted by such ellipsoids is shown in Fig. 5.24.

5.5.6 Difference Fourier Maps

Difference Fourier maps are useful as a pictorial indication of the state of the accuracy of the structure determination, as a probe of thermal motion, and as a method for locating light atoms.

A difference Fourier transform is computed in the following way. After a well-refined structure has been determined, the structure factors used to get it are labelled F^{obs}_{hkl}. Another set of structure factors are determined by Fourier transformation of the computed electron density map ($\rho(\mathbf{r})$ map). These structure factors are labelled F^{cal}_{hkl}. The difference Fourier transform is defined as

$$\rho_{obs} - \rho_{cal} = \frac{1}{V} \sum_h \sum_k \sum_l (F^{obs}_{hkl} - F^{cal}_{hkl}) e^{2\pi i(hX + kY + lZ)} \quad (5.32)$$

The Fourier difference map shows certain features quite clearly. If an atom has been located slightly out of position, the difference map will show a depression and adjacent positive hump near the atomic site. This indicates that the computed atom position is in error in the direction of the hump and needs to be moved back toward the depression. Second, ringlike depressions or elevations which are symmetrical about an atomic center indicate that the thermal motion is in need of refinement. This symmetrical defect will disappear if the atomic occupancy is appropriately spread out, via corrected thermal motion terms. Finally, light atoms, which would have appeared as only minor perturbations in a standard electron density map, can show up as well-defined entities in the difference map.

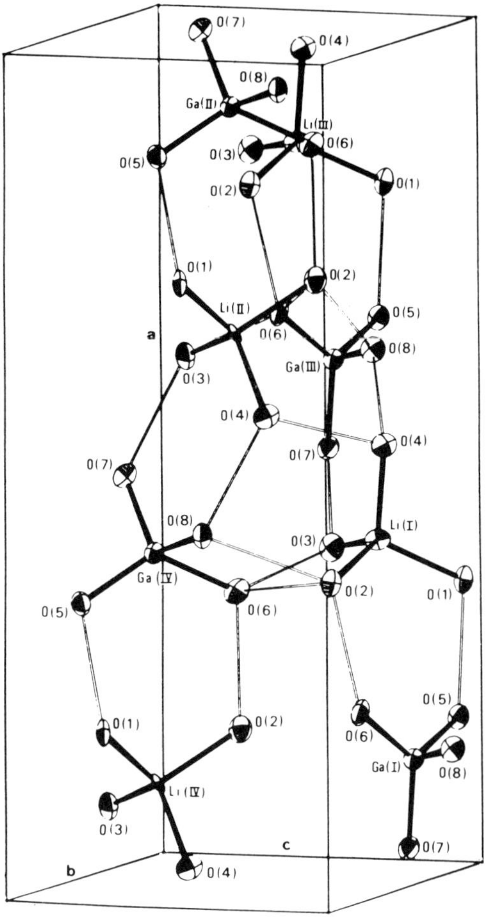

Figure 5.24. Crystal structure of lithium gallate hexahydrate, $LiGaO_2 \cdot 6H_2O$. Atomic positions depicted by thermal motion ellipsoids. Hydrogen atoms not shown. [C. Caranoni, G. Pepe, and L. Capella, *Acta Cryst.*, **B34**, 741 (1978)]

5.5.7 Refinement

It was indicated earlier that crystal structures are refined by iterative processes. Generally, the atomic positions found in the preceding iteration step are used to compute new structure factors. These structure factors are then used to compute a new electron density map. Provided that the atomic positions are approximately correct, this procedure always acts to more accurately position the atoms. This refinement occurs through sign changes of the structure factors which contribute little to the Fourier sum near the proposed peak position.

The iterative process is performed by a least squares treatment of both structure factors and thermal parameters. The standard by which refinement is judged is the discrepancy index R

$$R = \frac{\sum_h \sum_k \sum_l \left| |F_{hkl}^{\text{obs}}| - |F_{hkl}^{\text{cal}}| \right|}{\sum_h \sum_k \sum_l |F_{hkl}^{\text{obs}}|} \tag{5.33}$$

R values of 0.05 or 0.06 indicate a well-refined structure. For completely wrong structures, the expected values of R are 0.83 for centrosymmetric crystals and 0.59 for noncentrosymmetric crystals.

PROBLEMS

5.1. Show that base-centered tetragonal is not different from one of the 14 Bravais lattices.

5.2. Why is face-centered monoclinic not one of the 14 Bravais lattices?

5.3. Sketch and label the equivalent points for threefold and sixfold rotation and for threefold rotation-inversion.

5.4. Prove, using equivalent points, that twofold rotation-inversion is identical with a mirror plane.

5.5. Show that 3 combined with a mirror perpendicular to this axis is identically $\bar{3}$.

5.6. Sketch the 422 point group and give the positions of the equivalent points.

5.7. Draw diagrams to show a set of equivalent positions and the set of symmetry elements for the space groups

a) Pn b) $P\frac{2}{n}$ c) $P\frac{2_1}{c}$ d) $I4$

5.8. Derive the coordinates of the equivalent points for $P222$.

5.9. The diagrams shown in Fig. P5.9 show a set of equivalent positions in a unit cell. Find the crystal system and suggest a name for the space group.

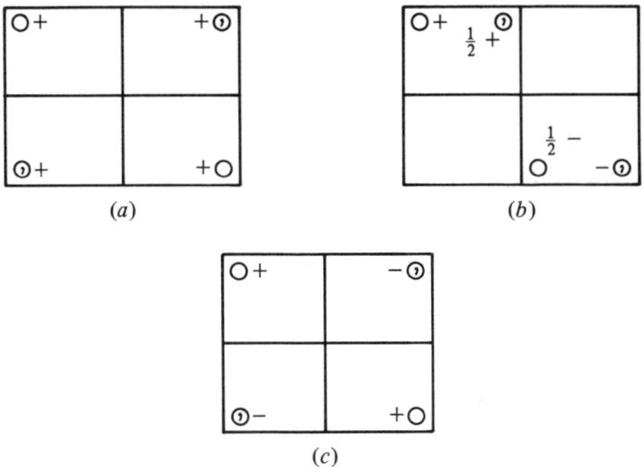

Figure P5.9.

5.10. In each of the following cases we are given the crystal class of a crystal and the indices of one face appearing on the crystal. Give the indices of the other faces belonging to the same form.
(a) $4/m$ (121) (d) 222 (121)
(b) $4/m$ (010) (e) 222 (021)
(c) $4m$ (121)

5.11. (a) Given one point at some position *(x,y,z)* relative to an origin, list the symmetry-related points (equivalent points) for point group $2mm$.
(b) List the general set of equivalent points for the space group $I2_12_12_1$. (Take the first point at *(x,y,z)*.)
(c) Suppose the initial equivalent point were located right on a 2_1 axis. What are the (special) equivalent points for this case?

5.12. (a) List the general set of equivalent points for space group $I\bar{4}2m$.
(b) Suppose an element crystal had atoms in these sites. Write an expression for the structure factor.
(c) What is the rule for systematic extinctions for the case?
(d) Would the answer to (c) be changed if there were fewer atoms and these were at *X*,0,0 and the symmetry-related positions?

5.13. The orthorhombic crystalline form of $CaCO_3$ has a density of 2.93 g/cm³. Rotation photographs around the *a,b,c* axes yield values of zeta (ζ) for the first layer line of 0.312, 0.194, and 0.270, respectively. (CuK_α radiation used; $\lambda = 1.54$ Å). Indexing the photographs yields reflections of the following types

200	011	110	111
400	012	130	112
	021	220	211
020		240	311
040	102	330	
	104		etc.
002			
004	202		
	302		
	304		

(a) What are cell dimensions?
(b) How many molecules per cell?
(c) What is the space group? (Use the *International Tables for X-ray Crystallography)*

5.14. The table below gives x-ray powder diffraction data for *d,l*-poly-(propylene oxide). This material is orthorhombic, with $a = 10.40$, $b = 4.64$, $c = 6.92$ Å.
(a) What is the Bravais lattice? *(P, C, I, F)*
(b) There is a screw axis associated with each crystallographic axis (**a**, **b**, and

c). For instance, along **b**, the screw axis is 2_1. What are the screw axes along the other crystallographic axes?

dÅ	I/I_1
5.18	45
4.21	100
3.89	6
3.61	4
3.46	6
3.24	<1
3.10	8
2.89	4
2.77	18
2.59	10
2.450	8
2.319	2
2.270	2
2.166	4
2.115	2
2.072	20
1.984	2
1.938	4
1.902	18
1.860	2
1.735	4
1.681	2
1.635	14

5.15. Figure P5.15 is the rotating crystal pattern of a cubic crystal, taken using CuK_α radiation ($\lambda = 1.54$ Å). The camera diameter was 57.3 mm.
(a) Find the lattice constant.
(b) This material has a density 6.100 g/cm³. Its composition is Cu_2O. How many atoms are there in the unit cell?
(c) Find the indices for each point on the film and label the points accordingly.
(d) Determine a rule for which reflections appear (or, conversely, a rule for systematic extinctions.) Can you learn anything about the atomic positions?

5.16. The structure of marcasite, FeS_2. The unit cell contains two "molecules" related by the space group symmetry *Pmnn* (listed as space group No. 58 in Vol. II of the *International Tables for Crystallography).*
(a) What is the position of the Fe atoms?
(b) What positions are possible for the S atoms? No systematic extinctions additional to those arising from the two *n* glide planes of *Pmnn* are observed.
(c) Show that this eliminates two of the three possible sets of positions for the S atoms.

Further information about the atomic coordinates can be obtained only by a comparison of observed and calculated intensities. Some experimental data are given in the following table.

hkl	f_{Fe}	f_S	*L.p.**	I_{obs}
200	15.6	9.0	4.45	Strong
400	10.2	6.1	1.84	Medium
020	17.9	10.2	6.05	Zero
040	12.3	7.3	2.68	Medium
002	19.4	11.0	7.42	Strong
004	13.8	8.1	3.40	Zero
110	19.4	11.0	7.35	Very strong
103	15.0	8.7	4.08	Very strong

* Lorentz-polarization factor

(d) The sulphur coordinates are (0*uv*) where *u* and *v* are two unknown parameters. List the values of *u* and *v* that are consistent with the observations that the (020) and (004) reflections are too weak to be observed.

(e) Which of the four values of *u* are consistent with (110) being very strong? (This will eliminate two values, the other two are related as *u* to $\bar{u}$; i.e. by the center of symmetry at the origin of the cell).

(f) Which of the eight values of *v* are consistent with (103) being very strong?

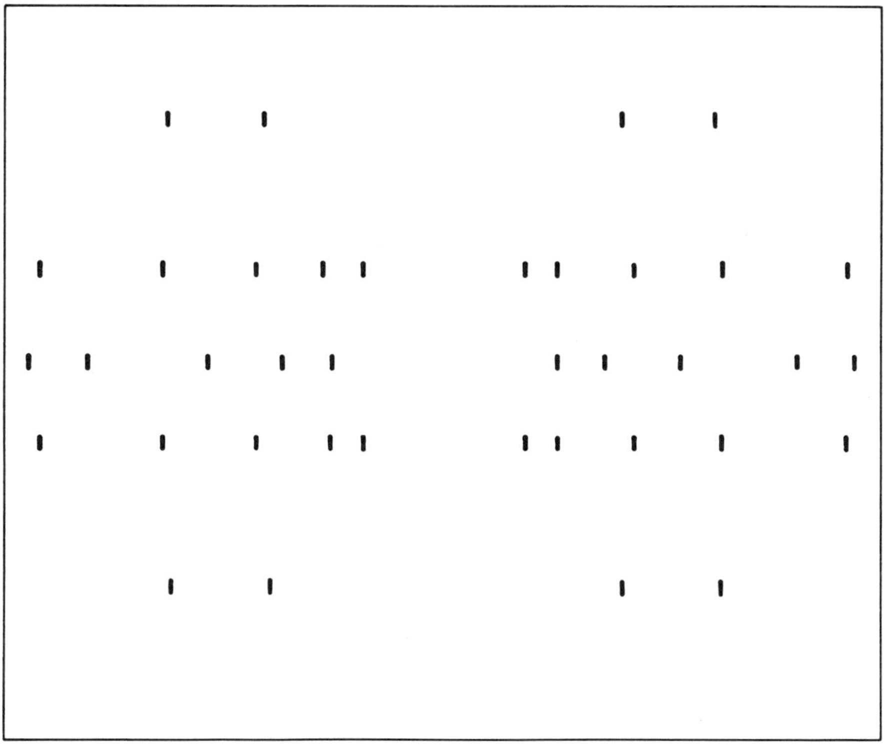

Figure P5.15.

(This will eliminate six and leave two which are related by the center of symmetry).

(g) Show that these values of $\pm u$ are $\pm v$ give good agreement for the other intensities listed above.

(h) It remains to determine which combination of $\pm u$, $\pm v$ corresponds to the S coordinates. To this end, the intensities of the *hkl* general reflection might be considered. Is it possible by examining some particular reflections or combination of reflections to select the correct sign combination for S parameters, and if so which?

(i) Make a drawing of the crystal structure.

BIBLIOGRAPHY

A. Crystallography

1. L. A. AZAROFF, *Elements of X-ray Crystallography,* Chs. 1–4, McGraw-Hill Book Company, N.Y. (1968).

2. M. G. BUERGER, *Introduction to Crystal Geometry,* McGraw-Hill Book Company, (1971).

3. G. BURNS and A. M. GLAZER, *Space Groups for Solid State Scientists,* Academic Press, Inc., N.Y. (1978).

4. N. F. M. HENRY and KATHLEEN LONSDALE, eds., *International Tables for X-ray Crystallography, 2nd Ed., Vol. I,* Ch. 5, The Kynoch Press, Birmingham, England (1965).

5. F. C. PHILLIPS, *An Introduction to Crystallography, 2nd Ed.,* Longmans, Green & Co., London (1956).

B. Crystal Structure Analysis

1. G. E. BACON, *Applications of Neutron Diffraction in Chemistry,* Pergamon Press, Inc., N.Y. (1963).

2. M. J. BUERGER, *X-ray Crystallography,* John Wiley & Sons, Inc., N.Y. (1942).

3. P. A. EGLESTAFF, *Thermal Neutron Scattering,* Chs. 10–14, Academic Press, Inc., N.Y. (1965).

4. JENNY P. GLUSKER and K. N. TRUEBLOOD, *Crystal Structure Analysis: A Primer,* Oxford, N.Y. (1972).

5. R. HOSEMANN and S. N. BAGCHI, *Direct Analysis of Diffraction by Matter,* Chs. 13–15, North-Holland, Amsterdam (1962).

6. M. F. C. LADD and R. A. PALMER, *Structure Determination by X-ray Crystallography,* Plenum Publishing Corporation, N.Y. (1977).

7. G. N. RAMACHANDRAN and R. SRINIVASAN, *Fourier Methods in Crystallography,* Wiley-Interscience, N.Y. (1970).

8. G. H. STOUT and L. H. JENSEN, *X-ray Structure Determination; a Practical Guide,* Macmillan, Inc., N.Y. (1968).

9. M. M. WOOLFSON, *An Introduction to X-ray Crystallography,* Chs. 7–9, Cambridge University Press, Cambridge (1970).

Disordered Crystals 6

6.1 GENERAL FORMALISM

This section shows how the general problem of scattering by a disordered crystal is to be treated. By disorder, we mean simply a loss in perfect periodicity. We may distinguish between chemical and physical disorders. By "chemical" we refer to local changes in atomic type. By "physical" we refer to the displacement of atoms from their perfect crystal sites. We shall see that the degree of coupling of disorder among unit cells becomes a critical factor and determines the disposition in resp of intensity removed from the Bragg peaks.

To begin, we first observe how disorder enters the intensity relation. We recall that for a unit cell composed of N_c atoms, the structure factor is written

$$F(\mathbf{s}) = \sum_{p=1}^{N_c} f_p e^{-2\pi i \mathbf{s}_{hkl} \cdot \mathbf{r}_p} \tag{6.1}$$

where $\mathbf{s}_{hkl}$ is the scattering vector associated with the hkl reflection. (However, it is not restricted to the rel point; in fact, we will now see that one effect of disorder is to delocalize some of the scattered intensity.) For a crystal containing regions of less than perfect order we must be careful to label each unit cell so that its specific composition (through the f_p's) and displacements (through $\mathbf{r}_p$'s) may be indicated. In this case, we label each unit cell by an integer n. Thus the structure factor of the nth cell is

$$(F_{hkl})_n = \sum_{p=1}^{N_c} f_{pn} e^{-2\pi i \mathbf{s}_{hkl} \cdot \mathbf{r}_{pn}} \tag{6.2}$$

where $\mathbf{r}_{pn}$ is the vector from the origin of real space to the pth atom of the nth cell. Inserting (6.2) into the general intensity formula, we have, for a system if N unit cells

$$\frac{I(\mathbf{s})}{I_e(s)}=\frac{A(\mathbf{s})A^*(\mathbf{s})}{I_e(s)}=\left(\sum_{n=1}^{N}\sum_{p=1}^{N_c} f_{pn}e^{-2\pi i\mathbf{s}_{hkl}\cdot\mathbf{r}_{pn}}\right)\left(\sum_{n'=1}^{N}\sum_{p=1}^{N_c} f_{pn'}e^{2\pi i\mathbf{s}_{hkl}\cdot\mathbf{r}_{pn'}}\right)$$

$$=\left[\sum_{n=1}^{N}(F_{hkl})_n e^{-2\pi i\mathbf{s}_{hkl}\cdot\mathbf{r}_n}\right]\left[\sum_{n'=1}^{N}(F^*_{hkl})_{n'}e^{2\pi i\mathbf{s}_{hkl}\cdot\mathbf{r}_{n'}}\right] \tag{6.3}$$

$$=\sum_{n=1}^{N}\sum_{n'=1}^{N}(F_{hkl})_n(F^*_{hkl})_{n'}e^{2\pi i\mathbf{s}_{hkl}\cdot(\mathbf{r}_{n'}-\mathbf{r}_n)}$$

where $\mathbf{r}_n$ is the vector from the origin of real space to a fiducial mark (e.g., the center of a specific atom) in the nth unit cell. Setting

$$m=n'-n \qquad \mathbf{r}_m=\mathbf{r}_{n'}-\mathbf{r}_n \tag{6.4}$$

we have

$$\frac{I(\mathbf{s})}{I_e(s)}=\sum_{m=1}^{N}\left[\sum_{n=1}^{N}(F_{hkl})_n(F^*_{hkl})_{n+m}\right]e^{-2\pi i\mathbf{s}_{hkl}\cdot\mathbf{r}_m} \tag{6.5}$$

Here the bracketed term is an autocorrelation function for unit cells. Since the summation over m takes place only after summation over n is completed, we may replace the n sum by the number of terms contained times the average value

$$\sum_{n=1}^{N}(F_{hkl})_n(F^*_{hkl})_{n+m}=N_m\overline{(F_{hkl})_n(F^*_{hkl})_{n+m}} \tag{6.6}$$

Here N_m is the number of nonzero terms in the sum. Clearly if the crystal is large and the value of m small, N_m will be very nearly the total number of unit cells. On the other hand if m is large (large $\mathbf{r}_{n'}-\mathbf{r}_n$) or if the crystal is very small, a number of potential pairs with $\mathbf{r}_n$ within the crystal will be nullified because $\mathbf{r}_{n'}$ will be outside the crystal limits. *For now,* let us take $N_m=N$. We shall return later to the case of small particle size effects.

We now proceed to deduce the dependence on intercell defect correlations. Let $\overline{F}$ be the average structure factor for the perturbed crystal

$$\overline{F}_{hkl}=\frac{1}{N}\sum_{n=1}^{N}(F_{hkl})_n \tag{6.7}$$

Then we may let a quantity ϕ^n_{hkl} represent the deviation of the nth cell from the average value

$$(F_{hkl})=\overline{F}_{hkl}+\phi^n_{hkl} \tag{6.8}$$

Using these definitions, we can now obtain a new expression for the average of (6.6)

$$\overline{(F_{hkl})_n\,(F^*_{hkl})_{n+m}} = \overline{[\bar{F}_{hkl} + \phi^n_{hkl}]\,[\bar{F}^*_{hkl} + (\phi^{n+m}_{hkl})^*]}$$
$$= |\bar{F}_{hkl}|^2 + \bar{F}_{hkl}\,\overline{(\phi^{n+m}_{hkl})^*} + \bar{F}^*_{hkl}\,\overline{\phi^n_{hkl}} + \overline{(\phi^n_{hkl})(\phi^{n+m}_{hkl})^*} \tag{6.9}$$

Since ϕ^n_{hkl} is a deviation from the average, its average value is zero and the second and third terms on the right disappear. Thus

$$\overline{(F_{hkl})_n\,(F^*_{hkl})_{n+m}} = |\bar{F}_{hkl}|^2 + \overline{(\phi^n_{hkl})(\phi^{n+m}_{hkl})^*} \tag{6.10}$$

Inserting (6.6) and (6.10) into the intensity expression, (6.5), we have

$$\frac{I(\mathbf{s})}{NI_e(s)} = \sum_{m=0}^{(N-1)} |\bar{F}_{hkl}|^2 e^{2\pi i\mathbf{s}_{hkl}\cdot\mathbf{r}_m} + \sum_{m=0}^{(N-1)} \overline{(\phi^n_{hkl})(\phi^{n+m}_{hkl})^*}\, e^{2\pi i\mathbf{s}_{hkl}\cdot\mathbf{r}_m} \tag{6.11}$$

$\bar{F}_{hkl}$, being an averaged quantity, bears no dependence on $\mathbf{r}_m$. Thus the first term on the right becomes

$$\sum_{m=0}^{(N-1)} |\bar{F}_{hkl}|^2 e^{2\pi i\mathbf{s}_{hkl}\cdot\mathbf{r}_m} = |\bar{F}_{hkl}|^2 \sum_{m=0}^{(N-1)} e^{2\pi i\mathbf{s}_{hkl}\cdot\mathbf{r}_m} = |\bar{F}_{hkl}|^2 \cdot Z(\mathbf{s}) \tag{6.12}$$

Finally we write

$$\frac{I_{hkl}}{NI_e(s)} = I_1^{hkl} + I_2^{hkl} \tag{6.13}$$

where

$$I_1^{hkl} = |\bar{F}_{hkl}|^2 \cdot Z(\mathbf{s}) \tag{6.14}$$

is just the ordinary diffraction spikes, with now a new intensity given by $|\bar{F}_{hkl}|^2$ and

$$I_2^{hkl}(\mathbf{s}) = \sum_{m=1}^{N} \overline{(\phi^n_{hkl})(\phi^{n+m}_{hkl})^*}\, e^{2\pi i\mathbf{s}_{hkl}\cdot\mathbf{r}_m} \tag{6.15}$$

is a new term which, as we shall see, directs scattered intensity into off-Bragg directions.

Let us now explore the behavior of $I_2(\mathbf{s})$. We begin with the case in which there is no correlation between different defective cells. This would be the case of a random solid solution or random point defects. In this case

$$\overline{(\phi^n_{hkl})(\phi^{n+m}_{hkl})^*} = \begin{cases} \overline{|\phi^{n+m}_{hkl}|^2} & \text{if } m = 0 \\ 0 & \text{if } m \neq 0 \end{cases} \tag{6.16}$$

By (6.10)

$$\overline{|\phi^n_{hkl}|^2} = \overline{|(F_{hkl})_n|^2} - |(\bar{F}_{hkl})_n|^2 \tag{6.17}$$

Thus

$$I_2^{hkl}(\mathbf{s}) = \sum_{m=0}^{(N-1)} \overline{|\phi^n_{hkl}|^2} e^{2\pi i\mathbf{s}_{hkl}\cdot\mathbf{r}_0} = \overline{|\phi^n_{hkl}|^2} = \overline{|(F_{hkl})_n|^2} - |(\bar{F}_{hkl})_n|^2 \tag{6.18}$$

We shall now see that $I_2(\mathbf{s})$ is a very diffuse term in such a case.

Consider a random, substitutional binary BCC alloy with atomic fractions X_A, $1 - X_A$ of the two components. For a given unit cell

$$F_n = f_1 + f_2 e^{\pi i(h+k+l)} = \begin{cases} f_1 + f_2 & \text{if } h+k+l = \text{even} \\ 0 & \text{if } h+k+l = \text{odd} \end{cases} \tag{6.19}$$

where f_1 and f_2 are the atomic-scattering factors associated with the two atomic positions in the unit cell. For this case—random placement of atoms of type A and B among all sites—$\bar{f_1} = \bar{f_2}$, and, hence, for the nonzero reflections

$$\begin{aligned} \overline{F_n} &= \overline{f_1 + f_2} = 2\bar{f} = 2[X_A f_A + (1 - X_A) f_B] \\ \overline{F_n^2} &= \overline{(f_1 + f_2)^2} = \overline{f_1^2 + 2f_1 f_2 + f_2^2} = 2\overline{f^2} + 2\bar{f}^2 \\ &= 2[X_A f_A^2 + (1 - X_A) f_B^2] + 2\bar{f}^2 \\ (\overline{F_n})^2 &= 4\bar{f}^2 \end{aligned} \tag{6.20}$$

Substituting (6.20) into (6.18), we have

$$\frac{I_2^{hkl}}{N} = \overline{|\phi_{hkl}|^2} = 2\overline{f^2} - 2\bar{f}^2 = 2X_A(1 - X_A)(f_A - f_B)^2 \tag{6.21}$$

We see from (6.21) that the intensity which has gone from the hkl diffraction spike now appears as a very broad background, whose shape is that of the square of the difference of the atomic scattering factors and whose intensity is controlled by the volume fraction of A. The intensity is strongest when $X_A = 0.5$.

The decrease in intensity of the sharp Bragg reflection can be evaluated for a random alloy as follows. If each atom in each cell contributed only to the diffraction spike, the intensity $(I_1^{hkl})^{\ddagger}$ would be given by

$$\frac{(I_1^{hkl})^{\ddagger}}{NI_e} = 4(X_A f_A^2 + X_B f_B^2) \tag{6.22}$$

The actual intensity of the spike would be

$$\frac{I_1^{hkl}}{NI_e} = 4\bar{f}^2 = 4(X_A f_A + X_B f_B)^2 \tag{6.23}$$

and the decrease in intensity due to the defective nature of the system would be

$$\frac{\Delta I_1^{hkl}}{NI_e} = \frac{(I_1^{hkl})^{\ddagger} - I_1^{hkl}}{NI_e} = 4X_A X_B (f_A - f_B)^2 \tag{6.24}$$

The ratio $\Delta I_1^{hkl}/I_1^{hkl}$ is then given by

$$\frac{\Delta I_1^{hkl}}{I_1^{hkl}} = \frac{X_A X_B (f_A - f_B)^2}{X_A f_A^2 + X_B f_B^2} \tag{6.25}$$

Figure 6.1 gives the computed ratio $\Delta I_1/I_1$ for the (111) reflection of two alloy

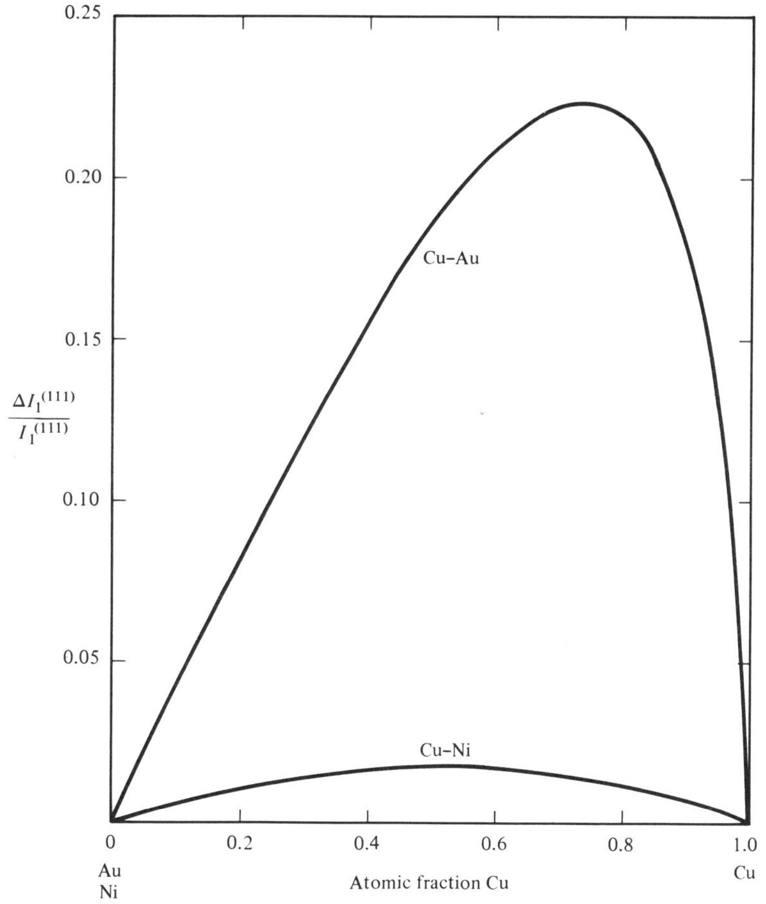

Figure 6.1. The computed ratio $\Delta I_l / I_l$ for the (111) reflections of the Cu-Au and Cu-Ni alloy series. Both alloy series have complete solid solubility.

series with complete substitutional solid solubility. In the Cu-Ni series, the atomic scattering factors of Cu and Ni are not much different. In this case the intensity decrease never exceeds 2 percent. For the Au-Cu series, in which f_{Au} is considerably greater than f_{Cu}, much larger intensity decreases should be observed.

Recall that for this case of a *random* solid solution, the $I_2(\mathbf{s})$ term has the very broad character of a difference in atomic-scattering factors. If ordering of any sort occurred, $I_2(\mathbf{s})$ would become more localized, and in specific ways. We shall treat these cases later.

Another instructive example of defect effects is that of a sinusoidal variation, of either composition or displacement in a crystal. We might think of a spinodal microstructure as typifying a sinusoidal composition distribution and of a lattice

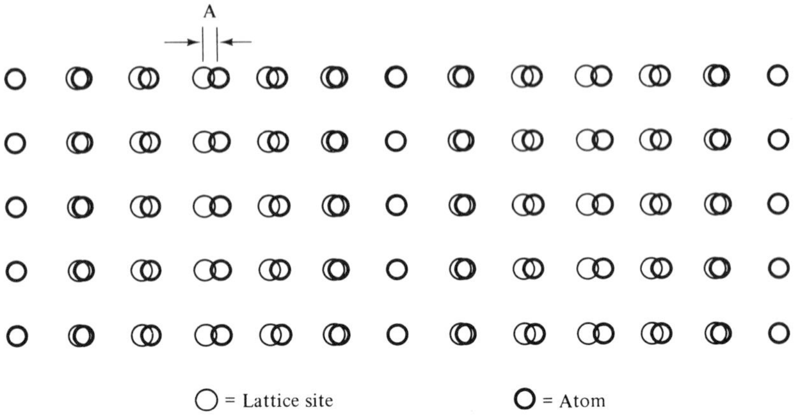

Figure 6.2. Longitudinal displacement wave.

phonon as typifying a sinusoidal displacement wave. We shall now treat the displacement wave.

Figure 6.2 illustrates a longitudinal displacement distribution of amplitude A and wavelength $\Lambda = 7a$. The displacement ΔZ_n associated with any level, Z_n, of the perfect crystal is, aside from a time dependence

$$\Delta Z_n = A \cos\left(\frac{2\pi Z_n}{\Lambda}\right) \tag{6.26}$$

Let us treat diffraction from the 001 planes. Let the structure factor of any unit cell, for the perfect crystal, be F. All atoms in the nth cell are considered to be displaced by an amount $\mathbf{\Delta Z}_n$ in <001> directions. Thus, the structure factor of the nth cell is

$$F_n = Fe^{2\pi i \mathbf{s}\cdot\mathbf{\Delta Z}_n} = Fe^{2\pi i s_{001}\Delta Z_n} \tag{6.27}$$

The average structure factor is

$$\overline{F} = \overline{Fe^{2\pi i s_{001}\Delta Z_n}} \tag{6.28}$$

If the amplitude A is always small, the exponential can be expanded to give

$$\begin{aligned} \overline{F} &= F\overline{[1 + 2\pi i s_{001}\Delta Z_n - 2\pi^2 s_{001}^2(\Delta Z_n)^2]} \\ &= F[1 + 2\pi i s_{001}\overline{\Delta\pi Z_n} - 2\pi^2 s_{001}^2\overline{(\Delta Z_n)^2}] \end{aligned} \tag{6.29}$$

Since $\overline{\Delta Z_n}$ is by definition zero

$$\overline{F} = F[1 - 2\pi^2 s^2\overline{(\Delta Z_n)^2}] = F[1 - \pi^2 s^2 A^2] \tag{6.30}$$

The deviation ϕ_n is, then

$$\phi_n = F_n - \overline{F} = F[e^{2\pi i s\Delta Z_n} - 1 + \pi^2 s^2 A^2] \tag{6.31}$$

In addition

$$\begin{aligned}\phi_n\phi^*_{n+m} &= F^2[e^{2\pi is\Delta Z_n} - 1 + \pi^2s^2A^2][e^{-2\pi is\Delta Z_{n+m}} - 1 + \pi^2s^2A^2] \\ &= F^2[2\pi is\Delta Z_n + \pi^2s^2A^2][-2\pi is\Delta Z_{n+m} + \pi^2s^2A^2] \qquad (6.32) \\ &= F^2\left[2\pi isA\cos\frac{2\pi Z_n}{\Lambda} + \pi^2s^2A^2\right]\left[-2\pi isA\cos\frac{2\pi Z_{n+m}}{\Lambda} + \pi^2s^2A^2\right]\end{aligned}$$

Ignoring sA terms of order higher than two, we have

$$\phi_n\phi^*_{n+m} = F^2\left[4\pi^2s^2A^2\cos\frac{2\pi Z_n}{\Lambda}\cos\frac{2\pi Z_{n+m}}{\Lambda}\right] \qquad (6.33)$$

Using the general law

$$2\cos[\tfrac{1}{2}(x+y)]\cos[\tfrac{1}{2}(x-y)] = \cos x + \cos y \qquad (6.34)$$

we find

$$\phi_n\phi^*_{n+m} = F^2 2\pi s^2A^2\left[\cos\frac{2\pi(Z_n + Z_{n+m})}{\Lambda} + \frac{\cos 2\pi(Z_{n+m} - Z_n)}{\Lambda}\right] \qquad (6.35)$$

Writing

$$Z_{n+m} = Z_n + Z_m \qquad (6.36)$$

we find

$$\phi_n\phi^*_{n+m} = 2\pi^2s^2A^2F^2\left[\cos\frac{2\pi(2Z_n + Z_m)}{\Lambda} + \cos\frac{2\pi Z_m}{\Lambda}\right. \qquad (6.37)$$

Recall now that what we need in our expression (6.15) for $I_2(s)$ is $\overline{\phi_n\phi^*_{n+m}}$, where the average is over all n. Thus

$$\overline{\phi_n\phi^*_{n+m}} = 2\pi^2s^2A^2F^2\cos\frac{2\pi Z_m}{\Lambda} \qquad (6.38)$$

Inserting (6.38) into (6.15), we have

$$\begin{aligned}I_2^{00l}(s) &= 2\pi^2s^2A^2F^2\sum_{m=-N/2}^{N/2}\cos\frac{2\pi Z_m}{\Lambda}e^{2\pi isZ_m} \\ &= \pi^2s^2A^2F^2\sum_{m=-N/2}^{N/2}\left[e^{2\pi i(s+\frac{1}{\Lambda})Z_m} + e^{2\pi i(s-\frac{1}{\Lambda})Z_m}\right] \qquad (6.39) \\ &= \pi^2s^2A^2F^2\left[\sum_{m=-N/2}^{N/2}e^{2\pi i(s+\frac{1}{\Lambda})ma} + \sum_{m=-N/2}^{N/2}e^{2\pi i(s-\frac{1}{\Lambda})ma}\right]\end{aligned}$$

The right side of (6.39) shows a series of pairs of spikes, each spike located at

$$s = \frac{p}{a} \pm \frac{1}{\Lambda} \qquad (6.40)$$

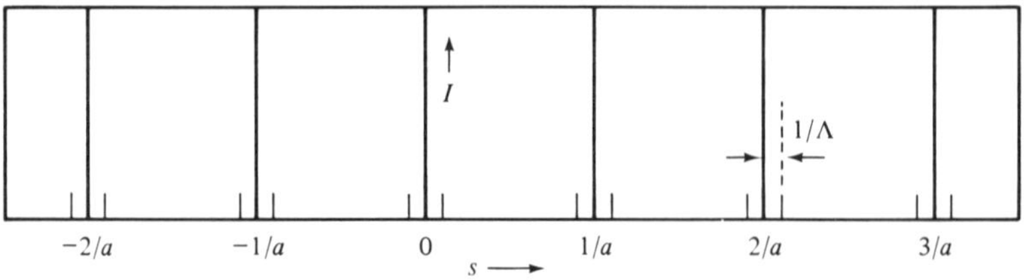

Figure 6.3. Satellite peaks, each located $1/\Lambda$ from the centers of $(00l)$ reflections.

where p is an integer. That is, each principal $00l$ diffraction (at p/a) has two satellites, located at $\pm 1/\Lambda$ from its center. This situation is shown in Fig. 6.3.

These results are extended easily to an arbitrary lattice wave in a crystal, for an arbitrary *hkl* reflection. To make the extension, we give the lattice wave a direction, by replacement of the scalar $1/\Lambda$ with a wave vector **K** (whose

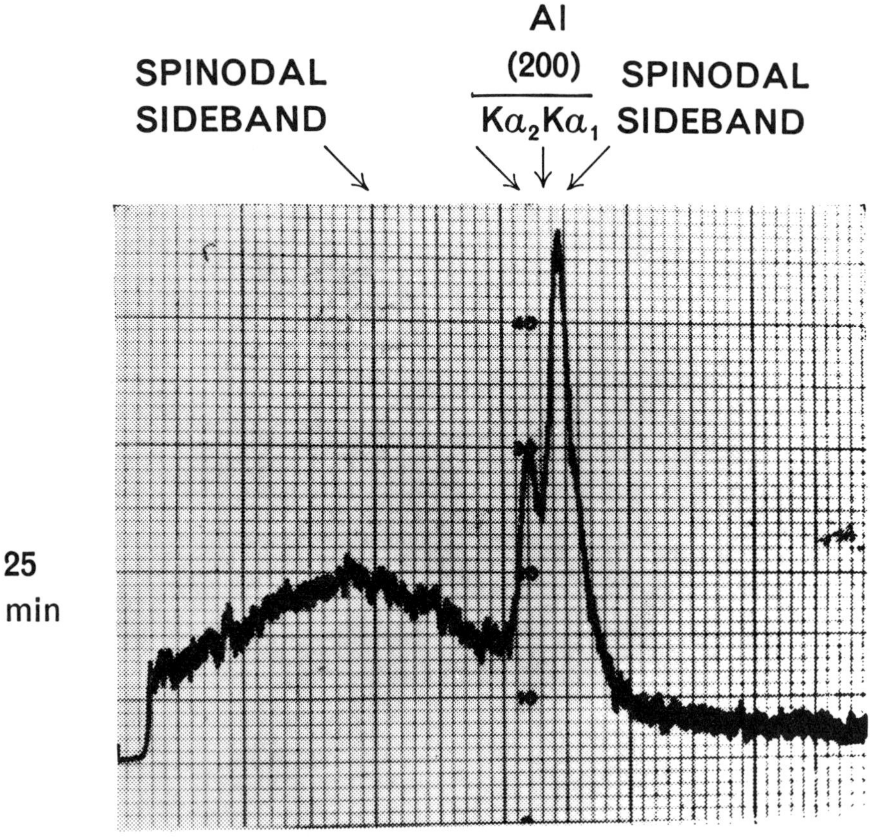

Figure 6.4. Spinodal sideband formed during the isothermal aging of quenched 50 Al-50 Zn at 22°C. [D. L. Douglass, *J. Mater. Sci.*, **4**, 130 (1969)]

magnitude is $1/\Lambda$). To indicate the vibration direction, we replace the scalar A with a vector $\mathbf{A}$. For a transverse wave $\mathbf{A} \cdot \mathbf{K} = 0$, for a longitudinal wave $\mathbf{A} \cdot \mathbf{K} = AK$. The displacement at the cell whose position vector is $\mathbf{r}_n$ is $\mathbf{A} \cos(2\pi\mathbf{K} \cdot \Delta\mathbf{r}_n)$ and the structure factor of the nth cell becomes

$$F_n = F\, e^{2\pi i(\mathbf{s} \cdot \mathbf{A}) \cos(2\pi\mathbf{K} \cdot \Delta\mathbf{r}_n)} \tag{6.41}$$

Going through precisely the same steps as before, we find that each relnode $\mathbf{r}^*_{hkl}$ has two satellites, at $\pm\mathbf{K}$ from it. The magnitudes of each satellite is $N\pi^2(\mathbf{s} \cdot \mathbf{A})^2F^2$.

The result can also be easily extended to sinusoidal composition fluctuations. Here, the position of the nth cell is fixed, but its structure factor magnitude varies from $F_{\min}$ to $F_{\max}$, with $F_{\text{ave}} = (F_{\min} + F_{\max})/2$. Thus

$$F_n = F_{ave} + \frac{(F_{\max} - F_{\min})}{2} \cos 2\pi\mathbf{K} \cdot \mathbf{r}_n \tag{6.42}$$

Going through the steps above, we could find again satellites at $\pm\mathbf{K}$. The magnitudes of the satellites would be

$$I_{\text{sat}} = \tfrac{1}{16}(F_{\max} - F_{\min})^2 \tag{6.43}$$

Examples of such satellites, observed by x-ray and by electron diffraction in spinodal alloys are shown in Figs. 6.4 and 6.5.

Figure 6.5. Satellite electron diffraction reflections about 022 in the athermal decomposition of Co_3Ti (M. N. Thompson, Ph.D. dissertation, University of Cambridge, 1971)

6.2 TWO GENERALIZATIONS

The following are two rules of thumb which can be used to visualize defect effects on diffraction patterns. First, displacement disorder, independent of its detail, lowers the intensity of the *hkl* peaks by an amount

$$\frac{\Delta I_{hkl}}{I_{hkl}} = \exp(-2\pi^2 s^2 A^2) \tag{6.44}$$

where A is a measure of the degree of disorder. Second, the narrower the function $\overline{\phi_n \phi^*_{n+m}}$ in real space, the broader the function $I_2(\mathbf{s})$ in resp.

The first of these generalizations can be seen from (6.27) to (6.30). In those, there was really no reliance on an explicit form for ΔZ_n, and the result, (6.30), can be rewritten as the exponential of (6.44), since it is taken that ΔZ_n is very small.

The proof of the second generalization is just the general reciprocity relating to Fourier transformation. That is, considering the transform

$$F(s) = \mathscr{F}[f(x)] \tag{6.45}$$

if $f(x)$ is narrow, $F(s)$ is broad, and if $f(x)$ is broad, $F(s)$ is narrow. The examples of Chap. 1 indicate this behavior. In our previous two examples here, we have seen this also. The random alloy case gave $|\overline{\phi_n \phi^*_{n+m}}| = \delta(m)|\overline{\phi_n}|^2$, a delta function. This produced a resp response that was flat in s, except for the variations in $(f_B - f_A)^2$. On the other hand, the broad sinusoidal variation gave a pair of spikes. (Note also that the distance of these satellite spikes goes as $1/\Lambda$). In addition to these two cases, it is instructive to look at a Cauchy function—a representation of a disorder which is locally strong, but which tails off away from its core. The Cauchy function

$$f(x) = \frac{2b}{b^2 + (2\pi x)^2} \tag{6.46}$$

and its transform

$$g(s) = \exp[-b|s|] \tag{6.47}$$

appear as in Fig. 6.6. The breadths of the real space distortion correlation

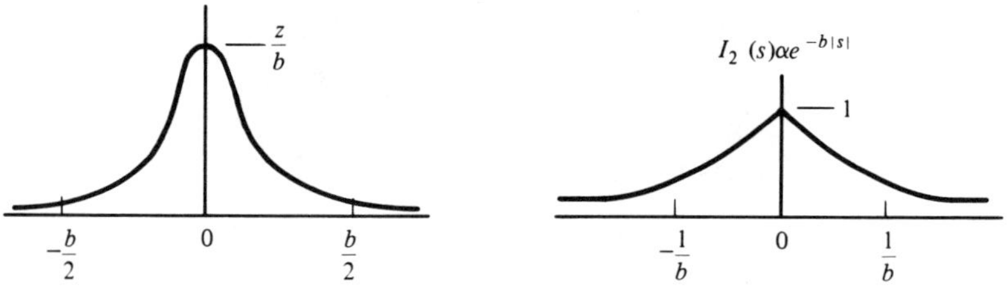

Figure 6.6. The Cauchy function and its Fourier transform.

and the $I_2(s)$ function go, respectively, as b and as b^{-1}. Thus, while the effect of a correlated defect on $I_1(s)$ depends only on the average displacement, the effect on $I_2(s)$ depends on the details of the spatial correlation.

We shall now look at three specific types of disorder: thermal vibrations, stacking faults, and short-range ordering.

6.3 STACKING FAULTS

6.3.1 General Formulation of the Problem

We first derive the pertinent expressions for the line broadening and peak shifting seen in materials with stacking faults. We consider only close-packed structures, illustrated in Fig. 6.7. Our unit cell is always hexagonal or pseudohexagonal.

The unit cell has a rhombus face in the closed-packed (CP) planes and has its other faces normal to the CP plane. For hexagonal closed-packed (HCP) the unit cell is $2c'$ thick, where c' is the spacing between CP planes. The FCC cell is $3c'$ thick. These are the true unit cells. We shall find it convenient, however, to often take unit cells only c' thick. The cell, in this case retains the same kind of shape, but it is one-half or one-third as tall as the conventional cells.

6.3.2 Growth Faults in HCP and FCC

Recall the expression for the intensity of radiation scattered by a crystal

$$\frac{I(\mathbf{s})}{I_e(s)} = \sum_n \sum_{n'} F_n F_{n'}^* \exp[2\pi i \mathbf{s} \cdot (\mathbf{r}_n - \mathbf{r}_{n'})] \tag{6.48}$$

We know that, for a crystal

$$\mathbf{s} = h\mathbf{a}^* + k\mathbf{b}^* + l\mathbf{c}^* \quad (h,k,l \text{ integers; } \mathbf{a}^*,\mathbf{b}^*,\mathbf{c}^* \text{ rel vectors}) \tag{6.49}$$

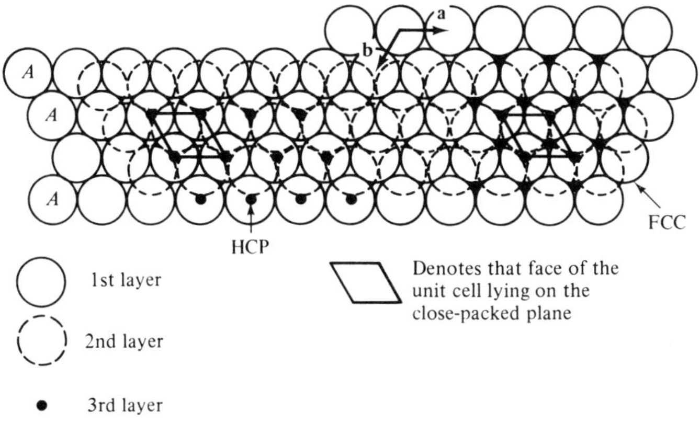

Figure 6.7. Packing geometry for close-packed structures.

We also make the substitution

$$\mathbf{r}_m = \mathbf{r}_n - \mathbf{r}_{n'} = m_h\mathbf{a} + m_k\mathbf{b} + m_l\mathbf{c}' \quad (m_h, m_k, m_l \text{ integers}) \tag{6.50}$$

Substituting (6.49) and (6.50) into (6.48), we obtain

$$\frac{I(\mathbf{s})}{I_e(\mathbf{s})} = \sum_n \sum_m F_n F^*_{n+m} \exp[2\pi i(hm_h + km_k + lm_l)] \tag{6.51}$$

Note that the two sums on the right each represent a triple sum (e.g., over m_h, m_k, and m_l).

Let us take the nth atom to be in an A-type plane. Then, if the $(n + m)$th plane is also A-type

$$F_n F^*_{n+m} = (FF^*) \exp[2\pi i(h\mathbf{a}^* + k\mathbf{b}^* + l\mathbf{c}^*) \cdot (m_h\mathbf{a} + m_k\mathbf{b} + m_l\mathbf{c}')] = |F|^2 \tag{6.52}$$

Call this quantity

$$F_A F_A^* = |F|^2 \tag{6.53a}$$

Similarly, if the nth plane is A-type and the $(n + m)$th B-type

$$\begin{aligned} F_A F_B^* &= |F|^2 e^{2\pi i(h\mathbf{a}^* + k\mathbf{b}^* + l\mathbf{c}^*) \cdot [(\frac{2}{3}\mathbf{a} + m_h\mathbf{a}) + (\frac{1}{3} + m_k)\mathbf{b} + m_l\mathbf{c}']} \\ &= |F|^2 e^{(\frac{2h}{3} + \frac{k}{3})2\pi i} = |F|^2 e^{4\pi i(\frac{h-k}{3})} \end{aligned} \tag{6.53b}$$

and, finally, if the nth plane is A type and the $(n + m)$th C-type

$$F_A F_C^* = |F|^2 e^{2\pi i(\frac{k-h}{3})} = F_C F_B^* \tag{6.53c}$$

Now, we rewrite (6.51) in the form

$$\frac{I(\mathbf{s})}{I_e(\mathbf{s})} = \sum_m \left\{ \left[\sum_n F_n F^*_{n+m} \right] e^{2\pi i(hm_h + km_k + lm_l)} \right\} \tag{6.54}$$

But $F_n F_{n+m}$ is independent of m_h and m_k [see (6.53)]; it depends only on m_l—i.e., on how the planes are stacked in the c-direction. Hence

$$\frac{I(\mathbf{s})}{I_e(s)} = \sum_{m_h} e^{2\pi i h m_h} \sum_{m_k} e^{2\pi i k m_k} \sum_{m_l} \left(\sum_n F_n F^*_{n+m} \right) e^{2\pi i l m_l} \tag{6.55}$$

Since h and m_h are both integers, $e^{2\pi i h m_h} = 1 = e^{2\pi i k m_k}$, and $\sum_{m_h} e^{2\pi i h m_h} = N_a$; $\sum_{m_k} e^{2\pi i k m_k} = N_b$, where N_a and N_b are the numbers of unit cells along the a and b axes, respectively. Then

$$\frac{I(\mathbf{s})}{I_e(s)} = N_a N_b \sum_{m_l} \left\{ \left[\sum_n F_n F^*_{n+m} \right] e^{2\pi i l m_l} \right\} \tag{6.56}$$

Now define

$$y_m \equiv \overline{F_n F_{n+m}} = \frac{\sum_{n=0}^{N_c-1} F_n F_{n+m}}{N_c} \tag{6.57}$$

where N_c is the number of unit cells along the c-axis. Then

$$\sum_{n=0}^{N_c-1} F_n F_{n+m} = N_c y_m \tag{6.58}$$

and

$$\frac{I(\mathbf{s})}{I_e(s)} = N_a N_b N_c \sum_{m_l} y_m \, e^{2\pi i l m_l} = \frac{V}{V_c} \sum y_m \, e^{2\pi i l m_l} \tag{6.59}$$

where V is the total irradiated volume of crystal and V_c is volume of the unit cell.

We must now evaluate y_m. This is done as follows. *A priori* we have equal probabilities that the nth plane will be A, B, or C. Let the probability of the $(n + m)$th plane being of the same type as the nth be P_m. For *growth* faulting, if the $(n + m)$th plane is not the same type as the nth, there are equal probabilities, $(1 - P_m)/2$, that it is either of the other two types. Hence

$$\begin{aligned} y_m = \overline{F_n F_{n+m}} &= \frac{1}{3} F_A \left[F_A^* P_m + \frac{(1 - P_m)}{2} (F_B^* + F_C^*)\right] \\ &+ \frac{1}{3} F_B \left[F_B^* P_m + \frac{(1 - P_m)}{2} (F_C^* + F_A^*)\right] \\ &+ \frac{1}{3} F_C \left[F_C^* P_m + \frac{(1 - P_m)}{2} (F_A^* + F_B^*)\right] \end{aligned} \tag{6.60}$$

Using (6.53), we find

$$y_m = P_m |F|^2 + $$

$$\frac{1}{6}\left[2e^{-2\pi i \frac{h-k}{3}} + e^{4\pi i \frac{h-k}{3}} + 2e^{2\pi i \frac{h-k}{3}} + e^{-4\pi i \frac{h-k}{3}}\right](1 - P_m)|F|^2 \tag{6.61}$$

Consider Fig. 6.8, a representation of the phase angles involved. It is recognized that

$$e^{2\pi i \frac{h-k}{3}} + e^{-4\pi i \frac{h-k}{3}} = 2 \cos\left(2\pi \frac{h-k}{3}\right) \tag{6.62}$$

Substituting (6.62) into (6.61), we have

$$\begin{aligned} y_m &= |F|^2 P_m + \frac{1}{2} |F|^2 \left(e^{2\pi i \frac{h-k}{3}} + e^{-2\pi i \frac{h-k}{3}}\right)(1 - P_m) \\ &= |F|^2 \left\{P_m + (1 - P_m) \cos\left[2\pi \left(\frac{h-k}{3}\right)\right]\right\} \end{aligned} \tag{6.63}$$

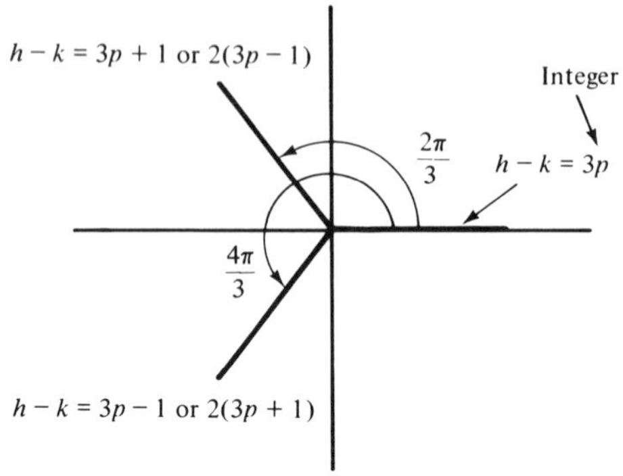

Figure 6.8. Representation of phase angles for (6.61).

Replacing P_m by

$$P_m = \frac{1}{3} + Q_m \tag{6.64}$$

and substituting in (6.63)

$$y_m = |F|^2 \left\{\frac{1}{3}\left[1 + 2\cos\frac{2\pi(h-k)}{3}\right] + Q_m\left[1 - \cos\frac{2\pi(h-k)}{3}\right]\right\} \tag{6.65}$$

Putting (6.65) into the intensity expression, (6.59) we obtain

$$\frac{I(\mathbf{s})}{NI_e(s)} = |F|^2 \left\{\frac{1}{3}\left[1 + 2\cos\frac{2\pi(h-k)}{3}\right] \sum_{m_l}^{N} e^{2\pi i l m_l} + \left[1 - \cos\frac{2\pi(h-k)}{3}\right] \sum_{m_l}^{N} Q_m\, e^{2\pi i l m_l}\right\} \tag{6.66}$$

We consider two cases now. If $h - k = 3p$, where p is an integer, then the first bracketed term is equal to $1 + 2\cos(2\pi) = 3$, while the second bracketed term is zero. Hence, *if* $h - k = 3p$, *then the intensity is* $N\,\delta\,(l)$ *for* $l =$ *integer, since* $\sum^{\infty} e^{2\pi i l m_l} = N\,\delta$ ($l =$ integer). However, if $h - k = 3p \pm 1$ (the only other possibilities), then the first bracketed term vanishes and we are left with

$$\left.\frac{I(\mathbf{s})}{I_e(s)}\right|_{h-k=3p\pm1} = \frac{V}{V_c}|F|^2 \frac{3}{2} \sum Q_m\, e^{2\pi i l m_l} \tag{6.67}$$

It is now pertinent to evolve an expression for P_m (so that (6.67) can be evaluated). To do so, we consider the several ways in which we can get an *A*-type plane at $n + m$, given an *A* at n. These are given in Table 6.1. Consider that the standard packing is hexagonal (*ABABAB* or *ACACAC*). The probabilities for 1 and 2 are together $P_{m-2}(1 - \alpha)$, where α is the probability of a stacking fault. The probabilities for 3 and 4 are less straightforward. Consider the follow-

TABLE 6.1
SEQUENCING LEADING TO A-PLANES AT n AND
$n + m$, ALLOWING GROWTH FAULTING

n . . .	$n + m - 2$	$n + m - 1$	$n + m$
1. A . . .	A	B	A
2. A . . .	A	C	A
3. A . . .	B	C	A
4. A . . .	C	B	A

ing sequences and their respective probabilities. The total number of ways of getting a stacking fault are represented by the sequences of Table 6.2.

TABLE 6.2
SEQUENCING LEADING TO
GROWTH FAULTS IN HCP

$$
\left.
\begin{array}{l}
\left.\begin{array}{l} A \ldots ABC \\ A \ldots ACB \end{array}\right\} (P_{m-2\alpha}) \\
\left.\begin{array}{l} A \ldots BAC \\ A \ldots CAB \end{array}\right\} (P_{m-1\alpha}) \\
A \ldots BCA \\
A \ldots CBA
\end{array}
\right\} (\alpha)
$$

Thus, the total probability for 3 and 4 is $\alpha(1 - P_{m-2} - P_{m-1})$, and

$$
P_m = \underbrace{(1 - \alpha) P_{m-2}}_{\substack{\text{by regular} \\ \text{sequence}}} + \underbrace{\alpha(1 - P_{m-2} - P_{m-1})}_{\substack{\text{by faulted} \\ \text{sequences}}}
\tag{6.68}
$$

Rearranging

$$
P_m + \alpha P_{m-1} - (1 - 2\alpha) P_{m-2} - \alpha = 0
\tag{6.69}
$$

Trying $P_m = \dfrac{1}{3} + q\rho^{|m|}$ as a solution to this linear difference equation, we have

$$
\frac{1}{3} + q\rho^{|m|} + \frac{\alpha}{3} + \alpha q\rho^{|m-1|} - \frac{1}{3} - q\rho^{|m-2|} + \frac{2\alpha}{3} + 2\alpha q\rho^{|m-2|} - \alpha
$$
$$
= q\rho^{|m|} + \alpha q\rho^{|m-1|} - (1 - 2\alpha) q\rho^{|m-2|}
\tag{6.70}
$$
$$
= 0
$$

Divide by $q\rho^{|m-2|}$ to get the following equation relating ρ and α (giving the value of ρ needed to use as a solution to (6.69))

$$
\rho^2 + \alpha\rho - (1 - 2\alpha) = 0
$$

Solving now for ρ

$$\rho = \frac{-\alpha \pm \sqrt{\alpha^2 + 4 - 8\alpha}}{2} \tag{6.72}$$

So we have found two solutions, called ρ_1 and ρ_2, for ρ, corresponding to the + and − signs before the radical. Hence the total solution for P_m is

$$P_m = \frac{1}{3} + q_1\rho_1^{|m|} + q_2\rho_2^{|m|} \tag{6.73}$$

We obtain q_1 and q_2 by using the initial conditions

$$\begin{aligned} P_0 &= 1 \\ P_1 &= 0 \end{aligned} \tag{6.74}$$

as follows

$$\begin{aligned} P_0 = 1 = \frac{1}{3} + q_1 + q_2; \quad q_1 + q_2 = \frac{2}{3} \\ P_1 = 0 = \frac{1}{3} + q_1\rho_1 + q_2\rho_2; \quad \rho_1 q_1 + \rho_2 q_2 = -\frac{1}{3} \end{aligned} \tag{6.75}$$

Solving by determinants

$$\begin{aligned} q_1 &= \frac{\begin{vmatrix} 2/3 & 1 \\ -1/3 & \rho_2 \end{vmatrix}}{\begin{vmatrix} 1 & 1 \\ \rho_1 & \rho_2 \end{vmatrix}} = \frac{\frac{2}{3}\rho_2 + \frac{1}{3}}{\rho_2 - \rho_1} = \frac{\frac{-\alpha}{3} - \frac{1}{3}\sqrt{4 - 8\alpha + \alpha^2} + \frac{1}{3}}{-\sqrt{4 - 8\alpha + \alpha^2}} \\ &= \frac{1}{3}\left[1 - \frac{(1-\alpha)}{\sqrt{4 - 8\alpha + \alpha^2}}\right] \end{aligned} \tag{6.76a}$$

$$q_2 = \frac{1}{3}\left[1 + \frac{(1-\alpha)}{\sqrt{4 - 8\alpha + \alpha^2}}\right] \tag{6.76b}$$

If we take $\alpha << 1$, then

$$q_1 \simeq \frac{1}{3}\left(1 - \frac{1}{2}\right) = \frac{1}{6} \tag{6.77a}$$

$$q_2 \simeq \frac{1}{3}\left(1 + \frac{1}{2}\right) = \frac{1}{2} \tag{6.77b}$$

And, using also $\sqrt{1+\epsilon} \simeq 1 + \frac{\epsilon}{2}$ for $\epsilon << 1$

$$\rho_1 = \frac{-\alpha + 2(1-\alpha)}{2} = 1 - \frac{3}{2}\alpha \tag{6.78a}$$

$$\rho_2 = \frac{-\alpha - 2(1-\alpha)}{2} = -\left(1 - \frac{\alpha}{2}\right) \tag{6.78b}$$

Substituting (6.77) and (6.78) into (6.73), we have

$$P_m = \frac{1}{3} + \frac{1}{6}\left(1 - \frac{3\alpha}{2}\right)^m + \frac{1}{2}\left[-\left(1 - \frac{\alpha}{2}\right)\right]^m \tag{6.79}$$

Inserting (6.79) into the intensity formula, (6.67), we find

$$\frac{I(\mathbf{s})}{I_e(s)}\bigg|_{h-k=3p\pm 1} = \frac{V}{V_c}|F|^2 \left\{\frac{1}{4}\sum_{-\infty}^{\infty}\left(1 - \frac{3}{2}\alpha\right)^{|m|} e^{2\pi ilm} + \frac{3}{4}\sum_{-\infty}^{\infty}\left[-\left(1 - \frac{\alpha}{2}\right)\right]^{|m|} e^{2\pi ilm}\right. \tag{6.80}$$

Consider the second term on the right. It is of the form

$$\sum_0^{\infty} (-1)^{|m|} a^{|m|} e^{imx} - 1 + \sum_0^{\infty} (-1)^{|m|} a^{|m|} e^{-imx}$$

Now

$$\sum_0^{\infty} (-1)^{|m|} a^{|m|} e^{imx} = 1 - ae^{ix} + a^2 e^{2ix} - a^3 e^{3ix} + \cdots \tag{6.81}$$

Multiplying and dividing (6.81) by $1 + ae^{ix}$, we have

$$\sum_0^{\infty} (-1)^{|m|} a^{|m|} e^{imx} = \frac{1}{1 + ae^{ix}} \tag{6.82a}$$

if $a < 1$. Similarly

$$\sum_0^{\infty} (-1)^{|m|} a^{|m|} e^{-imx} = \frac{1}{1 + ae^{-ix}} \tag{6.82b}$$

and

$$\sum_{-\infty}^{\infty} (-1)^m a^{|m|} e^{-imx} = \frac{1}{1 + ae^{ix}} - 1 + \frac{1}{1 + ae^{-ix}} = \frac{1 + 2a\cos x}{1 + a^2 + 2a\cos x} \tag{6.83}$$

In our case, we substitute $\left(1 - \frac{\alpha}{2}\right)$ for a and $2\pi l$ for x

$$\sum_{-\infty}^{\infty}\left[-\left(1 - \frac{\alpha}{2}\right)\right]^m e^{2\pi ilm}$$

$$= \frac{1 - 2\left(1 - \frac{\alpha}{2}\right)\cos 2\pi l}{1 + \left(1 - \frac{\alpha}{2}\right)^2 + 2\left(1 - \frac{\alpha}{2}\right)\cos 2\pi l} = \frac{4}{(V/V_c)|F|^2}\frac{I(\mathbf{s})}{I_e(s)} \tag{6.84}$$

This is then our intensity expression for $h - k = 3p \pm 1$.

The position, intensity, and breadth of the $h - k = 3p \pm 1$ reflections are found as follows. Let

$$l = p + \frac{1}{2} + \frac{\xi}{2\pi} \tag{6.85}$$

where ξ is a small number. Then

$$\cos 2\pi l = \cos [(2\pi p + \pi) + \xi] = -\cos \xi \tag{6.86}$$

Since $\xi << 1$

$$\cos 2\pi l \simeq -1 + \frac{\xi^2}{2} \tag{6.87}$$

and

$$\frac{4}{\left(\frac{V}{V_c}\right)|F|^2} \frac{I(\mathbf{s})}{I_e(s)} \simeq \frac{1 - (2-\alpha)\left(-1 + \frac{\xi^2}{2}\right)}{1 + \left(1 - \alpha + \frac{\alpha^2}{4}\right) + (2-\alpha)\left(-1 + \frac{\xi^2}{2}\right)} \simeq \frac{3 - \alpha - \xi^2}{\frac{\alpha^2}{4} + \xi^2} \tag{6.88}$$

Since α is constant, the peak value of intensity occurs very near $\xi = 0$—that is, right on $l = p + ½$. It can also be shown that the contribution of the first term on the right of (6.80) is negligible. The peak intensity is then

$$\frac{I(\mathbf{s})}{I_e(s)} = \frac{1}{4}\left(\frac{V}{V_c}\right)|F|^2(4)\frac{3-\alpha}{\alpha^2} = \left(\frac{V}{V_c}\right)|F|^2\frac{3-\alpha}{\alpha^2} \tag{6.89}$$

The value of $\xi = \xi^{\ddagger}$ needed to give half the peak intensity is found from

$$\frac{1}{2} = \frac{\frac{1}{4}\left(\frac{V}{V_c}\right)|F|^2\left(\frac{3 - \alpha - \xi^{\ddagger 2}}{\frac{\alpha^2}{4} + \xi^{\ddagger 2}}\right)}{\frac{1}{4}\left(\frac{V}{V_c}\right)|F|^2\left(\frac{3-\alpha}{\frac{\alpha^2}{4}}\right)} \tag{6.90}$$

Solving for $\alpha << 1$, we find

$$\xi^{\ddagger} \simeq \frac{\alpha}{2} \tag{6.91}$$

The full width of the peak at half maximum (FWHM) is then

$$\Delta l^{\ddagger}_{p+1/2} = 2\left(\frac{\xi^{\ddagger}}{2\pi}\right) = \frac{\alpha}{2\pi} \tag{6.92}$$

We similarly find that, for the first sum in (6.80), there are intensity peaks at $l = p$, with widths of approximately

$$\Delta l^{\ddagger}_{p} = \frac{3\alpha}{2\pi} \tag{6.93}$$

Therefore, resp appears as shown in Fig. 6.9. The half integral values come about because we have taken an unusual "unit cell"; the conventional one is twice as high.

Having gone through all this for HCP, the job is now much easier for FCC. What was regular stacking in HCP is now a fault in FCC, and conversely. Hence, we begin by replacing fault probabilities for HCP by the probability of finding no fault in FCC. We replace α in (6.69) by $(1 - \beta)$, where β is the fault probability in FCC

$$P_m + (1-\beta)P_{m-1} + (1-2\beta)P_{m-2} - (1-\beta) = 0 \tag{6.94}$$

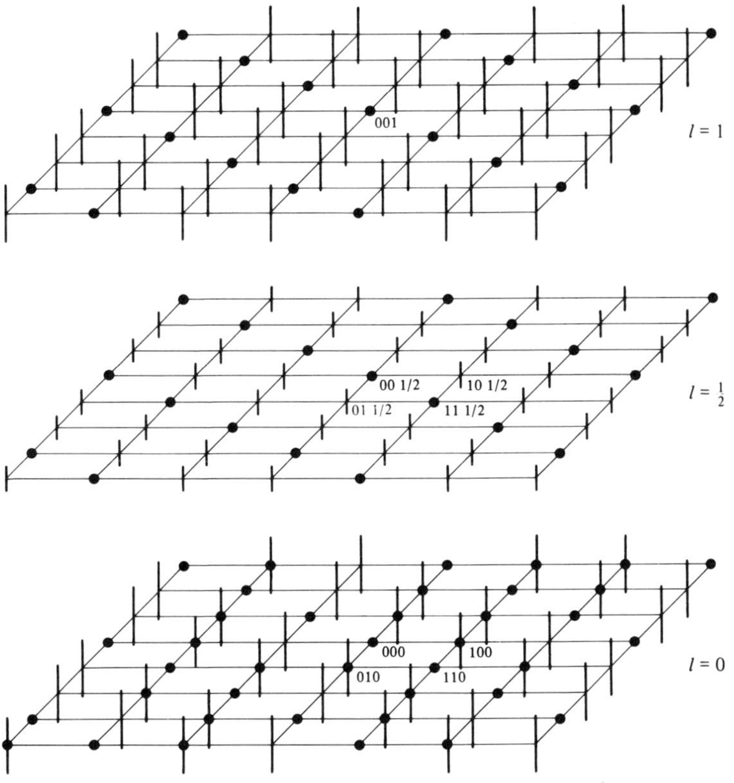

Figure 6.9. Resp for a FCC crystal containing growth faults.

Again we try

$$P_m = \frac{1}{3} + q\rho^{|m|}$$

$$\frac{1}{3} + q\rho^{|m|} + \frac{1}{3} + q\rho^{|m-1|} - \frac{\beta}{3} - \beta q\rho^{|m-1|} + \frac{1}{3} + q\rho^{|m-2|}$$
$$+ \frac{-2\beta}{3} - 2\beta q\rho^{|m-2|} - 1 + \beta = 0 \qquad (6.95)$$

and

$$\rho^2 + (1-\beta)\rho + (1-2\beta) = 0 \qquad (6.96)$$

Solving for ρ, we have

$$\rho = \frac{-(1-\beta) \pm \sqrt{-3 + 6\beta + \beta^2}}{2} \qquad (6.97)$$

Now ρ_1 and ρ_2 are complex numbers. The initial conditions are, again

$$P_0 = 1$$
$$P_1 = 0 \qquad (6.98)$$

giving, as before

$$q_1 = \frac{\frac{2}{3}\rho_2 + \frac{1}{3}}{\rho_- + \rho_+} \qquad (6.99a)$$

$$q_2 = \frac{-\frac{1}{3} - \frac{2}{3}\rho_1}{\rho_- + \rho_+} \qquad (6.99b)$$

Substituting (6.97) into (6.99), we have

$$q_1 = \frac{1}{3} - \gamma \qquad (6.100a)$$

$$q_2 = \frac{1}{3} + \gamma \qquad (6.100b)$$

where

$$\gamma = \frac{\beta}{3\sqrt{3 - 6\beta + \beta^2}} \qquad (6.101)$$

Since ρ_1, ρ_2 are complex numbers, we may write them as, e.g.,

$$\rho_1 = a + ib = \mathrm{r}e^{i\phi} = r(\cos\phi + i\sin\phi) \qquad (6.102a)$$

$$\rho_2 = \mathrm{r}(\cos\phi - i\sin\phi) \qquad (6.102b)$$

Using (6.97)

$$r^2 = a^2 + b^2 = \frac{1}{4}\{1 - 2\beta + \beta^2 + 3 - 6\beta + \beta^2\}$$
$$= \frac{1}{4}[4 - 8\beta + \beta^2] \simeq 1 - 2\beta \tag{6.103}$$

Since $\beta << 1$

$$r \simeq 1 - \beta \tag{6.104}$$

Also

$$\tan\phi = \frac{b}{a} = \frac{\sqrt{3 - 6\beta - \beta^2}}{-(1-\beta)} \simeq -\sqrt{3} \tag{6.105a}$$

$$\phi = \arctan(-\sqrt{3}) = \frac{2\pi}{3} \tag{6.105b}$$

Finally

$$P_m = \frac{1}{3} + \left(\frac{1}{3} - \gamma\right)(1-\beta)^{|m|}e^{i\phi m} + \left(\frac{1}{3} + \gamma\right)(1-\beta)^{|m|}e^{-i\phi m} \tag{6.106}$$

If $\gamma = 0$ (i.e., $\beta = 0$), then

$$P_m = \frac{1}{3} + \frac{2}{3}(1-\beta)^{|m|}\cos\frac{2\pi m}{3} \tag{6.107}$$

Substituting (6.107) into (6.67), we have

$$\frac{I(\mathbf{s})}{I_e(s)} = \frac{V|F|^2}{2V_c}\left\{\sum_{-\infty}^{\infty}(1-\beta)^{|m|}e^{2\pi i\left(l+\frac{1}{3}\right)m}\right.$$
$$\left. + \sum_{-\infty}^{\infty}(1-\beta)^{|m|}e^{2\pi i\left(l-\frac{1}{3}\right)m}\right\} \tag{6.108}$$

Similar to the evaluation of similar sums for the HCP case, these sums give maxima at $l = p \pm \frac{1}{3}$

If $\gamma \neq 0$, then the sine portions of $e^{i\phi m}$ in (6.108) do not disappear, and we have a term of the type

$$\sum_{-\infty}^{\infty}(1-\beta)^{|m|}\sin\frac{2\pi|m|}{3}\,e^{2\pi ilm}$$
$$= 2\sum_{0}^{\infty}(1-\beta)^{|m|}\sin\frac{2\pi m}{3}\cos(2\pi lm) \tag{6.109}$$
$$= 2\sum_{0}^{\infty}(1-\beta)^{m}\sin\left[2\pi\left(l+\frac{1}{3}\right)m\right] - \sum_{0}^{\infty}(1-\beta)^{m}\sin\left[2\pi\left(l-\frac{1}{3}\right)m\right]$$

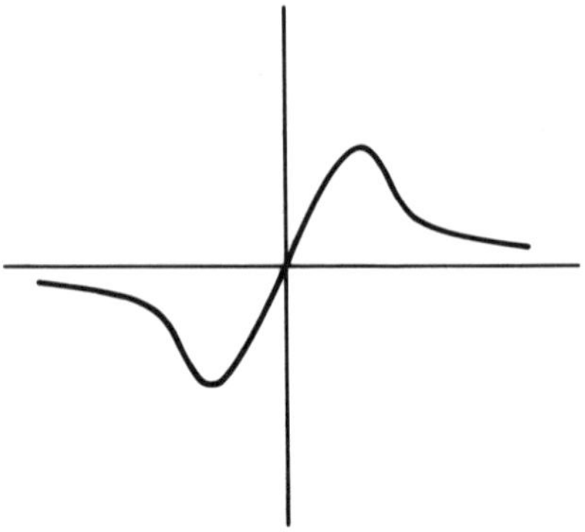

Figure 6.10. The form of the function of (6.109).

These are asymmetric functions of the form shown in Fig. 6.10. The *effect* of the sine terms is to skew the reflection. That is, adding the cosine terms and the sine terms pushes the peak a little off $l = p \pm ½$

6.3.3 Deformation Faults in FCC

Suppose that the perfect crystal arrangement is *ABCABC*. . . . Then the deformation-faulted arrangement is

$$ABCABCABCABABCABCABCABC\ldots$$

The ways of getting A at $n + m$, with A at n are given in Table 6.3.

TABLE 6.3
SEQUENCING LEADING TO A-PLANES AT n AND AT $n + m$, ALLOWING DEFORMATION FAULTING

$n\ldots$	$n+m-2$	$n+m-1$	$n+m$	*Probability*
$A\ldots$	A	B	A	$P^o_{m-2}(1-\alpha)\alpha$
$A\ldots$	A	C	A	$P^o_{m-2}\alpha(1-\alpha)$
$A\ldots$	B	C	A	$P^+_{m-2}(1-\alpha)(1-\alpha)$
$A\ldots$	C	B	A	$P^-_{m-2}\alpha\alpha$

Here P^o_m is the probability of having the same type plane at $n + m$ as was at n. P^+_m is the probability of having the plane next in the regular stacking order at $n + m$. P^-_m represents a probability for having the type plane preceding the one at n at $n + m$. Hence

$$P^o_m = 2\alpha(1-\alpha)\,P^o_{m-2} + (1-\alpha)^2\,P^+_{m-2} + \alpha^2 P^-_{m-2} \tag{6.110}$$

The ways of getting B at $n + m$ with A at n are shown in Table 6.4.

TABLE 6.4

SEQUENCING LEADING TO A-PLANE AT n AND B-PLANE AT $n+m$, ALLOWING DEFORMATION FAULTING

n . . .	$n+m-2$	$n+m-1$	$n+m$	*Probability*
A . . .	A	C	B	$P^o_{m-2}\alpha\alpha$
A . . .	B	C	B	$P^+_{m-2}(1-\alpha)\alpha$
A . . .	B	A	B	$P^+_{m-2}\alpha(1-\alpha)$
A . . .	C	A	B	$P^-_{m-2}(1-\alpha)(1-\alpha)$

Here

$$P^+_m = 2\alpha(1-\alpha)P^+_{m-2} + (1-\alpha)^2 P^-_{m-2} + \alpha^2 P^o_{m-2} \tag{6.111}$$

Both (6.110) and (6.111) are partial difference equations. But we can convert them into complete difference equations using the following identities

$$P^+_{m-2} + P^-_{m-2} + P^o_{m-2} = 1 \tag{6.112a}$$

$$P^o_{m-1} = (1-\alpha)P^-_{m-2} + \alpha P^+_{m-2} \tag{6.112b}$$

$$P^+_{m-1} = (1-\alpha)P^o_{m-2} + \alpha P^-_{m-2} \tag{6.112c}$$

Using determinants, (6.112a) and (6.112b) give

$$P^+_{m-2} = \frac{\begin{vmatrix} 1-P^o_{m-2} & 1 \\ P^o_{m-1} & 1-\alpha \end{vmatrix}}{\begin{vmatrix} 1 & 1 \\ \alpha & 1-\alpha \end{vmatrix}} = \frac{(1-\alpha) - P^o_{m-1} - (1-\alpha)P^o_{m-2}}{1-2\alpha} \tag{6.113a}$$

and

$$P^-_{m-2} = \frac{\begin{vmatrix} 1 & 1-P^o_{m-2} \\ \alpha & P^o_{m-1} \end{vmatrix}}{1-2\alpha} = \frac{P^o_{m-1} - \alpha + \alpha P^o_{m-2}}{1-2\alpha} \tag{6.113b}$$

Then, using (6.112a) and (6.112c)

$$P^o_{m-2} = \frac{\begin{vmatrix} 1-P^+_{m-2} & 1 \\ P^+_{m-1} & \alpha \end{vmatrix}}{\begin{vmatrix} 1 & 1 \\ 1-\alpha & \alpha \end{vmatrix}} = \frac{P^+_{m-1} + \alpha P^+_{m-2} - \alpha}{1-2\alpha} \tag{6.114a}$$

$$P^-_{m-2} = \frac{(1-\alpha) - P^+_{m-1} - (1-\alpha)P^+_{m-2}}{1-2\alpha} \tag{6.114b}$$

We are now in a position to determine P^o_m or P^+_m. We do P^o_m first. Substituting (6.114) into (6.110), we obtain

$$P_m^o = 2\alpha(1-\alpha)P_{m-2}^o + (1-\alpha)^2\left[\frac{(1-\alpha) - P_{m-2}^o - (1-\alpha)P_{m-2}^o}{1-2\alpha}\right] + \alpha^2\left[\frac{P_{m-1}^o - \alpha + \alpha P_{m-2}^o}{1-2\alpha}\right] \tag{6.115}$$

We now multiply through by $1 - 2\alpha$ and rearrange terms

$$(1-2\alpha)P_m^o = (1-2\alpha)(1-3\alpha+3\alpha^2)P_{m-2}^o - (1-2\alpha)P_{m-1}^o + (1-2\alpha)(1-\alpha+\alpha^2) \tag{6.116}$$

Finally

$$P_m^o + P_{m-1}^o + [1 - 3\alpha(1-\alpha)]\,P_{m-2}^o - [1-\alpha+\alpha^2] = 0 \tag{6.117}$$

We now try again a solution of the form

$$P_m^o = \frac{1}{3} + q\rho^{|m|} \tag{6.118}$$

Substituting (6.118) into (6.117) we have

$$\frac{1}{3} + q\rho^{|m|} + \frac{1}{3} + q\rho^{|m-1|} + \frac{1}{3} + q\rho^{|m-2|} - \alpha - 3\alpha q\rho^{|m-2|} + \alpha^2 + 3\alpha^2 q\rho^{|m-2|} - 1 + \alpha - \alpha^2 = 0 \tag{6.119}$$

or, dividing by $q\rho^{|m-2|}$

$$\rho^2 + \rho + (1 - 3\alpha + 3\alpha^2) = 0 \tag{6.120}$$

Solving (6.120), we have

$$\rho = \frac{-1 \pm \sqrt{1 - 4 + 12\alpha - 12\alpha^2}}{2} = -\frac{1}{2} \pm i\frac{\sqrt{3}}{2}(1-2\alpha) \tag{6.121}$$

We again make the substitution

$$\rho = re^{i\phi} = r\cos\phi + ir\sin\phi \tag{6.122}$$

We find, for the growth fault case

$$\tan\phi = \pm\sqrt{3}\,(1-2\alpha) \tag{6.123a}$$

$$r = \sqrt{1 - 3\alpha(1-\alpha)} \tag{6.123b}$$

To find q, we use the initial conditions

$$P_0^o = 1 \qquad P_1^o = 0 \tag{6.124}$$

Proceeding as in the growth fault analysis

$$P_0^o = 1 = \frac{1}{3} + q_1 + q_2 \tag{6.125a}$$

$$P_1^o = 0 = \frac{1}{3} + q_1\rho + q_2\rho^* \tag{6.125b}$$

Here ρ_1 and ρ_2 are complex conjugate. We have defined

$$\rho_1 = \rho_2^* = \rho \tag{6.126}$$

Solving (6.125) using determinants, we have

$$q_1 = \frac{\begin{vmatrix} \frac{2}{3} & 1 \\ -\frac{1}{3} & \rho^* \end{vmatrix}}{\begin{vmatrix} 1 & 1 \\ \rho & \rho^* \end{vmatrix}} = \frac{\frac{2}{3}\rho^* + \frac{1}{3}}{\rho^* - \rho} = \frac{-\frac{1}{3} - \frac{\sqrt{3}}{3}(1-2\alpha) + \frac{1}{3}}{-\sqrt{3}(1-2\alpha)} = \frac{1}{3} \tag{6.127a}$$

$$q_2 = \frac{\begin{vmatrix} 1 & \frac{2}{3} \\ \rho & -\frac{1}{3} \end{vmatrix}}{-\sqrt{3}(1-2\alpha)} = \frac{-\frac{1}{3} - \frac{2}{3}\rho}{-\sqrt{3}(1-2\alpha)} = \frac{-\frac{1}{3} + \frac{1}{3} - \frac{\sqrt{3}}{3}(1-2\alpha)}{-\sqrt{3}(1-2\alpha)} = \frac{1}{3} \tag{6.127b}$$

So,

$$\begin{aligned} P_m^o &= \frac{1}{3} + \frac{1}{3} r^{|m|} e^{i|m|\phi} + \frac{1}{3} r^{|m|} e^{-i|m|\phi} \\ &= \frac{1}{3} + \frac{2}{3} r^{|m|} \cos |m|\phi \\ &= \frac{1}{3} + \frac{2}{3} \{-[1 - 3\alpha(1-\alpha)]^{1/2}\}^m \cos\{|m| \arctan[\sqrt{3}(1-2\alpha)]\} \end{aligned} \tag{6.128}$$

It is now easy to determine P_m^+. The steps are similar to those just followed. First, substitute (6.114) into (6.111), to obtain

$$\begin{aligned} P_m^+ = 2\alpha(1-\alpha) + (1-\alpha)^2 \frac{(1-\alpha) - P_{m-1}^+ - (1-\alpha)P_{m-2}^+}{1-2\alpha} \\ + \alpha^2 \left[\frac{P_{m-1}^+ + \alpha P_{m-2}^+ - \alpha}{1-2\alpha}\right] \end{aligned} \tag{6.129}$$

Equation (6.129) is identical with (6.114), except that P^+ is substituted for P^o. Proceeding as before (steps (6.118)–(6.128)) we have

$$P_m^+ = \frac{i}{3} + q_1' \left[-\frac{1}{2} + i\frac{\sqrt{3}}{2}(1-2\alpha)\right]^{|m|} + q_2' \left[-\frac{1}{2} - i\frac{\sqrt{3}}{2}(1-2\alpha)\right]^{|m|} \tag{6.130}$$

But now the initial conditions are different

$$P_0^+ = 0 = \frac{1}{3} + q_1' + q_2' \tag{6.131a}$$

$$P_1^+ = (1-\alpha) = \frac{1}{3} + q_1'\rho + q_2'\rho^* \tag{6.131b}$$

Solving (6.131), we obtain

$$q_1' = \frac{\begin{vmatrix} -\dfrac{1}{3} & 1 \\ \dfrac{2-3\alpha}{3} & \rho^* \end{vmatrix}}{\begin{vmatrix} 1 & 1 \\ \rho & \rho^* \end{vmatrix}} = \frac{\dfrac{1}{6} + \dfrac{\sqrt{3}}{6}\, i(1-2\alpha) - \dfrac{2}{3} + \alpha}{-\sqrt{3}\, i(1-2\alpha)} = -\frac{1}{6} + i\frac{\sqrt{3}}{6} \tag{6.132a}$$

$$q_2' = \frac{\begin{vmatrix} 1 & -\dfrac{1}{3} \\ \rho & \dfrac{2-3\alpha}{3} \end{vmatrix}}{\begin{vmatrix} 1 & 1 \\ \rho & \rho^* \end{vmatrix}} = -\frac{1}{6} - i\frac{\sqrt{3}}{6} \tag{6.132b}$$

Note that

$$q_2' = q_1'^* \tag{6.133}$$

So,

$$P_m^+ = \frac{1}{3} + q_1'\, r^{|m|} e^{i|m|\phi} + q_1'^*\, r^{|m|} e^{-i|m|\phi} \tag{6.134}$$

Now let

$$q_1' = ae^{i\psi} \tag{6.135}$$

where

$$\tan\psi = -\sqrt{3} \tag{6.136a}$$

$$a = \frac{1}{6}\sqrt{1+3} = \frac{1}{3} \tag{6.136b}$$

Substituting (6.135) into (6.134), we get

$$\begin{aligned} P_m^+ &= \frac{1}{3} + ar^{|m|} e^{i(|m|\phi+\psi)} + ar^{|m|} e^{-i(|m|\phi+\psi)} \\ &= \frac{1}{3} + 2ar^{|m|} \cos(|m|\phi + \psi) \end{aligned}$$

$$= \frac{1}{3} + 2ar^{|m|}[\cos\psi \cos(|m|\phi) - \sin\psi \sin(|m|\phi) \qquad (6.137)$$

We found in (6.136) that $\tan\psi = -\sqrt{3}$. Hence, $\cos\psi = -\frac{1}{2}$ and $\sin\psi = \frac{\sqrt{3}}{2}$. So

$$P_m^+ = \frac{1}{3} + \frac{2}{3} r^{|m|} \left[-\frac{1}{2} \cos|m|\psi - \frac{\sqrt{3}}{2} \sin|m|\psi \right]$$

$$= \frac{1}{3} - \frac{1}{3} \{-[1 - 3\alpha(1-\alpha)]^{1/2}\}^m \{\cos[\sqrt{3}(1-2\alpha)|m|] \qquad (6.138)$$

$$+ \sqrt{3} \sin[\sqrt{3}(1-2\alpha)|m|]\}$$

Similarly, we would find

$$P_m^- = \frac{1}{3} - \frac{1}{3} \{-[1 - 3\alpha(1-\alpha)]^{1/2}\}^m \{\cos[\sqrt{3}(1-2\alpha)|m|]$$
$$- \sqrt{3} \sin[\sqrt{3}(1-2\alpha)|m|]\} \qquad (6.139)$$

As always

$$\frac{I(\mathbf{s})}{I_e(s)} = \sum_{m_l} y_m e^{2\pi i l m_l} \qquad (6.140)$$

In our case (see (6.60)

$$y_m = \overline{F_n F_{n+m}} = |F|^2 P_m^o + |F|^2 P_m^+ e^{\frac{2\pi}{3}(h-k)} \qquad (6.141)$$
$$+ |F|^2 p_m^- e^{-\frac{2\pi}{3}(h-k)}$$

If $h - k = 3p$, then the exponents are all reduced and $y_m = |F|^2$. There is no effect on these rel points. But if $h - k = 3p + 1$

$$y_m/|F|^2 = P_m^o + P_m^+ e^{i\frac{2\pi}{3}} + P_m^- e^{-i\frac{2\pi}{3}} \qquad (6.142)$$

And, for use in (6.140)

$$\frac{y_m}{|F|^2} e^{2\pi i l m_l} = P_m^o e^{i2\pi l m_l} + P_m^+ e^{i\frac{2\pi}{3}} e^{i2\pi l m_l} + P_m^- e^{-i\frac{2\pi}{3}} e^{i2\pi l m_l}$$

$$= P_m^o e^{i2\pi l m_l} + [P_m^+ + P_m^-] \cos\frac{2\pi}{3} e^{i2\pi m_l l} \qquad (6.143)$$

$$+ [P_m^+ - P_m^-] i \sin\frac{2\pi}{3} e^{i2\pi m_l l}$$

$$= \left\{ -\frac{1}{2} [(P_m^+ - P_m^-) + P_m^o] + i\frac{\sqrt{3}}{2} [P_m^+ - P_m^-] \right\} e^{i2\pi ml}$$

Now, (6.128), (6.138), and (6.139) can be expressed as

$$P_m^o = \frac{1}{3} + \frac{2}{3}\beta^{|m|}\cos|m|\gamma \tag{6.144a}$$

$$P_m^+ = \frac{1}{3} - \frac{1}{3}\beta^{|m|}\cos|m|\gamma - \frac{\sqrt{3}}{3}\beta^{|m|}\sin|m|\gamma \tag{6.144b}$$

$$P_m^- = \frac{1}{3} - \frac{1}{3}\beta^{|m|}\cos|m|\gamma + \frac{\sqrt{3}}{3}\beta^{|m|}\sin|m|\gamma \tag{6.144c}$$

Using (6.144), (6.143) can be written

$$\begin{aligned}\frac{y_m}{|F|^2} e^{2\pi ilm} l &= [\beta^{|m|}\cos|m|\gamma - i\beta^{|m|}\sin|m|\gamma]\, e^{i2\pi ml} \\ &= \beta^{|m|}[\cos(|m|\gamma)\cos(2\pi ml) + \sin(|m|\gamma)\sin(2\pi|m|l)]\end{aligned} \tag{6.145}$$

dropping the imaginary terms. Thus, using the trigonometric identity

$$\frac{y_m}{|F|^2} e^{2\pi ilm} = \beta^{|m|}\cos\left[2\pi|m|\left(l - \frac{\gamma}{2\pi}\right)\right] \tag{6.146a}$$

Similarly, for $h - k = 3p - 1$, we get

$$\frac{y_m}{|F|^2} e^{2\pi ilm} = \beta^{|m|}\cos\left[2\pi|m|\left(l + \frac{\gamma}{2\pi}\right)\right] \tag{6.146b}$$

And we now write the intensity expression, for $h - k = 3p \pm 1$

$$\frac{I(\mathbf{s})}{I_e(s)} = C\sum_{m=-\infty}^{\infty}\beta^{|m|}\cos\left[2\pi|m|\left(l \pm \frac{\gamma}{2\pi}\right)\right] \tag{6.147}$$

Here

$$\beta = -[1 - 3\alpha(1-\alpha)]^{1/2} = (-1)[1 - 3\alpha(1-\alpha)]^{1/2} \tag{6.148}$$

and

$$\beta^{|m|} = (-1)^{|m|}\{[1 - 3\alpha(1-\alpha)]^{1/2}\}^{|m|} = \{[1 - 3\alpha(1-\alpha)]^{1/2}\}^{|m|}\cos|m|\pi \tag{6.149}$$

Consider the product of cosines

$$\begin{aligned}\cos(|m|\pi)\cos\left[2\pi|m|\left(l - \frac{\gamma}{2\pi}\right)\right] &- \sin(|m|\pi)\sin\left[2\pi|m|\left(l + \frac{\gamma}{2\pi}\right)\right] \\ &= \cos\left[2\pi|m|\left(l + \frac{\gamma}{2\pi} + \frac{1}{2}\right)\right]\end{aligned} \tag{6.150}$$

Thus (6.147) can be rewritten as

$$\frac{I(\mathbf{s})}{I_e(s)} = C\left\{2\sum_{m=0}^{\infty}[(1 - 3\alpha(1-\alpha))^{1/2}]^m\cos 2\pi m\left(l + \frac{1}{2} \pm \frac{\gamma}{2\pi}\right) - 1\right\} \tag{6.151}$$

It can be shown that

$$\sum_{n=0}^{\infty} a^n \cos x = \frac{1 - a \cos x}{1 + a^2 - 2a \cos x} \qquad (\alpha \leqq 1) \tag{6.152}$$

Thus we have

$$\frac{I(\mathbf{s})}{I_e(s)} = C \left\{ 2 \left[\frac{1 - [1 - 3\alpha(1-\alpha)]^{1/2} \cos 2\pi \left(l + \frac{1}{2} \pm \frac{\gamma}{2\pi} \right)}{1 + 1 - 3\alpha(1-\alpha) - 2[1 - 3\alpha(1-\alpha)]^{1/2} \cos \left[2\pi \left(l + \frac{1}{2} \pm \frac{\gamma}{2\pi} \right) \right]} \right] - 1 \right\} \tag{6.153a}$$

Using $\alpha \ll 1$

$$\frac{I(\mathbf{s})}{I_e(s)} \simeq C \left\{ \frac{\frac{3}{2}\alpha(1-\alpha)}{1 - \frac{3}{2}\alpha(1-\alpha) - [1 - 3\alpha(1-\alpha)]^{1/2} \cos \left[2\pi \left(l + \frac{1}{2} \pm \frac{\gamma}{2\pi} \right) \right]} \right\} \tag{6.153b}$$

As in the growth fault cases, the intensity peaks sharply as the denominator of the intensity expression becomes very small. The denominator reaches its minimum when

$$l^* + \frac{1}{2} \pm \frac{\gamma}{2\pi} = p' \text{ (an integer)} \tag{6.154}$$

Especially note that the peak is shifted from a simple half integer by the amount $\frac{\gamma}{2\pi}$

$$l^* = p' - \frac{1}{2} + \frac{\gamma}{2\pi} \qquad \text{for } h - k = 3p + 1 \tag{6.155a}$$

$$l^* = p' - \frac{1}{2} - \frac{\gamma}{2\pi} \qquad \text{for } h - k = 3p - 1 \tag{6.155b}$$

The value of γ is arctan $\sqrt{3}(1 - 2\alpha)$. When $\alpha = 0$, $\gamma = \frac{\pi}{3}$; so

$$l = p' - \frac{1}{2} + \frac{1}{6} \qquad \text{for } h - k = 3p + 1 \tag{6.156}$$

But now recall that we should do this in terms of the real unit cell, in which case $l = \frac{l'}{3}$, where l' is the rel point for the real cell. Then

$$l' = 3p' - \frac{3}{2} - \frac{1}{2} = 3p' - 2 \tag{6.157}$$

If, however, $\alpha \neq 0$, then $\arctan \sqrt{3}(1-2\alpha) = \gamma = \frac{\pi}{3} - \delta$, where $\delta \ll 1$, giving

$$l' = 3p' - 2 + \delta \qquad \text{for } h - k = 3p + 1 \tag{6.158a}$$

$$l' = 3p' - 1 - \delta \qquad \text{for } h - k = 3p - 1 \tag{6.158b}$$

Resp now appears as in Fig. 6.11.

So far, we have been dealing in hexagonal coordinates. The transformation from hexagonal coordinates *(hkl)* to cubic coordinates *(HKL)* is accomplished as follows

$$H = \frac{-2(2h+k)+l}{3}$$
$$K = \frac{2(h-k)+l}{3} \tag{6.159}$$
$$L = \frac{2(h+2k)+l}{3}$$

Furthermore, we can easily determine which cubic reflections move which way, since

$$l' = H + K + L \tag{6.160}$$

i.e., l' is a <111>. What we find is shown in Fig. 6.12. Note that most reflections are broken up into two or more subpeaks.

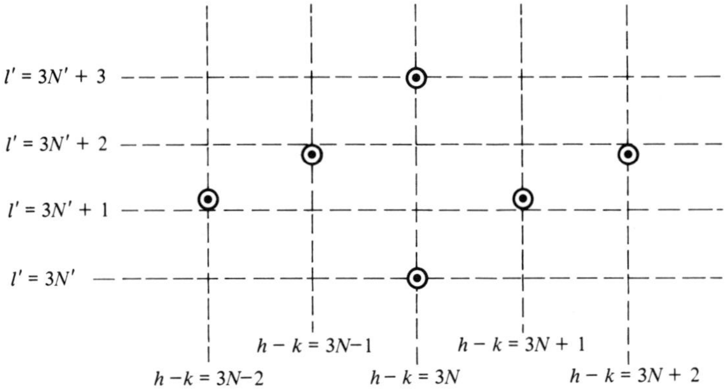

Figure 6.11. Projection, normal to **c**, of resp for a FCC crystal containing deformation faults.

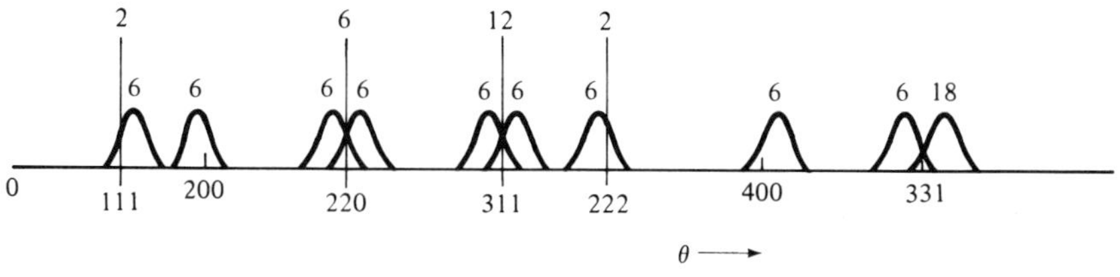

Figure 6.12. Shifting of Bragg peaks due to deformation faulting in FCC.

6.4 THERMAL DIFFUSE SCATTERING: THE DETERMINATION OF PHONON SPECTRA AND ELASTIC MODULI

We have seen that, when atomic displacements are not correlated, the value of $I_2(\mathbf{s})$ is

$$\frac{I_2(\mathbf{s})}{NI_e(s)} = \overline{|F|^2} - |\overline{F}|^2 \tag{6.161}$$

For a simple lattice this becomes $\overline{f^2} - \bar{f}^2$ and decreases smoothly with s. One might think that, to a first approximation, this result might apply to thermal vibrations within a crystal. However, these vibrations occur not as independent displacements about the lattice sites; the displacements are strongly spatially correlated as displacement waves. The effects of this strong correlation on the diffraction behavior were first observed by Laval in 1939. In this experiment, a KCl platelet was irradiated with MoK_α x-rays along (001). In this case, only the (211) lie on the Ewald sphere. In spite of this, a heavy exposure—seen in Fig. 6.13 shows spots near *all* the rel points. This behavior must be explained by the displacement correlation.

The displacement of the nth atom is a composite of wavelike motions in many lattice directions

$$\Delta\mathbf{r_n} = \sum_{\nu} \mathbf{A}_\nu \cos\left[2\pi(\nu t - \mathbf{k}_\nu - \mathbf{r_n})\right] \tag{6.162}$$

We have already seen that each one of these cosine waves produces a pair of satellite peaks located $\mathbf{k}_\nu$ away from the principal peak. The magnitude of each of the intensity satellites is $\pi^2 N|F_{hkl}|^2(\mathbf{s} \cdot \mathbf{A}_\nu)^2$. We now use α_k as the angle between $\mathbf{A}_\nu$ and $\mathbf{s}$. We also note that the wave amplitude $\mathbf{A}_\nu$ lies at some arbitrary angle to $\mathbf{k}_\nu$. We can think of one longitudinal ($\mathbf{A}_{\nu 1}$) and two transverse ($\mathbf{A}_{\nu 2}$, $\mathbf{A}_{\nu 3}$) components. Thus, the intensity at each point $\mathbf{s}$ is the sum of all three components of all waves ν.

$$\frac{I_2(\mathbf{s})}{NI_e(s)} = \pi^2|F_{hkl}|^2 s^2 \sum_{i=1,2,3} A_{\nu i}^2 \cos^2 \alpha_i(k_\nu) \tag{6.163}$$

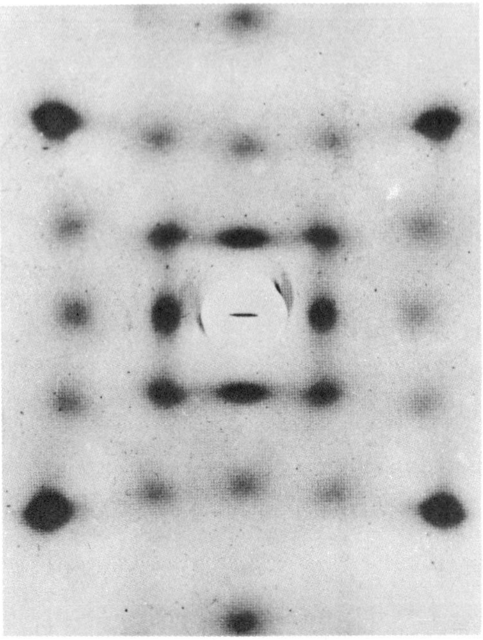

Figure 6.13. Thermal diffuse scattering pattern from a KCl crystal. Monochromatic MoK_α radiation was used. (100) was normal to both incident beam and plane film. None of the observable spots come from relnodes lying precisely on the Ewald sphere; in all cases, the relnode centers lie slightly off the sphere. [A. Guinier, *X-ray Diffraction in Crystals, Imperfect Crystals, and Amorphous Bodies* (W. H. Freeman, San Francisco, 1963)]

From (6.163) we see that to each position **s** in resp there correspond three displacement amplitude terms. However, in certain cases, two of these terms are zero. Consider the $I_2(\mathbf{s})$ (thermal diffuse scattering) about 200 for a cubic material. The situation is sketched in Fig. 6.14. Here we compare a point $\mathbf{k}_{\nu_\perp}$ from 200 with a point $\mathbf{k}_{\nu\|}$ from 200. For $k_{\nu_\perp}$ the longitudinal vibration cosine terms, cos $\alpha_1(k_\nu)$ and cos $\alpha_3(k_\nu)$ are zero, while the transverse component, cos $\alpha_2(k_\nu)$ is equal to unity. Conversely, for $k_{\nu\|}$, cos $\alpha_1(k_\nu) = 1$ and cos $\alpha_2(k_\nu) =$ cos $\alpha_3(k_\nu) = 0$. Thus the intensities at $k_{\nu_\perp}$, and $k_{\nu\|}$ provide the $A_{\nu i}$ terms for the case in which $\mathbf{k}_\nu$ is along a (100) direction.

$$(A_{\nu1}^{100})^2 = \frac{I_2(\mathbf{r}_{200}^* + \mathbf{k}_{\nu\|})}{NI_e(s)\pi^2|F_{200}|^2 s_{200}^2}$$

$$(A_{\nu2}^{100})^2 = (A_{\nu3})^2 = \frac{I_2(\mathbf{r}_{200}^* + \mathbf{k}_{\nu_\perp})}{NI_e(s)\pi^2|F_{200}|^2 s_{200}^2} \tag{6.164}$$

Using similar reasoning, the amplitude spectrum at any $\mathbf{k}_\nu$ can be found, although for directions not lying on symmetry elements the trigonometry is a little more complex. It is clear that the complete phonon spectrum should be available from such thermal-diffuse scattering (TDS) measurements.

This same approach can also provide the full matrix of elastic compliances, **S**, or stiffnesses, **C**. This comes about because the displacements $\mathbf{A}_\nu$ depend directly on the moduli (low moduli providing large displacements). A short

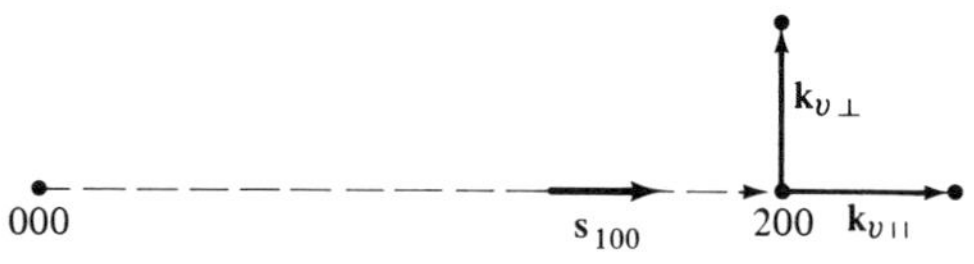

Figure 6.14. Reciprocal space geometry associated with the analysis of diffuse scattering.

review of some elements of formal elasticity should make this clear and enable us to establish the necessary connections between TDS measurements and moduli. While the methodology should provide information for cases of arbitrary crystal symmetry, we will here treat only the cubic case, in the interest of time and space.

For an arbitrary Hookean elastic solid, conditions of force equilibria and displacement compatibility must always be met. The force equilibria in the three cartesian directions are given by

$$\rho \frac{\partial^2 u_1}{\partial t^2} = \frac{\partial \sigma_{11}}{\partial x_1} + \frac{\partial \sigma_{12}}{\partial x_2} + \frac{\partial \sigma_{13}}{\partial x_3} \tag{6.165a}$$

$$\rho \frac{\partial^2 u_2}{\partial t^2} = \frac{\partial \sigma_{21}}{\partial x_1} + \frac{\partial \sigma_{22}}{\partial x_2} + \frac{\partial \sigma_{23}}{\partial x_3} \tag{6.165b}$$

$$\rho \frac{\partial^2 u_3}{\partial t^2} = \frac{\partial \sigma_{31}}{\partial x_1} + \frac{\partial \sigma_{32}}{\partial x_2} + \frac{\partial \sigma_{33}}{\partial x_3} \tag{6.165c}$$

where ρ is the density of the material and (u_1, u_2, u_3) are the displacements along (x_1, x_2, x_3). Also, in general, the stress and strain vectors $\boldsymbol{\sigma}$ and $\boldsymbol{\epsilon}$ are related through

$$\boldsymbol{\sigma} = \mathbf{C}\boldsymbol{\epsilon} \tag{6.166}$$

where the stiffness tensor matrix $\mathbf{C}$ is written

$$\mathbf{C} = \begin{vmatrix} c_{11} & c_{12} & c_{13} & c_{14} & c_{15} & c_{16} \\ c_{21} & c_{22} & c_{23} & c_{24} & c_{25} & c_{26} \\ c_{31} & c_{32} & c_{33} & c_{34} & c_{35} & c_{36} \\ c_{41} & c_{42} & c_{43} & c_{44} & c_{45} & c_{46} \\ c_{51} & c_{52} & c_{53} & c_{54} & c_{55} & c_{56} \\ c_{61} & c_{62} & c_{63} & c_{64} & c_{65} & c_{66} \end{vmatrix} \tag{6.167}$$

For cubic symmetry, this becomes

$$\mathbf{C}_c = \begin{vmatrix} c_{11} & c_{12} & c_{12} & 0 & 0 & 0 \\ c_{12} & c_{11} & c_{12} & 0 & 0 & 0 \\ c_{12} & c_{12} & c_{11} & 0 & 0 & 0 \\ 0 & 0 & 0 & c_{44} & 0 & 0 \\ 0 & 0 & 0 & 0 & c_{44} & 0 \\ 0 & 0 & 0 & 0 & 0 & c_{44} \end{vmatrix} \tag{6.168}$$

Using (6.168), the terms on the right of (6.165a) become

$$\begin{aligned}
\frac{\partial \sigma_{11}}{\partial x_1} &= c_{11}\frac{\partial \epsilon_{11}}{\partial x_1} + c_{12}\frac{\partial \epsilon_{22}}{\partial x_1} + c_{12}\frac{\partial \epsilon_{33}}{\partial x_1} \\
\frac{\partial \sigma_{12}}{\partial x_2} &= c_{44}\frac{\partial \epsilon_{12}}{\partial x_2} \\
\frac{\partial \sigma_{13}}{\partial x_3} &= c_{44}\frac{\partial \epsilon_{12}}{\partial x_3}
\end{aligned} \tag{6.169}$$

Using the general relation between strains and displacements

$$\epsilon_{ij} = \frac{1}{2}\left(\frac{\partial u_i}{\partial x_j} + \frac{\partial u_j}{\partial x_i}\right) \tag{6.170}$$

and inserting (6.169) into (6.165a), we have

$$\begin{aligned}
\rho\frac{\partial^2 u_1}{\partial t^2} &= c_{11}\frac{\partial^2 u_1}{\partial x_1^2} + c_{12}\frac{\partial^2 u_2}{\partial x_1 \partial x_2} + c_{12}\frac{\partial^2 u_3}{\partial x_1 \partial x_3} \\
&\quad + c_{44}\left(\frac{\partial^2 u_1}{\partial x_1 \partial x_2} + \frac{\partial^2 u_1}{\partial x_1 \partial x_3} + \frac{\partial^2 u_2}{\partial x_1^2} + \frac{\partial^2 u_3}{\partial x_1^2}\right)
\end{aligned} \tag{6.171}$$

Similarly, for the other two equilibrium equations we have

$$\begin{aligned}
\rho\frac{\partial^2 u_2}{\partial t^2} &= c_{11}\frac{\partial^2 u_2}{\partial x_2^2} + c_{12}\frac{\partial^2 u_1}{\partial x_1 \partial x_2} + c_{12}\frac{\partial^2 u_3}{\partial x_2 \partial x_3} \\
&\quad + c_{44}\left(\frac{\partial^2 u_2}{\partial x_1 \partial x_2} + \frac{\partial^2 u_2}{\partial x_2 \partial x_3} + \frac{\partial^2 u_1}{\partial x_2^2} + \frac{\partial^2 u_3}{\partial x_2^2}\right)
\end{aligned} \tag{6.172}$$

$$\begin{aligned}
\rho\frac{\partial^2 u_3}{\partial t^2} &= c_{11}\frac{\partial u_3}{\partial x_3^2} + c_{12}\frac{\partial^2 u_1}{\partial x_1 \partial x_3} + c_{12}\frac{\partial^2 u_2}{\partial x_2 \partial x_3} \\
&\quad + c_{44}\left(\frac{\partial^2 u_3}{\partial x_1 \partial x_3} + \frac{\partial^2 u_3}{\partial x_2 \partial x_3} + \frac{\partial^2 u_1}{\partial x_3^2} + \frac{\partial^2 u_2}{\partial x_3^2}\right)
\end{aligned} \tag{6.173}$$

These equilibrium equations are satisfied by waveform displacements:

$$\begin{aligned}
u_1 &= A\cos\alpha_1 \exp\left[2\pi i\left(\frac{v_1}{\Lambda}t - \mathbf{k}\cdot\mathbf{r}\right)\right] \\
u_2 &= A\cos\alpha_2 \exp\left[2\pi i\left(\frac{v_2}{\Lambda}t - \mathbf{k}\cdot\mathbf{r}\right)\right] \\
u_3 &= A\cos\alpha_3 \exp\left[2\pi i\left(\frac{v_3}{\Lambda}t - \mathbf{k}\cdot\mathbf{r}\right)\right]
\end{aligned} \tag{6.174}$$

where α_1, α_2, and α_3 are the angles between displacement $\mathbf{A}$ and the coordinate directions; v_1, v_2, v_3 are the wave velocity components in the three coordinate directions; $|\mathbf{k}| = 1/\Lambda$ is the displacement wavelength; and the scalar product $\mathbf{k} \cdot \mathbf{r}$ is written

$$\mathbf{k} \cdot \mathbf{r} = k(lx_1 + mx_2 + nx_3) \tag{6.175}$$

Here the l, m, and n are direction cosines between $\mathbf{k}$ and the coordinate axes. Insertion of (6.174) into (6.171)–(6.173) yields

$$\begin{aligned}
&[c_{11}l^2 + c_{44}(m^2 + n^2) - \rho v^2] \cos \alpha_1 + (c_{12} + c_{44})lm \cos \alpha_2 \\
&\qquad\qquad + (c_{12} + c_{44})ln \cos \alpha_3 = 0 \\
&[c_{11}m^2 + c_{44}(l^2 + n^2) - \rho v^2] \cos \alpha_2 + (c_{12} + c_{44})lm \cos \alpha_1 \\
&\qquad\qquad + (c_{12} + c_{44})mn \cos \alpha_3 = 0 \\
&[c_{11}n^2 + c_{44}(l^2 + m^2) - \rho v^2] \cos \alpha_3 + (c_{12} + c_{44})ln \cos \alpha_1 \\
&\qquad\qquad + (c_{12} + c_{44})mn \cos \alpha_2 = 0
\end{aligned} \tag{6.176}$$

Suppose now that we set x_1 along [100] and x_2, x_3 along [010], [001]. Let us consider the case in which $\mathbf{k}$ is parallel to [100]. Here $l = 1$, $m = n = 0$, and (6.176) becomes

$$\begin{aligned}
(c_{11} - \rho v^2) \cos \alpha_1 &= 0 \\
(c_{44} - \rho v^2) \cos \alpha_2 &= 0 \\
(c_{44} - \rho v^2) \cos \alpha_3 &= 0
\end{aligned} \tag{6.177}$$

As previously, if we consider scanning about (200), the case for $\mathbf{k}$ parallel to [100] provides information about only the longitudinal component, v_1, of v. Thus, here, $\cos \alpha_1 = 1$ and $\cos \alpha_2 = \cos \alpha_3 = 0$. Then

$$c_{11} = \rho v_1^2 \tag{6.178}$$

We will see below that v_1 can be related to the intensity at this value of $\mathbf{k}$. In the meantime, we shall note the experiments to get c_{12} and c_{44}.

Consider $\mathbf{k}$ along [110]. In this case, $l = m = 1/\sqrt{2}$ and $n = 0$. The three equations (6.176) become here

$$\begin{aligned}
&\left(\frac{c_{11}}{2} + \frac{c_{44}}{2} - \rho v^2\right) \cos \alpha_1 + \frac{(c_{12} + c_{44})}{2} \cos \alpha_2 = 0 \\
&\left(\frac{c_{11}}{2} + \frac{c_{44}}{2} - \rho v^2\right) \cos \alpha_2 + \frac{(c_{12} + c_{44})}{2} \cos \alpha_1 = 0 \\
&(c_{44} - \rho v^2) \cos \alpha_3 = 0
\end{aligned} \tag{6.179}$$

Combining the first two equations of (6.179), we have

$$\begin{aligned}
&(c_{11} + c_{12} + 2c_{44} - 2\rho v^2)(\cos \alpha_1 + \cos \alpha_2) = 0 \\
&(c_{11} - c_{12} - 2\rho v^2)(\cos \alpha_1 - \cos \alpha_2) = 0 \\
&(c_{44} - \rho v^2) \cos \alpha_3 = 0
\end{aligned} \tag{6.180}$$

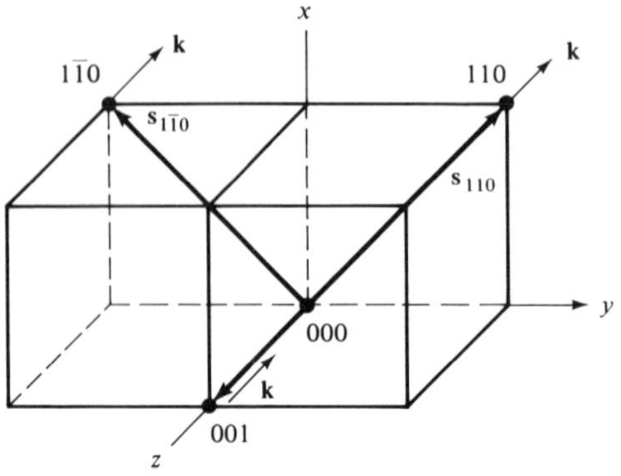

Figure 6.15. Resp geometry required for complete analysis of the elastic constants for a cubic material.

Suppose now that **s** is also along [110]. In this case $\cos\alpha_1 = \cos\alpha_2 = 1/\sqrt{2}$ and $\cos\alpha_3 = 0$. This gives

$$c_{11} + c_{12} + 2c_{44} = \rho(v_1^{110})^2 \tag{6.181}$$

If **s** is along [$1\bar{1}0$], with **k** still along [110], then $\cos\alpha_1 = -\cos\alpha_2 = 1/\sqrt{2}$ and $\cos\alpha_3 = 0$. In this case, (6.179) gives

$$c_{11} - c_{12} = 2\rho(v_2^{110})^2 \tag{6.182}$$

Finally, if **s** is along [001], then $\cos\alpha_1 = \cos\alpha_2 = 0$ and $\cos\alpha_3 = 1$. Here

$$c_{44} = \rho(v_3^{110})^2 \tag{6.183}$$

In resp, these three positions look as shown in Fig. 6.15. Clearly, inspection of these three positions in resp must, according to (6.180) – (6.182), provide a full determination of c_{11}, c_{12}, and c_{44}. The c_{11} so obtained should be the same as that found from the TDS about (100).

What remains then is a relation between $I_2(\mathbf{s})$ and the velocities v_i. This is obtained as follows. Equation (6.162) gave the displacement of the nth cell. From this sinusoidal form, we see that the average displacement for the wave of frequency ν must be $A^2/2$. One expression for $A(\nu)$ is obtained from kinetic energy considerations. The kinetic energy of the cell n is

$$e(\nu) = \frac{m\left(\dfrac{\partial \Delta r_\nu}{\partial t}\right)^2}{2} = 2\pi^2 m\nu^2 A^2(\nu) \tag{6.184}$$

where m is the mass of that cell. For the entire N cells of the crystal, the energy due to the ν waves is

$$\overline{E}(\nu) = Ne(\nu) = 2\pi^2 N m \nu^2 A^2(\nu) \tag{6.185}$$

A second expression for $\overline{E}(\nu)$ is obtained from the quantum description of lattice vibrations. In this case, the allowable energy states for a frequency ν are

$$E_n(\nu) = \left(\frac{1}{2} + n\right) h\nu \qquad n = 0,1,2,3, \ldots . \tag{6.186}$$

The average energy is then, using Boltzmann statistics,

$$\bar{E}(\nu) = \frac{\sum_{n=0}^{\infty} E_n e^{-E_n/kT}}{\sum_{n=0}^{\infty} e^{-E_n/kT}} \tag{6.187}$$

Substituting (6.186) into (6.187) and after some algebraic juggling, we find

$$\bar{E}(\nu) = \frac{1}{2} + \frac{h\nu}{e^{h\nu/kT} - 1} \tag{6.188}$$

Equating our two expressions, (6.185) and (6.188), for the energy stored by lattice waves of frequency ν, we have

$$A^2(\nu) = \frac{h}{2\pi m N \nu}\left[\frac{1}{2} + \frac{1}{e^{h\nu/kT} - 1}\right] \tag{6.189}$$

Recall that the lattice wave velocity v is given by $v = \nu/k$. Since we have chosen **k**'s to look at, the velocity v is defined if we deduce ν at **k.** This is accomplished via (6.189); $A^2(\nu)$ is related to $I_2(\mathbf{S})$ in (6.163), for instance. Having this value of $A^2(\nu)$, ν is found from (6.189) and v is computed from $v = \nu/k$. With an absolute value of v, the elastic constants can now be determined via (6.178) and (6.181)–(6.183).

PROBLEMS

6.1. We have seen that energy is conserved in x-ray scattering. That is, the total intensity integrated over resp will be a constant quantity for a given collection of atoms, independent of their state of aggregation (crystal, liquid, or gas). Many polymeric solids are semicrystalline; they are composed of fairly well defined crystalline and amorphous domains. In the mid 1950s, polymer scientists, using the conservation of energy principle, began to compute apparent degrees of cyrstallinity from the ratio of the integrated intensity of all scattering into sharp, crystalline peaks to the total integrated scattered intensity. However, it was soon recognized that this could be only an approximation, since intracrystalline defects would displace scattering from the crystalline peaks to the "background." In 1962, W. Ruland proposed that a "true" degree of crystallinity (i.e., the actual ratio of crystal domain to amorphous domain) could be obtained by accounting for the decrement of crystalline peak integrated intensities via a term $\exp(-ks^2)$, where k is a constant for the solid and is to be found by fitting this modified intensity formula to several peaks.

What is the theoretical justification for the use of the $\exp(-ks^2)$ factor and what must be its limitations? Show that a factor of this form can account for any type defect under the limitations you mention.

6.2. Bolling, Massalski, and McHargue measured, at 4.2 K, peak shifts which occurred during the cold-working of zone-refined Pb at 4.2 K. The peak shifts were defined by the angular separation between pairs of x-ray diffraction peaks. Specifically, they measured the angles $\Delta 2\theta_1$ between (200) and (111) and $\Delta 2\theta_2$ between (220) and (200). Their results are as follows

Specimen	$\Delta 2\theta_1$ (°)	$\Delta 2\theta_2$ (°)
Not cold-worked	5.05	16.12
Cold-worked	5.02	16.15

What is the stacking fault probability for the cold-worked material? Assume the virgin material to contain no stacking faults.

6.3. You have just made a new fibrous composite, by the solid state precipitation of a cubic spinel structure phase in a metal matrix. (A "spinel" is just a specific type of complex cubic crystal.) The composite is not giving the modulus you think it should, in the fiber direction. You are not sure whether the problem is in the finite fiber length (which is hard to measure in this metal matrix) or if the fiber phase is simply not as hard as you had expected. An initial set of x-ray diffraction scans has indicated that, when you precipitate the spinel in a single crystal of the matrix, the [100] axes of crystal and (cubic) matrix are parallel. You have been unable to separate out the very fine fibers for examination and testing. Outline in as much detail as you can the minimum x-ray experiments you will need to run on the fiber-precipitated single crystals to get a value of the fiber-axis modulus of the spinel phase.

6.4. (a) Solid solutions can decompose to two-phase systems by either of two modes: nucleation and growth or spinodal decomposition. In the latter case, spatially sinusoidal compositional variations of wavelength λ occur

$$\rho = \frac{\rho_{max} + \rho_{min}}{2} + (\rho_{max} - \rho_{min}) \sin\left(\frac{2\pi x}{\lambda}\right)$$

Show that in this case, a pair of satellite peaks develops about each Bragg reflection, at a resp distance $1/\lambda$ from each lattice point. Will these satellites also appear in the small-angle regime (about the incident beam direction)?

(b) In most systems, the two satellites are not of equal intensity. This is due to the combined effect of composition and displacement fluctuations; since the two substituent species will not be of exactly the same size, there must also be a displacement wave of wavelength λ. Derive an expression to show that unequal satellite peak intensities are to be expected in this case.

BIBLIOGRAPHY

1. J. L. Amoros and M. L. Canut de Amoros, *La Difraccion Difusa de los Cristales Moleculares,* C.S.I.C., Madrid (1965).

2. J. M. Cowley, *Diffraction Physics,* Chs. 7, 12, 17, and 18, North-Holland, Amsterdam (1975).

3. O. S. Edwards and H. Lipson, *Proceedings of the Royal Society (London), Section A, Vol. 180,* p. 277 (1940).

4. A. Guinier, *X-ray Crystallographic Technology,* Ch. 5, Hilger and Watts, London (1952).

5. A. Guinier, *X-ray Diffraction in Crystals, Imperfect Crystals, and Amorphous Bodies,* Chs. 6–9, W. H. Freeman & Company Publishers, San Francisco (1963).

6. R. Hoseman and S. N. Bagchi, *Direct Analysis of Diffraction by Matter,* Ch. 5, North-Holland, Amsterdam (1962).

7. R. W. James, *The Optical Principles of Diffraction of X-rays,* Chs. 5 and 10, G. Bell & Sons, Ltd., London (1954).

8. H. P. Klug and L. E. Alexander, *X-ray Diffraction Procedures for Polycrystalline and Amorphous Materials, 2nd Ed.,* Ch. 9, Wiley-Interscience, N.Y. (1959).

9. C. A. Johnson, *Acta Crystallographica, Vol. 16,* p. 490 (1963).

10. M. T. F. von Laue, *Röntgenstrahlinterferenzen, 3. Ausg.,* Ch. 4, Akademische Verlagsgesellschaft, Frankfurt (1960).

11. M. S. Paterson, *Journal of Applied Physics, Vol. 23,* p. 805 (1952).

12. L. H. Schwartz and J. B. Cohen, *Diffraction from Materials,* Ch. 7, Academic Press, Inc., N.Y. (1977).

13. A. Taylor, *X-ray Metallography,* Chs. 14 and 15, Wiley, N.Y. (1945).

14. B. E. Warren, *X-ray Diffraction,* Chs. 11–13, Addison-Wesley Publishing Company, Inc. Reading, MA (1969).

15. W. A. Wooster, *Diffuse X-ray Reflections from Crystals,* Clarendon, Oxford (1962).

Small Particulate Systems 7

7.1 INTRODUCTION

We saw briefly in Sect. 3.5 that the effect of having a small particle size is to broaden diffraction peaks. In fact, such *particle-size broadening* affords a useful measure of the dimensions of small crystallites. We shall see here how one quantitatively obtains such information. An extension of this method to the broadening of the forward-scattered beam permits one to measure the dimensions of noncrystalline, as well as crystalline domains. This technique is termed *small-angle scattering.* Small-angle scattering will be treated as a special case of particle-size broadening.

Physically, particle-size broadening arises because of incomplete destructive interference. Consider the infinite crystal of Fig. 7.1*a*. Suppose that this crystal is set just barely off the exact Bragg angle. In that case, the rays scattered from adjacent planes have a phase angle displacement of $\Delta\phi$. If the crystal is large enough, there is another plane, N away from the original plane, such that the wave scattered from it is just out of phase with the wave from the original plane. Thus for *every* plane there is another, N planes distant, which will nullify a wave scattered from it. This is why the Bragg condition is so sharp. If, on the other hand, the crystal is small enough that most planes do not have another N away, then the conditions of destructive interference are not met and a finite intensity will be observed at this off-Bragg setting. This situation is shown in Fig. 7.1*b*.

The critical crystal size at which finite intensity can be observed at an angle $\Delta(2\theta)$ off the Bragg angle θ_B can be estimated. Suppose that the number of

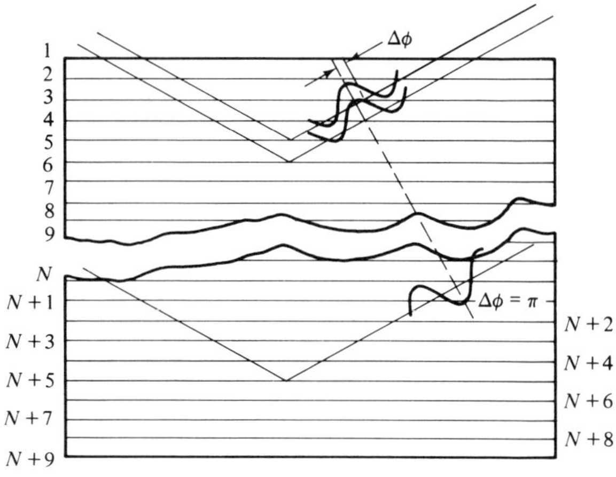

(a)

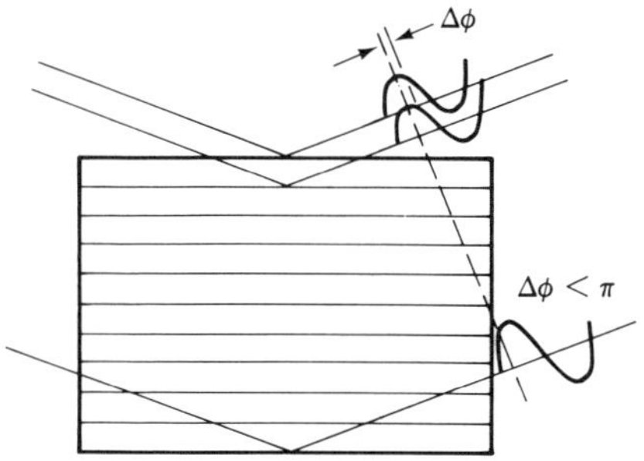

(b)

Figure 7.1. Representation of the maximum phase change $\Delta\phi$ in *(a)* a large crystal and *(b)* a small crystal. In *(a)* there are sufficiently many atomic planes for $\Delta\phi = \pi$, in *(b)* this is not possible.

planes needed for complete destructive interference is N. Then a crystal N planes thick will exhibit complete destruction for only its end planes. The intensity from such a crystal, when $\Delta(2\theta)$ from the Bragg condition, will be finite but not complete. Let us then take N as an approximate critical crystal thickness. For this angle setting, $N\Delta\phi = \pi$, by definition, where $\Delta\phi$ is the phase difference for waves scattered from adjacent planes. $\Delta\phi$ is related to the setting angle θ by $\Delta\phi = (2\pi/\lambda)(2d \sin\theta - 2d \sin\theta_B)$. For a relatively small missetting $\Delta(2\theta)$, $2(\sin\theta - \sin\theta_B) \simeq (\cos\theta_B)\Delta(2\theta)$. Combining our expressions above for $\Delta\phi$, we find for the critical crystal thickness D, $D = Nd \simeq \lambda/[4\cos\theta_B\Delta(2\theta)]$. A more precise criterion gives a constant closer to unity, rather than 4, in the denominator, as we shall see. The interpretation of this result is as follows.

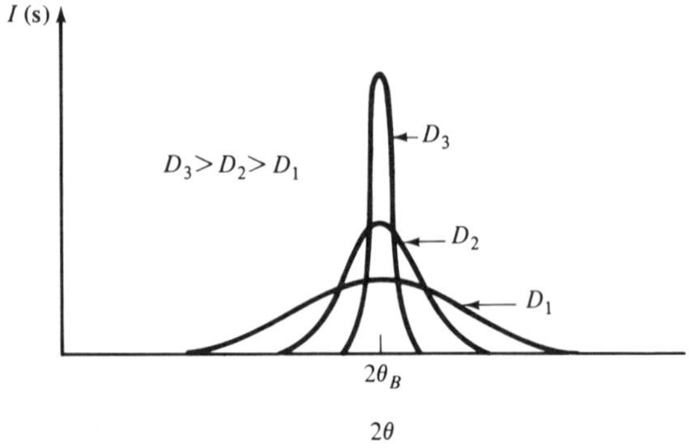

Figure 7.2. Sketch of Bragg peak at $2\theta_B$ for three different crystal diameters $D_1 < D_2 < D_3$.

For the present particle size *D*, the given setting results in finite intensity, but at larger $\Delta(2\theta)$ the intensity vanishes. For larger *D*, the intensity at this given $\Delta(2\theta)$ vanishes, while for a smaller particle the intensity will be finite, at the given $2(\theta)$ and also at larger $\Delta(2\theta)$. This situation is sketched, for a given *hkl* reflection, in Fig. 7.2. The FWHM breadth $\Delta_o(2\theta)$ for a given peak will be some $\lambda/(D \cos \theta)$.

In the next section we obtain a more general and precise relationship, using the Fourier transform scattering formalism.

7.2 THE FORMALISM OF PARTICLE-SIZE BROADENING

We begin by considering the formal scattering equations again, but now retaining a term, $\sigma(\mathbf{r})$, which is sensitive to the particle dimensions. Figure 7.3*a* represents

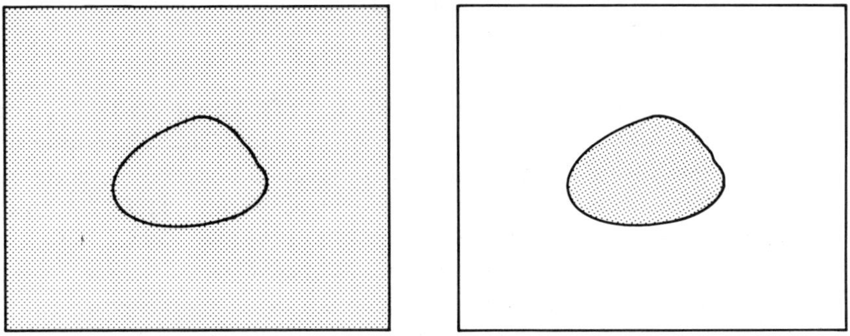

(*a*) (*b*)

Figure 7.3. *(a)* A very large system of uniform density ρ_o; *(b)* removal of all but the volume outlined in *(a)*.

a very large, homogeneous system whose density function is $\rho_o(\mathbf{r})$. A portion of that system is outlined. We now define a function $\sigma(\mathbf{r})$ which has value 1 within the outlined portion and is zero everywhere else. Figure 7.3*b* show the same outlined portion, but now separated from the larger matrix. The density function of this particle can be written as

$$\rho(\mathbf{r}) = \rho_o(\mathbf{r})\,\sigma(\mathbf{r}) \tag{7.1}$$

The amplitude of scattering from such a small particle is

$$\frac{A(\mathbf{s})}{A_e(s)} = \int \rho_o(\mathbf{r})\sigma(\mathbf{r})e^{2\pi i\mathbf{s}\cdot\mathbf{r}}\,dv_{\mathbf{r}} \tag{7.2}$$

Using the convolution theorem in the form of (1.63), we find

$$\frac{A(\mathbf{s})}{A_e(s)} = \mathscr{F}[\rho_o(\mathbf{r})] * \mathscr{F}[\sigma(\mathbf{r})] \tag{7.3}$$

Since $\mathscr{F}[\rho_o(r)]$ is a set of delta functions over a lattice, the effect of the convolution is to replicate the broader function $\mathscr{F}[\sigma(\mathbf{r})]$ at each lattice point.

As an example, let us consider a crystalline sphere of radius R. Here

$$\mathscr{F}[\sigma(\mathbf{r})] = \int \sigma(\mathbf{r})\,e^{2\pi i\mathbf{s}\cdot\mathbf{r}}\,dv_{\mathbf{r}} \tag{7.4}$$

$$= \int_{\substack{\text{volume of}\\ \text{sphere}}} e^{2\pi i\mathbf{s}\cdot\mathbf{r}}\,dv_{\mathbf{r}}$$

Taking spherical coordinates about the vector $\mathbf{s}$, we have $\mathbf{s}\cdot\mathbf{r} = sr\cos\phi$, $dv_{\mathbf{r}} = r^2 \sin\phi\, d\theta\, d\phi\, dr$. Thus

$$\mathscr{F}[\sigma(\mathbf{r})] = \int_{r=0}^{R}\int_{\phi=0}^{\pi}\int_{\theta=0}^{2\pi} e^{2\pi i\mathbf{s}\cdot\mathbf{r}}\,r^2\,d\theta\,d\phi\,dr \tag{7.5}$$

As in the case of equations (1.69)–(1.72), we find here

$$\mathscr{F}[\sigma(\mathbf{r})] = 2\pi\int_0^R r\frac{\sin 2\pi sr}{2\pi s}\,dr \tag{7.6}$$

Integrating over $\mathbf{r}$

$$\mathscr{F}[\sigma(r)] = 4\pi\,\frac{\sin(2\pi sR) - 2\pi sR\cos(2\pi sR)}{(2\pi s)^3} \tag{7.7}$$

The intensity of scattering from the *hkl* plane within the sphere will then be

$$\frac{I(\mathbf{s})}{I_e(s)} = \frac{1}{V_c^2}\,|F_{hkl}|^2\,|\mathscr{F}(\sigma_{hkl})|^2$$

$$= \frac{1}{V_c^2}|F_{hkl}|^2 \frac{1}{8\pi^3}\left[\frac{4\pi R^2}{s^{*4}} + \frac{1}{\pi s^{*6}} - \frac{4R}{s^{*5}} \sin 4\pi Rs + \left(\frac{4\pi R^2}{s^{*4}} - \frac{1}{\pi s^{*6}}\right) \cos 4\pi Rs\right] \tag{7.8}$$

where $\mathbf{s}^* = \mathbf{s} - \mathbf{r}^*_{hkl}$. (The V_c^{-2} term arises from the convolution of the real lattice function, $\Sigma\Sigma\Sigma\delta[\mathbf{r} - (u\mathbf{a} + v\mathbf{b} + w\mathbf{c})]$. The bracketed term on the right of (7.8) is shown in Fig. 7.4. The peak has been broadened by the finite particle size. The breadth of the peak at half maximum is some $(\Delta s)_{1/2} = 0.9/R$. (When one integrates over $\mathbf{s}$, in three dimensions, one finds still, for the integrated intensity, $(V/V_c^2)I_e(s)|F_{hkl}|^2$, where $V = \frac{4}{3}\pi R^3$.) This result leads to a breadth of the Bragg peak of some $\Delta(2\theta) = \lambda/(2R\cos\theta)$, using $(\Delta s)_{1/2} = \cos\theta\Delta(2\theta)/\lambda$.

In the above, we have computed an amplitude function and from this derived the intensity relationship. An alternate scheme is to derive the intensity relationship from an autocorrelation function. These two alternative paths are illustrated in Fig. 7.5. In some cases, one path is easier; in other cases, the other path may be easier.

Using the autocorrelation function approach, one proceeds as follows. As before, the local density function for a small particle is written $\rho(\mathbf{r}) = \rho_o(\mathbf{r})\sigma(\mathbf{r})$. The autocorrelation function is then

$$P(\mathbf{u}) = \int \rho_o(\mathbf{r})\rho_o(\mathbf{r}+\mathbf{u})\sigma(\mathbf{r})\sigma(\mathbf{r}+\mathbf{u})\,dv_{\mathbf{r}} \tag{7.9}$$

The product $\sigma(\mathbf{r})\sigma(\mathbf{r}+\mathbf{u})$ has value unity if both $\mathbf{r}$ and $\mathbf{r}+\mathbf{u}$ lie within the particle. Otherwise the product is zero. The function of this product is then

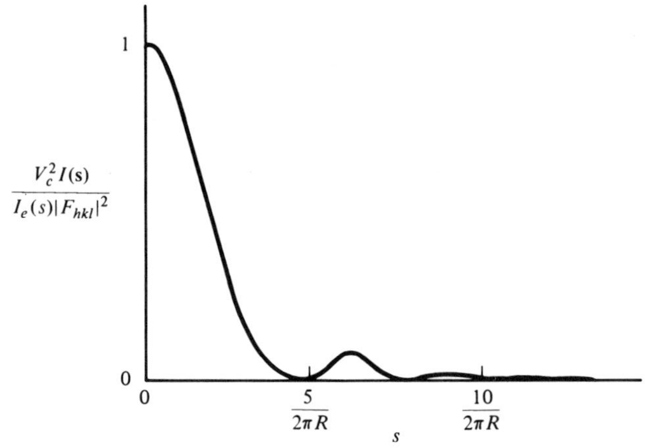

Figure 7.4. The variation of the intensity of scattering from a uniform sphere with the scattering vector **s**.

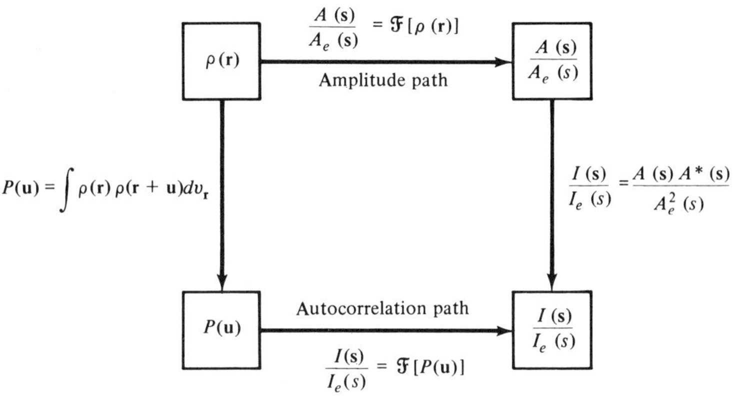

Figure 7.5. Alternative paths for structure analysis.

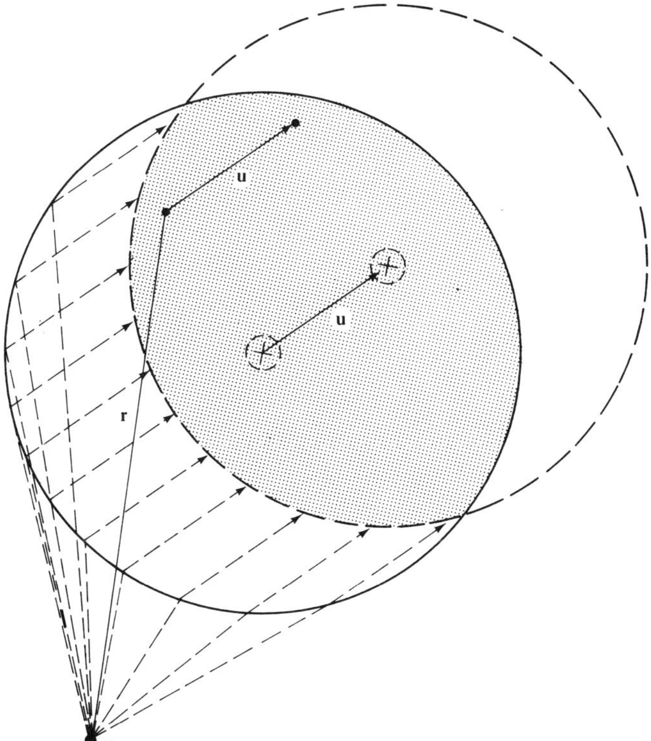

Figure 7.6. "Ghost particle" construction for dealing with scattering from small particles. The real particle (here a sphere) is drawn with a solid line. The "ghost particle" is drawn with a broken line. The origin of real space is 0.

only to limit the values of **r** which can be used for the product $\rho(\mathbf{r})\rho(\mathbf{r} + \mathbf{u})$. There is a useful geometrical construct for dealing with the allowable values of **r**. The construct is shown in Fig. 7.6. A sphere with solid border represents the particle. A given vector **u** is shown drawn from the center of the sphere. A second, imaginary sphere (dashed line surface) whose center is displaced by **u** from the real sphere is also drawn. A shaded area common to real and virtual particles, is shown, as is a **u**-vector whose head and tail lie entirely in that common volume. In fact, all **u**-vectors whose heads lie in common volume give unit value of $\sigma(\mathbf{r})\sigma(\mathbf{r} + \mathbf{u})$. The **u**-vectors whose heads lie outside the common volume do not contribute to $\sigma(\mathbf{r})\sigma(\mathbf{r} + \mathbf{u})$. This is indicated at the left of the real sphere; there a vector **u** whose head just touches the common volume still has its tail at the surface of the real sphere, but any head lying outside the common volume has its tail outside the real particle and the $\sigma(\mathbf{r})\sigma(\mathbf{r} + \mathbf{u})$ product is zero. Thus the volume over which **r** can be used is just that common volume. This common volume can be written as a fraction $\phi(\mathbf{u})$ of the total volume V of the particle. Thus

$$P(\mathbf{u}) = V\phi(\mathbf{u}) \int \rho_o(\mathbf{r})\rho_o(\mathbf{r} + \mathbf{u})\, dv_{\mathbf{r}} \tag{7.10}$$

Inserting (7.10) into the general intensity formula, we have

$$\frac{I(\mathbf{s})}{I_e(s)} = V \int \phi(\mathbf{u}) P_o(\mathbf{u}) e^{2\pi i \mathbf{s}\cdot\mathbf{r}}\, dv_{\mathbf{r}} \tag{7.11}$$

$$= V\mathscr{F}[P_o(\mathbf{u})] * \mathscr{F}[\phi(\mathbf{u})]$$

where

$$P_o(\mathbf{u}) = \int \rho_o(\mathbf{r})\rho_o(\mathbf{r} + \mathbf{u})\, dv_{\mathbf{r}} \tag{7.12}$$

The Fourier transform of $P_o(\mathbf{u})$ is the transform of the electron density within an infinitely large crystal. This quantity is, for a crystalline solid

$$\mathscr{F}[P_o(\mathbf{u})] = [|F(\mathbf{s})|^2 \cdot |Z(\mathbf{s})|^2] \tag{7.13}$$

as before. Thus

$$\frac{I_{hkl}}{(I_e)_{hkl}} = \frac{V}{V_c^2} \mathscr{F}[\phi_{hkl}(\mathbf{u})] |F_{hkl}|^2 \tag{7.14}$$

Let us, as an example, consider again the spherical particle. Here the common volume is

$$V\phi(\mathbf{u}) = 2\pi \left(\frac{2}{3} R^3 - \frac{R^2 u}{2} + \frac{u^3}{24}\right) \tag{7.15}$$

The Fourier transform of $\phi(\mathbf{u})$ is

$$\mathscr{F}[\phi(\mathbf{u})] = \frac{3}{32\pi^4 R^3} \left[\frac{4\pi R^2}{s^4} + \frac{1}{\pi s^6} - \frac{4R}{s^5} \sin 4\pi Rs + \left(\frac{4\pi R^2}{s^4} - \frac{1}{\pi s^6}\right) \cos 4\pi Rs\right] \quad (7.16)$$

Substituting (7.16) into (7.14) yields the expected result: (7.8).

An inverse relationship exists between the dimension of a particle in real space and the degree of line broadening. This is most easily seen in the example of a squat rectangular parallelopiped, as in Fig. 7.7. The thickness of the plate is h and the width is w. Shown also are diffracting vectors $\mathbf{s}_x$, $\mathbf{s}_y$, and $\mathbf{s}_z$ and their associated diffracting planes. The shape factor, $\mathscr{F}[\sigma(\mathbf{r})]$ for this particle is

$$\mathscr{F}[\sigma(\mathbf{r})] = \int_{-h/2}^{h/2} \int_{-w/2}^{w/2} \int_{-w/2}^{w/2} e^{2\pi i \mathbf{s}\cdot\mathbf{r}}\, dx\, dy\, dz \quad (7.17)$$

For diffraction from planes normal to the thin direction, $\mathbf{s} \cdot \mathbf{r} = s_z z$. Thus

$$\mathscr{F}_z[\sigma(\mathbf{r})] = \int_{-h/2}^{h/2} e^{2\pi i s_z z} dz = w^2 z \frac{\sin (\pi h s_z)}{(\pi h s_z)} \quad (7.18)$$

The intensity of each diffraction node is then spread in the z-direction as $|\mathscr{F}_z[\sigma(\mathbf{r})]|^2$

$$|\mathscr{F}_z[\sigma(\mathbf{r})]|^2 = z^2 \frac{\sin^2 (\pi h s_z)}{(\pi h s_z)^2} = z^2 \frac{\sin^2 (\pi h s_z)}{(\pi h s_z)^2} \quad (7.19)$$

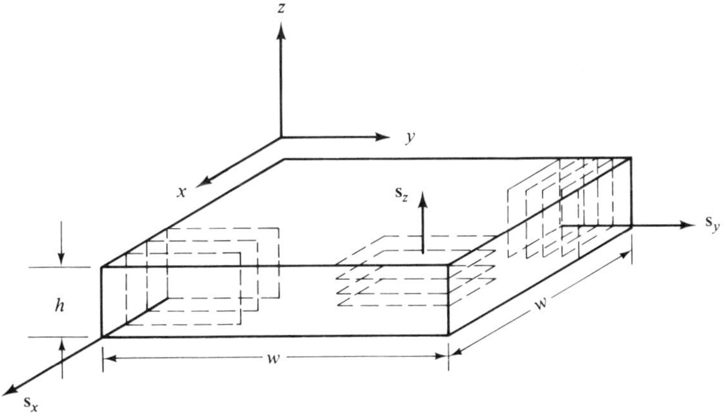

Figure 7.7. Relationships between real and reciprocal space geometries for a flat, square plate of width w and thickness h.

Similarly, diffracting from planes normal to x or y yields

$$|\mathscr{F}_x[\sigma(\mathbf{r})]|^2 = w^2 \frac{\sin^2(\pi w s_x)}{(\pi w s_x)^2}$$
$$|\mathscr{F}_y[\sigma(\mathbf{r})]|^2 = w^2 \frac{\sin^2(\pi w s_y)}{(\pi w s_y)^2} \tag{7.20}$$

These functions are plotted in Fig. 7.8. In this figure the w/h ratio is given as 10. The full width at half maximum (FWHM) breadths of these peaks are found via

$$\frac{\sin^2(\pi w(s_x)_{1/2}]}{[\pi w(s_x)_{1/2}]^2} = \frac{1}{2} = \frac{\sin^2[\pi h(s_z)_{1/2}]}{[\pi h(s_z)_{1/2}]^2} \tag{7.21}$$

Solving (7.21), we have for the FWHM's

$$2(s_x)_{1/2} = \frac{2.78}{\pi w}$$
$$2(s_z)_{1/2} = \frac{2.78}{\pi h} \tag{7.22}$$

Recall now that these $|\mathscr{F}[\sigma(\mathbf{r})]^2$ curves are convoluted into $\mathscr{F}[\rho_o(\mathbf{r})]$, which, for a "perfect" crystalline solid is set of delta functions arrayed in resp. Thus, each node in resp is broadened equally. If we now represent the intensity distribution about each node by the FWHM in each direction, the intensity distribution of a plane in resp appears as in Fig. 7.9. The relnodes are elongated in the direction parallel to the thin direction of the crystals. In general, the breadth of the relnode in any resp direction is inversely proportional to the average real space thickness in that direction. (This average is an area average over a plane perpendicular to the **s** chosen.)

Suppose we had an equiaxed particle, $w = h$. Then the relnode has breadth $2.78/(\pi h)$ in all directions. In order to translate this breadth into an experimental line broadening, we must convert $2s_{1/2}$ into a $\Delta(2\theta)$. As above, we note that $2s_{1/2} = \cos\theta\Delta(2\theta)/\lambda$. Substituting into (7.22), we find

$$\Delta(2\theta) = \frac{2.78\lambda}{\pi w \cos\theta} = \frac{0.88\lambda}{w\cos\theta} \tag{7.23}$$

This is the Scherrer formula, which is widely used for determining the average dimension of roughly equiaxed particles. In resp, the situation is as depicted in Fig. 7.10; a relnode of breadth $0.88/w$ subtends an arc of length $0.88/(w \cos\theta)$ on the Ewald sphere, and this is equivalent to an angular spread of $0.88\lambda/(w\cos\theta)$.

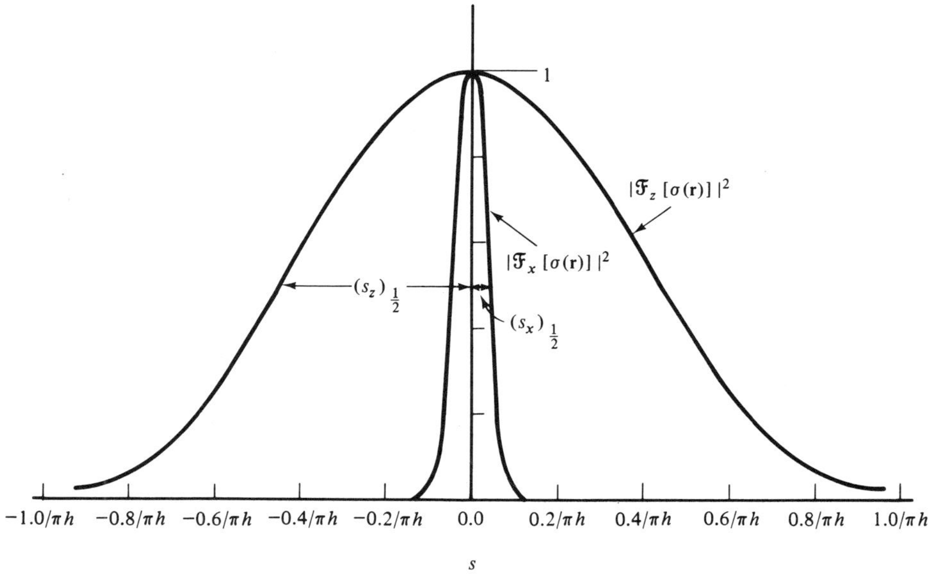

Figure 7.8. The shape factor plotted against s for scattering vectors $\mathbf{s}_x$ and $\mathbf{s}_z$, respectively, in the plane of the sheet and normal to it.

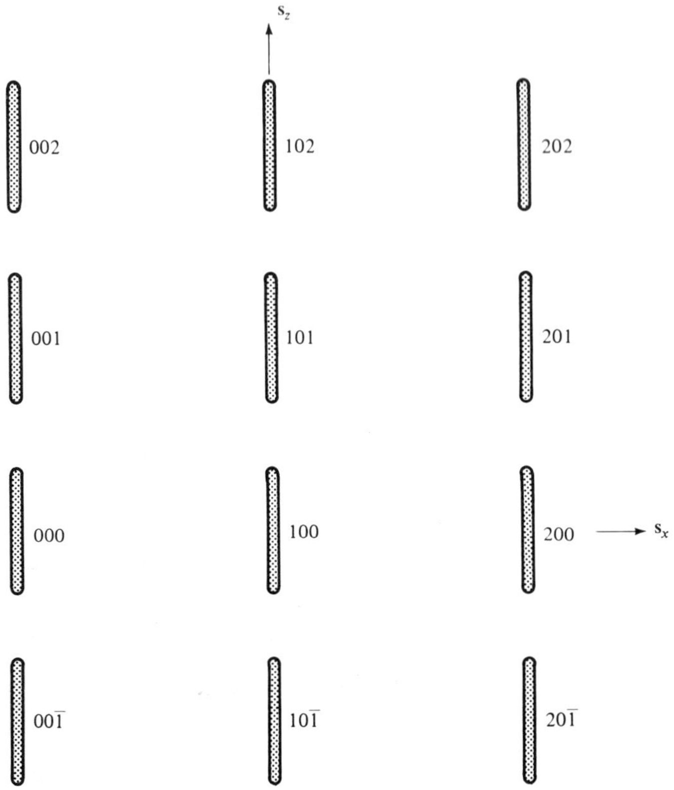

Figure 7.9. $\mathbf{s}_x\mathbf{s}_z$ plane in resp.

In order to get a feeling for the magnitude of the line broadening for different size equiaxed particles, Table 7.1 is presented. This table shows the value of $\Delta(2\theta)$ (the FWHM breadth) for CuK_α x-radiation ($\lambda = 1.54$ Å) and a Bragg angle of 45°. With a good deal of care, one can measure intensity with an angular accuracy of some 0.005°, but more commonly an accuracy of some 0.05° is considered excellent. Thus a reasonable upper limit to measurable particle size is several hundred angstroms.

While the absolute line broadening $\Delta(2\theta)$ depends on the wavelength used, the *relative* broadening is independent of λ. By relative broadening is meant the fraction of the total angle between diffraction lines or spots. This relative angle is given by the relative broadening of relnodes with respect to the resp spacing. The ratio $\Delta s/a^*$ of relnode breadth to rel spacing is 0.88 a/w. In changing from one wavelength to another, only the size of the Ewald sphere changes, and this has little effect on the relative angular broadening, as indicated in Fig. 7.11. There we see the breadths of the Ewald sphere intersections for Ewald spheres of different wavelengths. We observe that the ratio of breadth

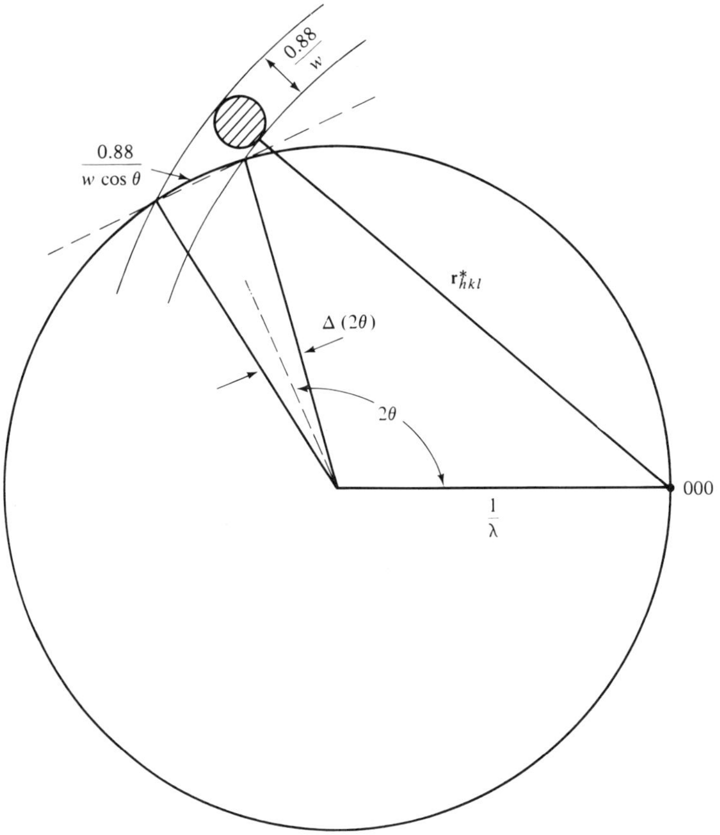

Figure 7.10. Resp geometry relating to the Scherrer formula.

TABLE 7.1
LINE BROADENING FOR AN EQUIAXED PARTICLE ($\lambda = 1.54$ Å; $\theta_B = 45°$)

w	$\Delta(2\theta)(°)$
0.1 cm	5.5×10^{-6}
0.01 cm	5.5 10^{-5}
0.001 cm	5.5 10^{-4}
1μ	5.5 10^{-3}
$0.1\mu = 1000$ Å	0.055
100 Å	0.55
10 Å	5.5

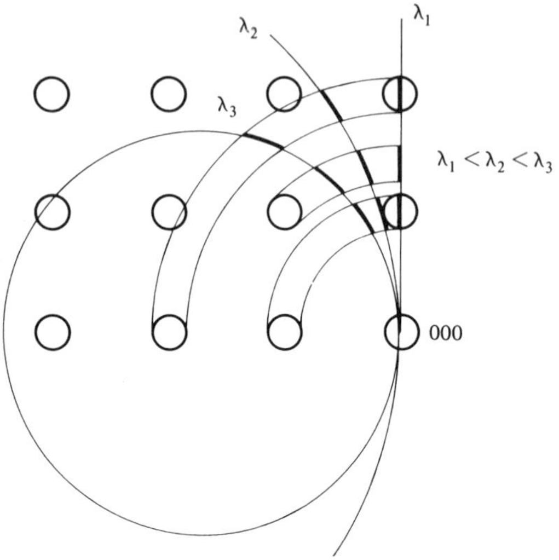

Figure 7.11. Description of the geometry leading to the weakness of the wavelength dependence of relative particle-size broadening.

of intersection on the Ewald sphere to the spacing between intersections is not much affected by the radius of the Ewald sphere. Thus relative line broadening in electron diffraction is similar to that in x-ray diffraction.

If the particle is not equiaxed, then the line broadening relates to a specific dimension of the particle—the dimension parallel to $\mathbf{r}^*_{hkl}$. This is seen in Fig. 7.12. Here the plane AA', perpendicular to $\mathbf{r}^*_{hkl}$ within the relnode, intersects the Ewald sphere at point B in a powder experiment. Therefore each point on the trace of the relnode on the Ewald sphere represents the intensity of a plane lying normal to $\mathbf{r}^*_{hkl}$. Thus the Ewald sphere samples intensity vectors parallel to $\mathbf{r}^*_{hkl}$. The intensity at each point on the Ewald sphere is then the integral of the function $\mathscr{F}[\sigma(\mathbf{r})]$ over a plane normal to $\mathbf{r}^*_{hkl}$ in the relnode. Let us

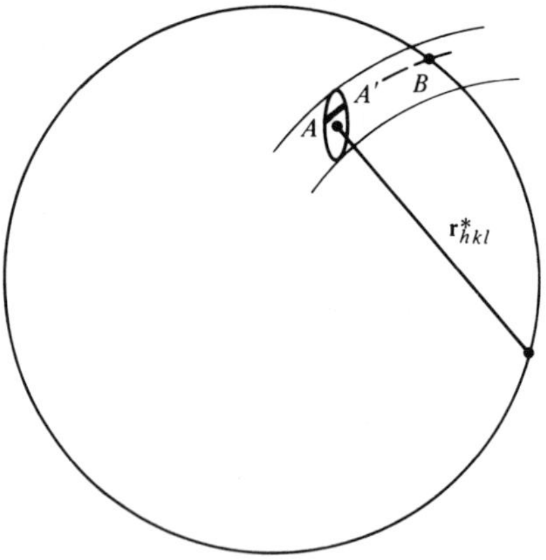

Figure 7.12. Dependence of the magnitude of particle-size broadening on the shape of the particles.

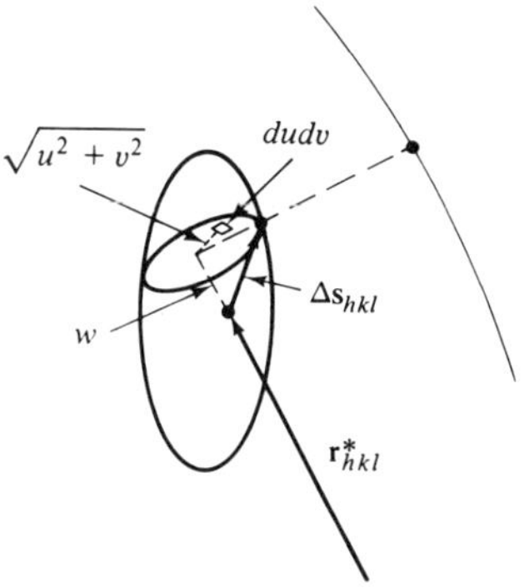

Figure 7.13. The separation of the vector $\mathbf{s}_{hkl}$ from the center of the relnode into components u and v normal to $\mathbf{r}^*_{hkl}$.

separate the vector $\Delta\mathbf{s}_{hkl}$ in the relnode to components u and v normal to $\mathbf{r}^*_{hkl}$ and w parallel to $\mathbf{r}^*_{hkl}$, as in Fig. 7.13. Then the intensity at an Ewald sphere point $\mathbf{r}^*_{hkl} + \mathbf{w}$ from 000 is, for a given value of w, $\int \mathscr{F}[\sigma(\mathbf{r})]du\,dv$.

A possible ambiguity exists regarding line broadening. We saw in the preceding chapter that lattice distortions can also lead to line broadening. However, such defect broadening does not affect all relnodes uniformly. The 000 node is unbroadened and the broadening becomes larger as $|\mathbf{r}^*_{hkl}|$ increases. If an observed line broadening is due solely to particle-size effects, the broadening must obey the Scherrer formula, $\Delta(2\theta) = 0.88\lambda/(w \cos \theta)$. If it is not known a priori whether the particles are equiaxed or not, it is best to test for compliance with the Scherrer relation, using several orders of the *same* reflection, as these will all affect $\Delta(2\theta)$ in the same way. If it is known that the particles are equiaxed, then all reflections can be used to test for compliance. It is not simple to decouple combined particle-size and distortion effects, unless one is a much larger effect than the other.

7.3 SMALL-ANGLE SCATTERING

7.3.1 The Nature of Small-Angle Scattering

When radiation is scattered from a scattering center, the wave travels outward in all directions, including forward. There is, then, a forward-scattered beam. In the forward direction, the paths of the rays are all identical and the interference is perfectly constructive. In directions slightly deviant from the forward direction, the paths of the rays scattered from the several centers differ and partial or complete destructive interference occurs. This situation is precisely that already described for particle-size broadening of an *hkl* diffraction node and precisely the same formalism applies.

There are, however, two important differences between particle-size broadening of an *hkl* node and small-angle scattering. First, in order to observe *hkl* broadening, one must have a crystalline solid. Scattering toward 000, however,

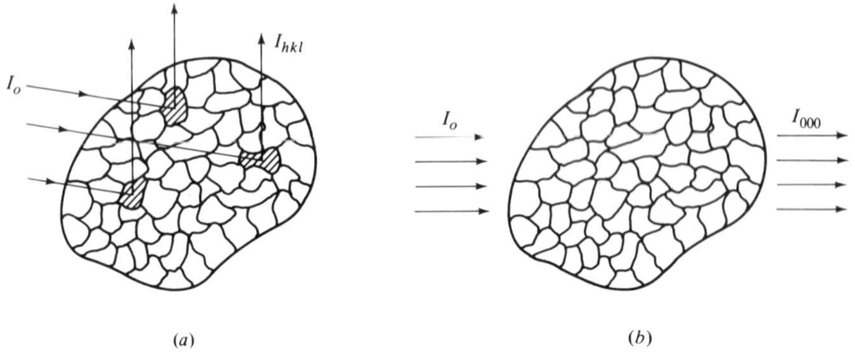

Figure 7.14. Illustration of the difference in the information given by *hkl* line broadening and small-angle scattering: *(a)* line broadening is sensitive to orientation and consequently would give the size of the component crystals in a small polycrystalline particle; *(b)* small-angle scattering is sensitive to density differences and consequently would measure the total diameter of such a polycrystalline particle.

is present for all materials—crystalline, amorphous, liquid, or gaseous. The method is in fact used to determine the size and shape of macromolecules in solution. Second, a density difference between the particle and its surroundings is required in order to observe small-angle scattering. This is illustrated in Fig. 7.14.

In Fig. 7.14*a* we note that only a few grains of a polycrystalline solid will be in position to reflect, for a given incident direction. Only those grains, acting independently, give rise to the *hkl* diffraction node. These grains are then the particles whose sizes and shapes are measured by particle-size broadening of the *hkl* reflection. In the forward direction (Fig. 7.14*b*), *all* grains simultaneously contribute to the scattering; their relative orientation is unimportant. The only means by which one region can be differentiated from another in their forward-directed scattering is through an electron density difference, since the electron density determines the magnitude of the scattered amplitude. This situation is shown in Fig. 7.15.

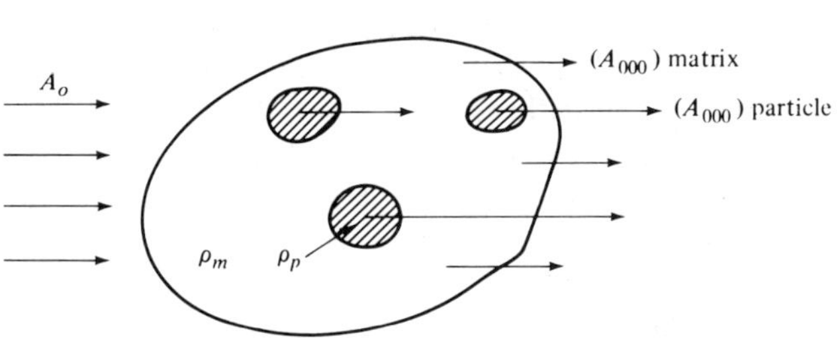

Figure 7.15. The situation determining the amplitude of scattering from a two-phase system. The amplitude of forward scattering from each phase is proportional to the density and volume of that phase.

Since scattering in the forward direction is not affected by defect effects, one can with more confidence attempt to measure the details of the small-angle scattering curve and to deduce structure information from this. The next sections indicate some of the information which can be found via a detailed analysis of small-angle scattering data.

Before going on to discuss the analyses of small-angle scattering curves, it is useful to mention two unusual, but frequent, uses of small-angle scattering. First, recall that x-ray scattering near the forward direction is dependent upon a density difference between regions of two (or more) kinds. In fact, one of the regions can be void. Small-angle scattering is therefore frequently used to study small voids in materials. Figure 7.16 shows a thermal neutron scattering curve due to voids in aluminum which had been exposed to high energy neutrons in a reactor. Second, think of the microstructure of an organic polymer melt. This melt consists of a collection of nearly randomly coiled carbon/hydrogen chains. The details of this microstructure had eluded scientists for years. However, it was finally realized that the dimensions of the individual coils could be measured using neutron small-angle scattering. The trick is to mix in a small fraction of chains in which deuterium has been substituted for hydrogen. From Appendix Table B.4 we note that the neutron-scattering factor for hydro-

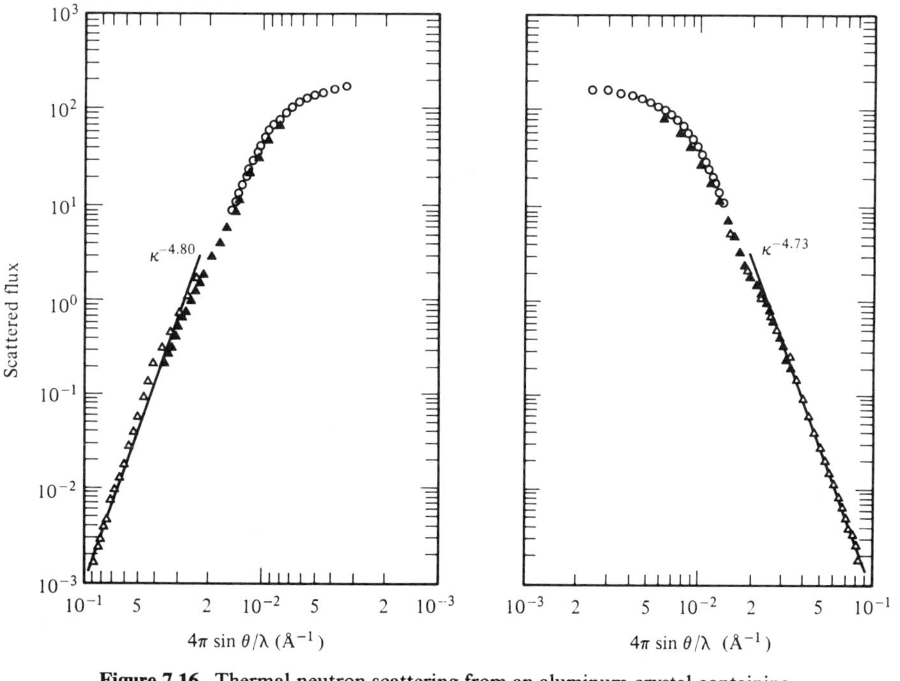

Figure 7.16. Thermal neutron scattering from an aluminum crystal containing radiation-induced small voids. The two sides of this figure represent scattering in the positive and negative directions about zero degrees. [R. W. Hendricks, J. Schelten, and W. Schmatz, *Phil. Mag.*, **30**, 819 (1974)]

gen is -0.38×10^{-12} cm, whereas that of deuterium is $+0.65 \times 10^{-12}$ cm. Thus the deuterium-containing chains will scatter neutrons significantly differently from the matrix. Figure 7.17 shows neutron small-angle scattering from hydrogen-bearing polyethylene chains in an otherwise deuterated polyethylene melt.

7.3.2 The Guinier Approximation

The Guinier approximation provides a relation that allows small-angle scattering data to be used to evaluate particle sizes. The relation is

$$\frac{I(\mathbf{s})}{I_e(s)} \propto \exp\left(-\frac{4}{3}\pi^2 R_g^2 s^2\right) \tag{7.24}$$

where R_g is the radius of gyration of the particles. This relation is nearly exact for a very dilute suspension of spherical particles and becomes more and more inaccurate as the suspension becomes denser and the particles less equiaxed.

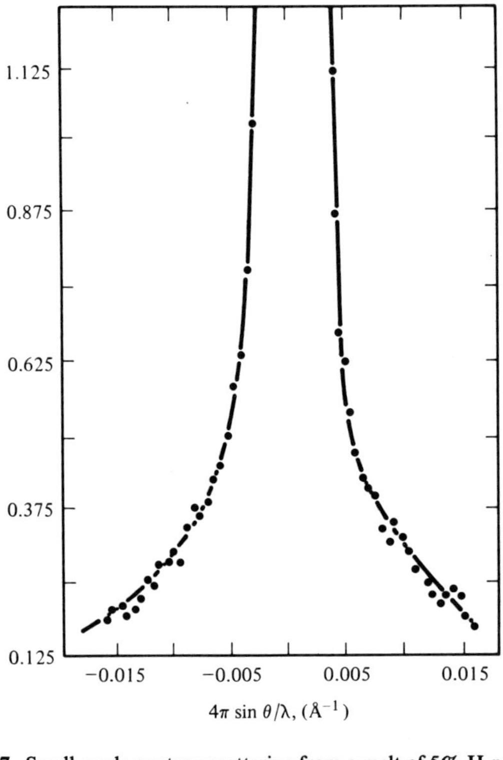

Figure 7.17. Small-angle neutron scattering from a melt of 5% H-polyethylene in a D-polyethylene matrix. (J. Schelten, G. D. Wignall, and D. G. H. Ballard, *Polymer,* **15,** 682 (1974), by permission of the publishers, IPC Business Press, Ltd. ©)

We shall look at the derivation, first for a dilute suspension of spherical particles.

If the suspension is sufficiently dilute, the scattering from the several particles do not mutually interfere. Thus we need only consider scattering by one sphere; later we can multiply by the number N of spheres per unit volume to get the scattering per unit volume.

We must also observe that *interatomic*-level effects show up at tens of degrees and have no effect in the small-angle regime. Thus we may (rigorously, it can be shown) replace the real electron density $\rho(x)$ within the sphere by the average density ρ_1. Now suppose that the suspension matrix has an average electron density ρ_2. For a whole space containing only one sphere of radius R, the amplitude of scattering is given by

$$\begin{aligned}\frac{A_1(\mathbf{s})}{A_e(s)} &= \rho_2 \int_{-\infty}^{\infty} e^{-2\pi i \mathbf{s}\cdot\mathbf{r}}\, dv_{\mathbf{r}} \\ &- \rho_2 \int_0^{2\pi}\int_0^{\pi}\int_0^{R} e^{2\pi i \mathbf{s}\cdot\mathbf{r}}\, r^2 \sin\phi\, dr\, d\phi\, d\psi \\ &+ \rho_1 \int_0^{2\pi}\int_0^{\pi}\int_0^{R} e^{2\pi i \mathbf{s}\cdot\mathbf{r}} \sin\phi\, dr\, d\phi\, d\psi\end{aligned}$$

Here the first term on the right represents all of space with an average density ρ_2. In the second term we take out a spherical hole, and in the third we fill it to a density ρ_1. The first term above is a delta function at the origin and cannot be observed. Thus

$$\frac{A_1(\mathbf{s})}{A_e(s)} = (\rho_1 - \rho_2) \int_0^{2\pi}\int_0^{\pi}\int_0^{R} e^{2\pi i \mathbf{s}\cdot\mathbf{r}}\, r^2 \sin\phi\, dr\, d\phi\, d\psi \tag{7.25}$$

Integrating over ψ and ϕ, we have

$$\frac{A_1(\mathbf{s})}{A_e(s)} = 4\pi(\rho_1 - \rho_2) \int_0^{R} r^2 \frac{\sin 2\pi sr}{2\pi sr}\, dr \tag{7.26}$$

Let us now assume that $2\pi sr << 1$. We shall justify this later. Then

$$\frac{\sin 2\pi sr}{2\pi sr} \simeq 1 - \frac{(2\pi sr)^2}{3!} \tag{7.27}$$

and

$$\begin{aligned}\frac{A_1(\mathbf{s})}{A_e(s)} &= 4\pi(\rho_1 - \rho_2) \int_0^{R} \left(r^2 - \frac{4\pi^2 s^2 r^4}{6}\right) dr \\ &= \frac{4}{3}\pi(\rho_1 - \rho_2) R^3 \left(1 - \frac{2\pi^2 s^2 R^2}{5}\right)\end{aligned} \tag{7.28}$$

For small arguments, $e^{-x} \simeq 1 - x$. Therefore

$$\frac{A_1(\mathbf{s})}{A_e(s)} \simeq \frac{4}{3}\pi R^3(\rho_1 - \rho_2)\exp(-2\pi^2 s^2 R^2/5) \tag{7.29}$$

The intensity of scattering by this single particle is

$$\frac{I_1(\mathbf{s})}{I_e(s)} = \left(\frac{4}{3}\pi R^3\right)^2 (\rho_1 - \rho_2)^2 \exp(-4\pi^2 s^2 R^2/5) \tag{7.30}$$

For N particles per unit volume, $I_1(\mathbf{s})/I_e(s)$ would be multiplied by N and

$$\begin{aligned}\frac{I(\mathbf{s})}{I_e(s)} &= N\left(\frac{4}{3}\pi R^3\right)^2 (\rho_1 - \rho_2)^2 \exp(-4\pi^2 s^2 R^2/5) \\ &= N V_p^2 (\rho_1 - \rho_2)^2 \exp(-4\pi^2 s^2 R^2/5)\end{aligned} \tag{7.31}$$

where V_p is the volume of the particle.

An alternate expression of Guinier's law is obtained by squaring (7.28). Subsequent inversion of both sides of the equation and use of the small argument approximation $1/(1-\epsilon) \simeq 1 + \epsilon$ leads to

$$I^{-1}(s) \propto [N V_p^2 (\rho_1 - \rho_2)^2]^{-1}(1 + 4\pi^2 s^2 R^2/5) \tag{7.32}$$

Particle sizes are therefore found by measuring $I(\mathbf{s})/I_e(s)$ in the small angle regime and plotting the data appropriately. At small angles $I_e(s)$ is essentially independent of s. Therefore one can write

$$\log I(s) = \log(\text{constant}) - \frac{4\pi^2}{(2.303)(5)} R^2 s^2 \tag{7.33}$$

The slope of a plot of $\log I(s)$ against s^2 must be

$$\text{Slope} = -\frac{4\pi^2 R^2}{(2.303)(5)} \tag{7.34}$$

and the sphere's radius is found from this slope.

Figure 7.18 is a set of small-angle x-ray scattering data from spherical metal (catalyst) particles in a ceramic matrix. These data are plotted according to Guinier's law, $\log I(s)$ against $(2\theta)^2$. A straight line results, as predicted. Likewise, Fig. 7.19 is a Guinier plot of the molten polyethylene small-angle neutron scattering data of Fig. 7.17.

The radius of gyration of a sphere of radius R is

$$R_g = \sqrt{\frac{3}{5}}\, R \tag{7.35}$$

Substituting this into (7.31), we have

$$\frac{I(\mathbf{s})}{I_e(s)} = N(4\pi R^3)^2(\rho_1 - \rho_2)^2\, e^{-\frac{4}{3}\pi^2 R_g^2 s^2} \tag{7.36}$$

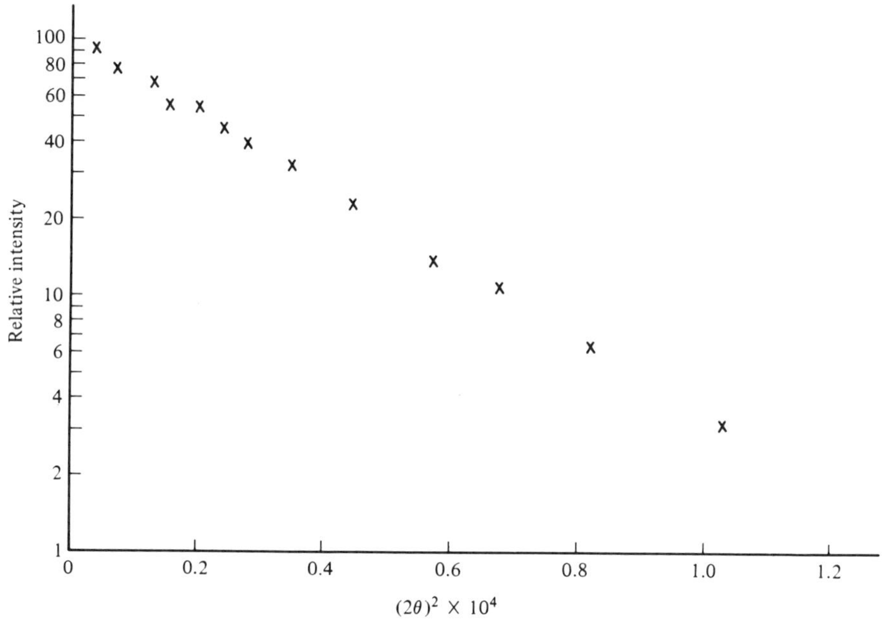

Figure 7.18. Guinier plot of small-angle x-ray scattering from noble metal catalyst particles in a ceramic matrix (support).

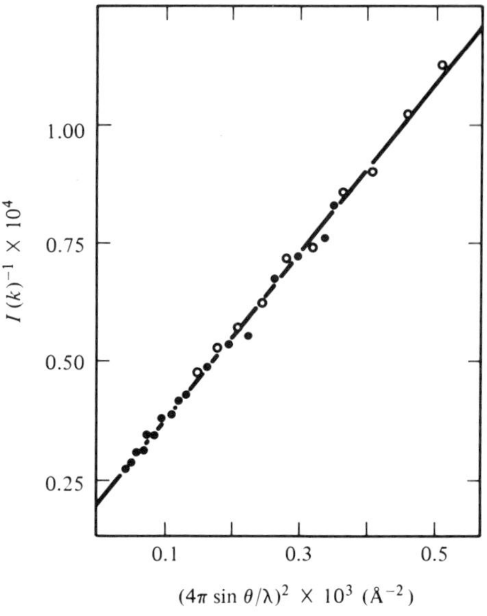

Figure 7.19. Guinier plot for 5% H-polyethylene molecules in a D-polyethylene melt. Open and closed circles denote scattering on both sides of $s = 0$. (J. Schelten, G. D. Wignall, and D. G. H. Ballard, *Polymer,* **15,** 682 (1974), by permission of the publishers, IPC Business Press, Ltd. ©)

To the first approximation this form is applicable to shapes other than spheres. A discussion of the applicability of this type of analysis to other shapes and to *distributions* of sizes can be found in the book by Guinier and Fournet.[1]

In the Guinier formula the intensity is scaled according to $(\rho_1 - \rho_2)^2$. The same scattered intensity will occur for ρ_1 as matrix or as particle, so long as the other phase is of electron density ρ_2. Thus electron deficient phases (or voids!) in more electron-dense materials can be very efficient scatterers, just as can heavy particles in electron-sparse materials.

In some cases, a Lorentz correction must be made in order to extract dimensional information from small-angle scattering data. These are the cases in which a set of anisotropic particles is distributed with some range of orientations. As before, the Lorentz correction is used to correct for our sampling of the orientation distribution. As an example, let us look at the scattering from platelets. First recall that platelets will elongate relnodes parallel to the thin direction of the platelets. For a situation in which the platelets are all similarly oriented, resp will appear as in Fig. 7.9. The distribution of intensity in the *z*-direction goes as $\sin^2 (\pi h s_z)/(\pi h s_z)^2$. This expression can be converted to a Guinier format by using the small-argument approximation, $\sin \epsilon = \epsilon - \epsilon^3/3!$. In this case

$$|\mathscr{F}_z[\sigma(\mathbf{r})]|^2 \simeq V^2\left[1 - \frac{(\pi h s_z)^2}{3}\right] \simeq V^2 e^{-\frac{\pi^2 h^2 s_z^2}{3}} \tag{7.37}$$

Guinier-Preston zones (platelike clusters) in alloy single crystals give such scattering. Figure 7.20 shows a small-angle x-ray pattern from such a system. Here the thin copper-rich platelets lie on {100} planes. In this experiment, [010] is parallel to the incident beam. The streaks along the *x*- and *y*-directions of the film are from platelets lying on (100) and (001). Suppose now that there were a collection of thin platelets, but the platelets were oriented randomly in the x-ray beam. In this case the relrods are rotated uniformly about the center

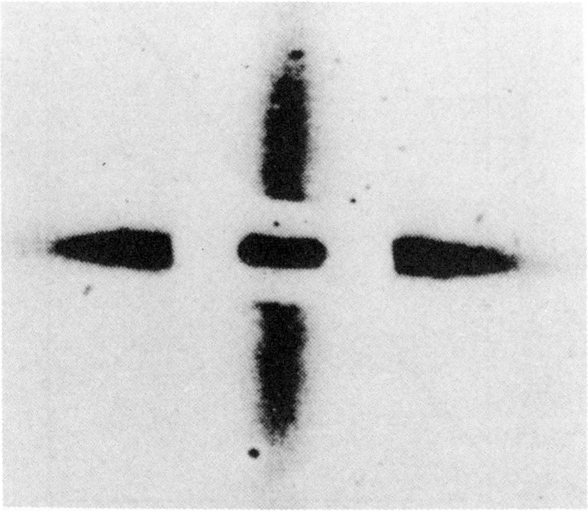

Figure 7.20. Small-angle x-ray scattering pattern from a single crystal of a cubic 96 Al-4Cu alloy. The [100] axis is parallel to the incident beam and normal to the film and [010] is vertical. MoK_α radiation was used, with a 4 cm specimen-to-film distance. [A. Guinier and G. Fournet, *Small-Angle Scattering of X-rays* (John Wiley & Sons, NY, 1955)]

[1] A. Guinier and G. Fournet, *Small-Angle Scattering of X-rays* (John Wiley & Sons, NY, 1955).

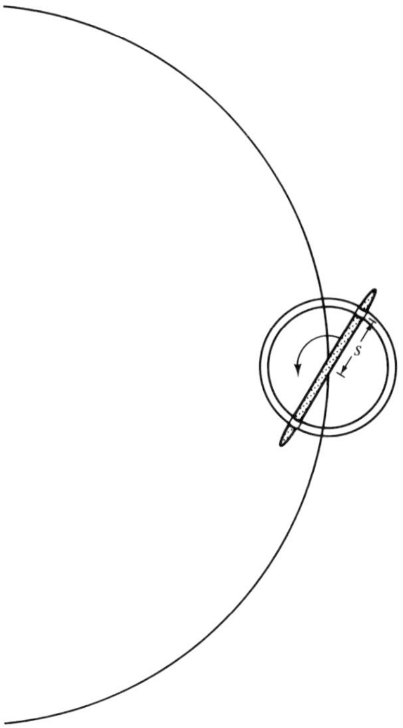

Figure 7.21. Relrods representing scattering from thin plates assume all orientations in resp.

of 000, as indicated in Fig. 7.21. Here a volume element distant s from 000 is distributed uniformly about a spherical shell. The Ewald sphere samples only a fraction of this shell. It was shown in Chap. 3 that such sampling resulted in a correction factor of $\lambda^3/(\sin\ \theta\ \sin\ 2\theta)$. In the small-angle regime, $\sin\theta \sin 2\theta \simeq \lambda^2 s^2/2$. For this random arrangement of plates, then, the measured intensity becomes

$$I(s) \propto s^{-2}\ e^{-\pi^2 h^2 s^2/3} \tag{7.38}$$

Taking logarithms

$$\log\,[s^2 I(s)] = \text{Constant} - \frac{\pi^2 h^2}{(3)(2.303)}\, s^2 \tag{7.39}$$

In this case, the intensity is multiplied by s^2 prior to plotting semilogarithmically. Figure 7.22 is a set of small-angle x-ray scattering curves taken at different times during the crystallization of a randomly oriented set of platelike polyethylene crystallites. The data are plotted according to (7.39). The unvarying slope of the curves shows that the crystal growth occurred at constant crystal thickness.

If a collection of plates exhibited cylindrical symmetry, then an element of the relrod of Fig. 7.22 intersects the Ewald sphere in only two places, not over a continuous ring. In this case, the sampling of orientations gives rise to

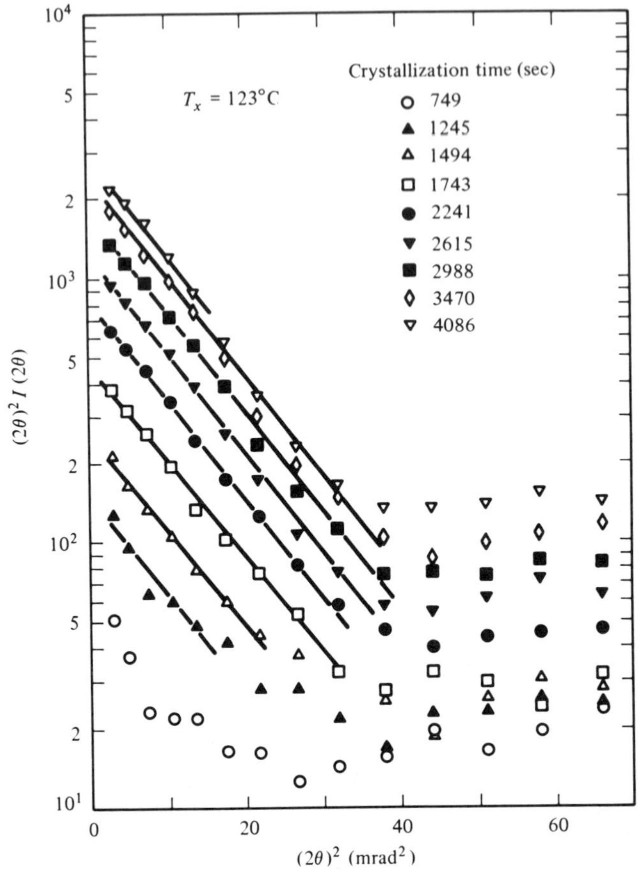

Figure 7.22. Small-angle x-ray scattering Guinier plots from scans taken during the isothermal crystallization of polyethylene from the melt at 123°C. Crystallization times associated with each scan are given in the figure legend. [J. M. Schultz, J. S. Lin, and R. W. Hendricks, *J. Appl. Cryst.*, **11,** 551 (1978)]

TABLE 7.2
INTENSITY RELATIONSHIPS FOR SPHERICALLY RANDOM DISTRIBUTIONS

Shape	$V^{-2}\lvert\mathscr{F}[\sigma(r)]\rvert^2$
Sphere of radius R	$\phi^2(2\pi sR) = \left[3\,\dfrac{\sin(2\pi sR) - 2\pi sR\cos(2\pi sR)}{(2\pi sR)^3}\right]^2$
Ellipsoid of revolution, with axes $2a$, $2a$, and $2\beta a$	$\displaystyle\int_0^{\pi/2} \phi^2(2\pi sa\sqrt{\cos^2\epsilon + \beta^2\sin^2\epsilon})\cos\epsilon\, d\epsilon$
Thin disks of thickness h and diameter $2R$	$\dfrac{2}{(2\pi sR)^2}\,\dfrac{\sin^2(\pi sh)}{(\pi sh)^2}$

an s^{-1} Lorentz correction, and the correct form of a Guinier plot would be log $[sI(s)]$ versus s^2.

Platelets and rods turn out to be the simplest cases. Many other shapes can be approximated by ellipsoids. Spherical averaging (Lorentz correcting) over random orientations of rods and ellipsoids has been computed. The results are tabulated in Table 7.2.

7.3.3 The Zero-Angle Extrapolation

The general form for the scattering amplitude for a single particle is given by

$$\frac{A(\mathbf{s})}{A_e(s)} = (\rho_1 - \rho_2) \int_{\substack{\text{one} \\ \text{particle}}} e^{2\pi i \mathbf{s}\cdot\mathbf{r}}\, dv_{\mathbf{r}} \tag{7.40}$$

At $\mathbf{s} = 0$ (zero angle)

$$\frac{A(0)}{A_e(0)} = (\rho_1 - \rho_2)\, V_p \tag{7.41}$$

where V_p is the volume per particle. If N_p represents the number of particles in one cubic centimeter, then $N_p V_p$ is the volume fraction V_f of particles. Hence

$$\frac{I(0)}{I_e(0)} = N_p(\rho_1 - \rho_2)^2 V_p^2 = V_f V_p(\rho_1 - \rho_2)^2 \tag{7.42}$$

Thus the extrapolation of a plot of, say, log $[I(\mathbf{s})/I_e(s)]$ versus s^2 to zero angle provides $I(0)/I_e(0)$ and this quantity gives $V_f V_p(\rho_1 - \rho_2)^2$. V_f can be measured by other means, for example, by dilatometry, and the densities ρ_1, ρ_2 are usually known. Thus V_p is found from the extrapolation and can be used as a crosscheck on the particle dimension obtained from the slope.

We must observe, however, that here we need a good measurement of $I_e(0)$. Since $I_e(0) = r_e^2 I_o$, this is tantamount to a measurement of the primary beam. Generally the primary beam is too intense to be measured directly; the number of photons per unit area per second is too great for the photon counting electronics. The obtaining of I_o turns out to be very difficult and usually introduces errors in excess of 10%.

7.3.4 The Porod Region

Another type of useful information can be extracted from the tails of a small-angle curve. To see this, we begin with the example of scattering by a sphere. The form of the intensity distribution from scattering by spheres is given in (7.8). For small-angle scattering this becomes

$$\frac{I(s)}{I_e(s)} = N[(\rho_1 - \rho_2)^2/(8\pi^3)]\left[\frac{4\pi R^2}{s^4} + \frac{1}{\pi s^6} - \frac{4R}{s^5}\sin 4\pi Rs + \left(\frac{4\pi R^2}{s^4} - \frac{1}{\pi s^6}\right)\cos 4\pi Rs\right] \tag{7.43}$$

where N is the number of particles. In the tails of the intensity curve, the values of s are large enough that the s^{-5} and s^{-6} terms can be ignored. The cosine term will also be near zero in this regime. Thus

$$\frac{I(s)}{I_e(s)} \simeq [(\rho_1 - \rho_2)^2/(16\pi^3)]\frac{A}{s^4} \tag{7.44}$$

where $A = N \cdot 4\pi R^2$ is the total surface area of the particles. This total surface is obtained from plots of $I(s)/I_e(s)$ versus s^{-4}. The slopes of the straight lines are $(\rho_1 - \rho_2)^2 A/(16\pi^3)$.

Figure 7.23 shows a set of log (Intensity) versus log $(2\theta)^2$ curves for several alumina catalysts. The curves each have a slope of -2, as predicted by (7.44). The specific area A was found for these specimens by nitrogen adsorption, as well as by x-ray small-angle scattering. The agreement, shown in Fig. 7.24, is very good.

7.3.5 The Small-Angle Invariant

A very useful quantity can be determined by integration of the small-angle scattering intensity over resp. For a two-phase system, this quantity, called the invariant, Q, is related to the volume fractions ϕ_1, $(1 - \phi_1)$ of the phases present and the densities ρ_1 and ρ_2 of the phases by

$$Q = \int [I(s)/I_e(s)]\,dv_s = V^2\phi_1(1 - \phi_1)(\rho_1 - \rho_2)^2 \tag{7.45}$$

We shall derive this relation below. What is to be noted here is that this measurement enables one to determine either ϕ_1 or $(\rho_1 - \rho_2)$, if the other is known. Generally, $\rho_1 - \rho_2$ is known or can be estimated and thus measurement of the invariant permits a determination of the volume fraction of the dispersed phase.

The above form of the invariant is derived as follows. We begin by writing Q in Fourier transform notation

$$Q = \iiint \rho(\mathbf{r})\rho(\mathbf{r} + \mathbf{u})e^{2\pi i\mathbf{s}\cdot\mathbf{u}}\,dv_{\mathbf{r}}\,dv_{\mathbf{u}}\,dv_{\mathbf{s}} \tag{7.46}$$

Writing the $\rho(\mathbf{r})$ in terms of deviations $\eta(\mathbf{r})$ from the average electron density ρ_o we have

$$Q = \iiint [\rho_o + \eta(\mathbf{r})][\rho_o + \eta(\mathbf{r}+\mathbf{u})] \exp(2\pi i \mathbf{s} \cdot \mathbf{u}) dv_{\mathbf{r}}\, dv_{\mathbf{u}}\, dv_{\mathbf{s}} \tag{7.47}$$

Expanding, the invariant becomes

$$\begin{aligned} Q = \rho_o^2 \iiint e^{2\pi i \mathbf{s}\cdot\mathbf{u}}\, dv_{\mathbf{r}}\, dv_{\mathbf{u}}\, dv_{\mathbf{s}} &+ \rho_o \iint \left[\int \eta(\mathbf{r}) dv_{\mathbf{r}}\right] e^{2\pi i \mathbf{s}\cdot\mathbf{u}} dv_{\mathbf{u}} dv_{\mathbf{s}} \\ &+ \rho_o \iint \left[\int \eta(\mathbf{r}+\mathbf{u}) dv_r\right] e^{2\pi i \mathbf{s}\cdot\mathbf{u}}\, dv_{\mathbf{u}} dv_{\mathbf{s}} \\ &+ \iiint \eta(\mathbf{r})\eta(\mathbf{r}+\mathbf{u}) e^{2\pi i \mathbf{s}\cdot\mathbf{u}} dv_{\mathbf{r}} dv_{\mathbf{u}} dv_{\mathbf{s}} \end{aligned} \tag{7.48}$$

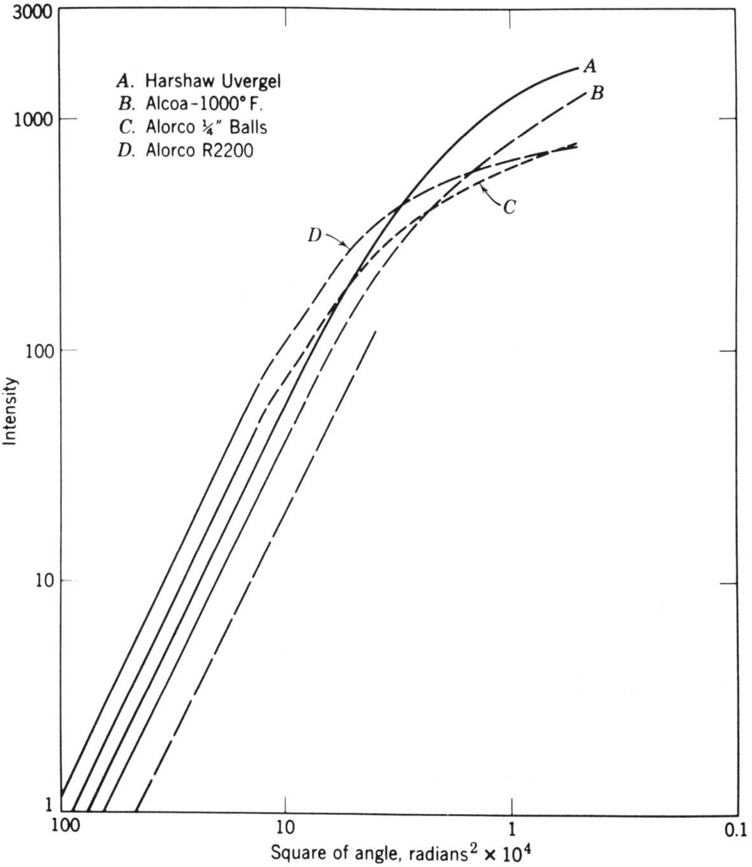

Figure 7.23. Log *I* versus log $(2\pi s)^2$ for different samples of alumina catalysts. The dashed line shows the theoretical $(2\pi s)^{-4}$ dependence. [R. A. Van Nordstrand and K. M. Hach, cited in A. Guinier and G. Fournet, *Small-Angle Scattering of X-rays* (John Wiley & Sons, NY, 1955)]

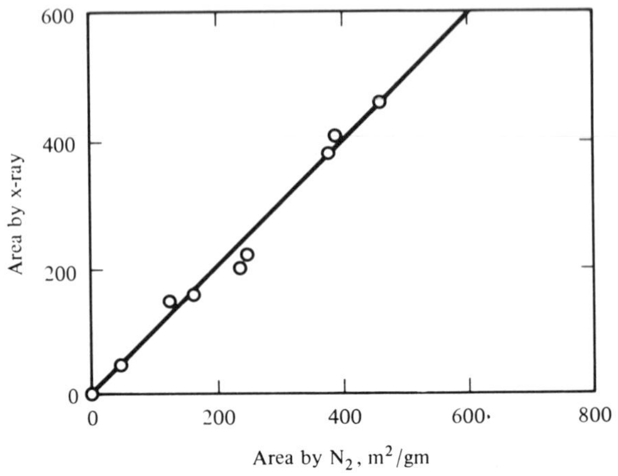

Figure 7.24. Specific surface areas as measured by x-rays and by nitrogen adsorption. [R. A. Van Nordstrand and K. M. Hach, cited in A. Guinier and G. Fournet, *Small-Angle Scattering of X-rays* (John Wiley & Sons, NY, 1955)]

The first term on the right is a delta function at the origin of resp and cannot be measured. Since the average value of $\eta(\mathbf{r})$ is zero, the integrals over $\mathbf{r}$ of the second and third terms are zero. We are left with

$$Q = \iiint \eta(\mathbf{r})\eta(\mathbf{r}+\mathbf{u}) \exp(2\pi i \mathbf{s}\cdot\mathbf{u})\, dv_{\mathbf{r}}\, dv_{\mathbf{u}}\, dv_{\mathbf{s}} \tag{7.49}$$

Integration over $\mathbf{s}$ yields a delta function at $\mathbf{u} = 0$

$$Q = \iint \eta(\mathbf{r})\eta(\mathbf{r}+\mathbf{u})\delta(\mathbf{u})dv_{\mathbf{r}}\, dv_{\mathbf{u}} \tag{7.50}$$

Subsequent integration over u gives

$$Q = \int \eta^2(\mathbf{r})dv_{\mathbf{r}}$$

This can be rewritten as

$$Q = V\overline{\eta^2(\mathbf{r})} \tag{7.51}$$

This is a fundamental equation and holds whatever the nature of the material.

For a two-phase material the average of $\eta^2(\mathbf{r})$ can now be written in terms of ϕ_1, ρ_1, ρ_2, and the average electron density ρ_o

$$Q/V = \overline{\eta^2(\mathbf{r})} = \phi_1(\rho_1 - \rho_o)^2 + (1-\phi_1)(\rho_o - \rho_2)^2 \tag{7.52}$$

Here ρ_o is related to the problem variables via

$$\rho_o = \phi_1\rho_1 + (1-\phi_1)\rho_2 \tag{7.53}$$

Substituting, we have

$$\frac{Q}{V} = \phi_1[\rho_1 - \phi_1\rho_1 - (1-\phi_1)\rho_2]^2 + (1-\phi_1)[\phi_1\rho_1 + (1-\phi_1)\rho_2 - \rho_2]^2$$
$$= \phi_1(1-\phi_1)(\rho_1-\rho_2)^2 \tag{7.54}$$

Note that this expression was evolved with no assumptions other than that only two phases exist and that those phases are each homogeneous. The small-angle invariant can be used to find ϕ_1 if $\rho_1 - \rho_2$ is known or $\rho_1 - \rho_2$ if ϕ_1 is known. Usually the densities of the phases are known and one seeks the relative amounts of the phases.

As an example of the use of the small-angle invariant, Fig. 7.25 is derived from the data of Fig. 7.22. We see here the change of Q during the isothermal crystallization of linear polyethylene. From such data, one can compute the transformation isotherm, ϕ_1 versus time.

7.3.6 Fourier Analysis

In general, any pattern of density variation can be represented as a Fourier series

$$\rho(\mathbf{r}) = \Sigma B(\mathbf{K}) \cos 2\pi\mathbf{K}\cdot\mathbf{r} \tag{7.55}$$

where $B(\mathbf{K})$ is the amplitude of the density wave of wavelength K^{-1}. These Fourier components $B(\mathbf{K})$ are related directly to the experimentally determined amplitudes. More specifically, we shall see that

$$VB(\mathbf{s}) = \left[\frac{I(\mathbf{s})}{I_e(s)}\right]^{1/2} \tag{7.56}$$

where V is the irradiated volume. One can sometimes use (7.56) to qualitatively interpret a set of data. An example of this use will be shown below.

We derive (7.56) as follows. In general, the intensity is written

$$\frac{I(\mathbf{s})}{I_e(s)} = \iint \rho(\mathbf{r})\rho(\mathbf{r}+\mathbf{u})\, e^{2\pi i\mathbf{s}\cdot\mathbf{u}}\, dv_{\mathbf{r}}\, dv_{\mathbf{u}} \tag{7.57}$$

Inserting (7.55) in (7.57), we have

$$\frac{I(\mathbf{s})}{I_e(s)} = \iiiint\limits_{\mathbf{LKur}} B(\mathbf{K})B(\mathbf{L})[\cos 2\pi\mathbf{K}\cdot\mathbf{r}]\,[\cos 2\pi\mathbf{L}\cdot(\mathbf{r}+\mathbf{u})]e^{2\pi i\mathbf{s}\cdot\mathbf{u}}\; dv_{\mathbf{r}}dv_{\mathbf{u}}\, dv_{\mathbf{K}}dv_{\mathbf{L}} \tag{7.58}$$

Integration over $\mathbf{u}$ yields

$$\frac{I(\mathbf{s})}{I_e(s)} = \frac{1}{2}V\iiint\limits_{\mathbf{LKr}} B(\mathbf{K})B(\mathbf{L})[e^{2\pi i\mathbf{L}\cdot\mathbf{r}}\,\delta(\mathbf{s}+\mathbf{L}) + e^{-2\pi i\mathbf{L}\cdot\mathbf{r}}\,\delta(\mathbf{s}-\mathbf{L})] \times \cos 2\pi\mathbf{K}\cdot\mathbf{r}\; dv_{\mathbf{r}}\, dv_{\mathbf{K}}\, dv_{\mathbf{L}} \tag{7.59}$$

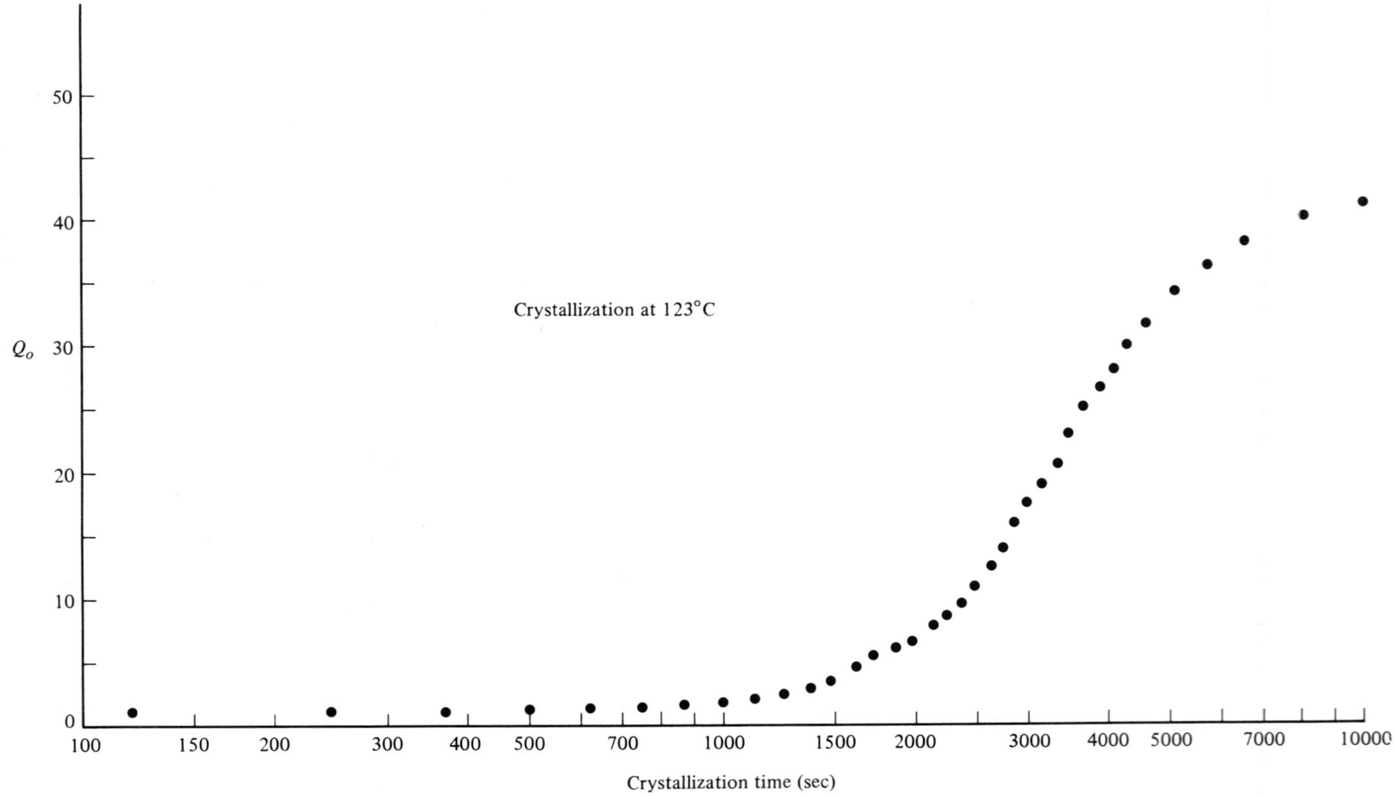

Figure 7.25. Change in the integrated intensity of a polyethylene specimen during crystallization at 123°C.

Subsequent integration over **r** produces

$$\frac{I(\mathbf{s})}{I_e(s)} = \frac{V^2}{4} \iint_{\mathbf{KL}} B(\mathbf{K})B(\mathbf{L})\{[\delta(\mathbf{L}+\mathbf{K}) + \delta(\mathbf{L}-\mathbf{K})][\delta(\mathbf{s}+\mathbf{L}) + \delta(\mathbf{s}-\mathbf{L})]dv_{\mathbf{K}}\, dv_{\mathbf{L}}$$

$$= \frac{V^2}{4}[B(-\mathbf{s})B(\mathbf{s}) + B(-\mathbf{s})B(-\mathbf{s}) + B(\mathbf{s})B(-\mathbf{s}) + B(\mathbf{s})B(\mathbf{s})] \tag{7.60}$$

It is difficult to envision a long-range density pattern that is not centrosymmetric. For a centrosymmetric microstructure, $B(\mathbf{s}) = B(-\mathbf{s})$ and, finally, rearranging (7.56)

$$\frac{I(\mathbf{s})}{I_e(s)} = V^2 B^2(\mathbf{s}) \tag{7.61}$$

This relation between Fourier density components $B(\mathbf{s})$ and the scattered intensity shows up particularly clearly in systems undergoing spinodal decomposition. Spinodal decomposition is a mode of phase transformation where the initial phase becomes unstable to density fluctuations. Fourier density waves above a critical wavelength grow continuously. A maximum growth rate is predicted for waves of a specific wave number $K_{\max}$. The dominant microstructure is then a density wave of that wave number. Since there is only one dominant wave number (Fourier component), the small-angle scattering pattern should exhibit only spikes at $\pm K$. These spikes should show no higher orders. Figures 7.26 and 7.27 show small-angle scattering curves taken at several times during the spinodal transformation of an Al-Zn alloy (Fig. 7.26) and during the heat treatment of a film of polypropylene which had solidified during rapid drawing from the melt (Fig. 7.27). In the latter case, it is suggested that the density wave represents a redistribution of defects along the axis of fibrillar crystals.

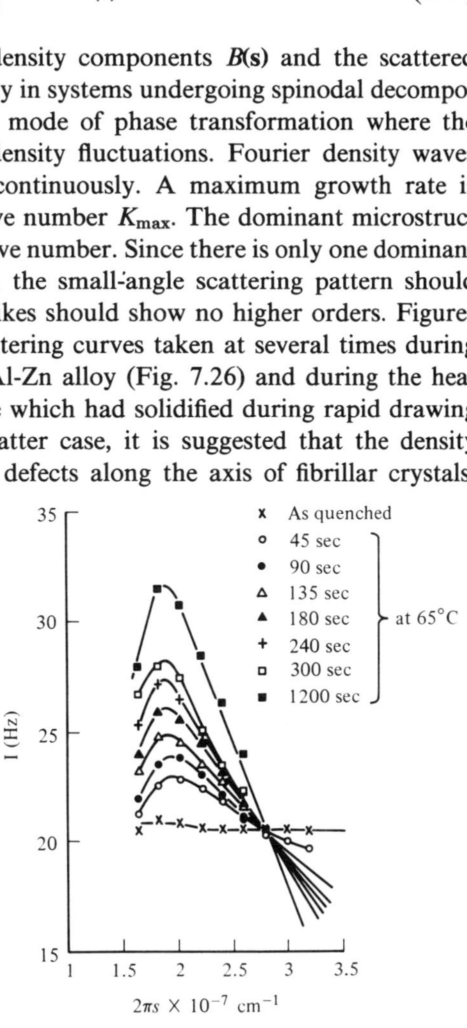

Figure 7.26. Intensity versus $2\pi s$ for 78Al/22Zn film quenched from 150°C and subsequently annealed at 65°C for the times indicated. [S. Agarwal and H. Herman, *Scripta Met.*, **7**, 503 (1973)]

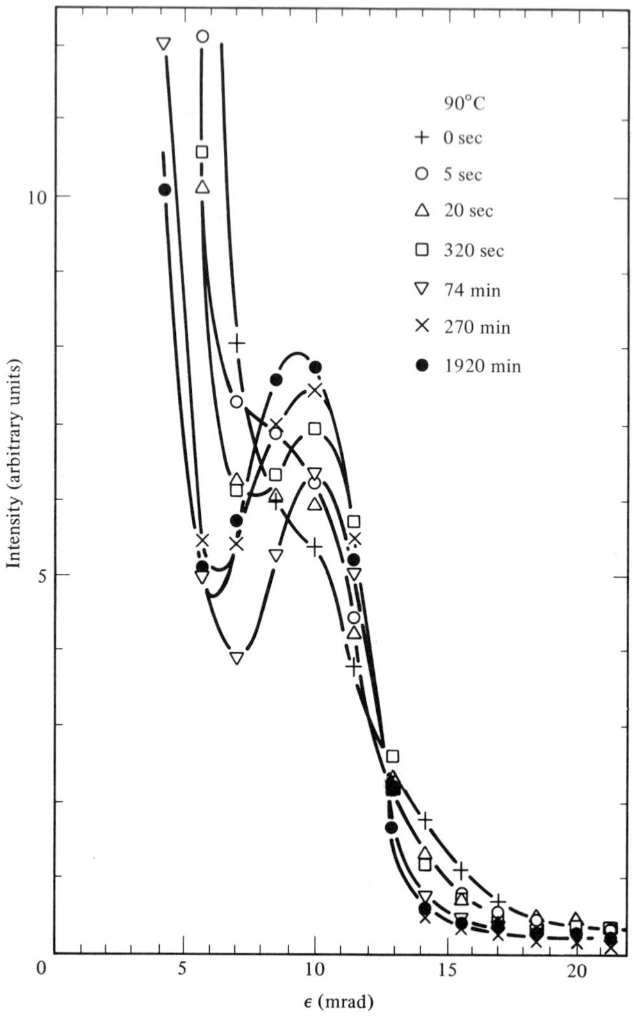

Figure 7.27. Intensity versus 2θ for an isotactic polypropylene film which had been crystallized during the extension of a melt at a strain rate in excess of $10^4 s^{-1}$. The film was subsequently heat-treated at 90°C for the times indicated. The data was taken at room temperature during interruptions of the heat treatment. (J. Petermann, J. M. Schultz, R. M. Gohil, R. W. Hendricks, and J. S. Lin, *Journal of Polymer Science, Polymer Physics Edition,* in press)

PROBLEMS

7.1. We have seen something of the effect of particle size on the breadth of a diffraction node. For a large crystal, the intensity peaks can be represented in resp as the points of a rel. For smaller crystals, these points become diffuse. We have also seen that a graphical description of the diffraction condition can be provided using the Ewald construction. Explain the following two effects, using the Ewald construction: (a) the particle size broadening of all relnodes is identical (half-height breadths are all the same), and (b) the lines of a powder pattern are always broadened by small particles more at higher diffracting angles than at lower diffracting angles (the relationship is that the breadth is proportional to $1/\cos \theta$).

7.2. Figure P7.2 is a set of powder x-ray diffraction peaks from a noble metal oxide prepared under different physical conditions. Using the $2\theta = 35.3°$ peak, compute the crystallite size for each sample. CuK_α radiation was used ($\lambda = 1.54$ Å).

7.3. Rybnikář [F. Rybnikář, *Collection of Czechoslovak Chemical Communications,* **31,** pp. 4080–4094 (1966)] took x-ray diffraction data from melt-crystallized polyoxymethylene, $(–CH_2–O–)_n$, after isothermal crystallization at various temperatures. He used CuK_α radiation ($\lambda = 1.54$ Å). Polyoxymethylene has a hexagonal crystal structure, with $a = 4.45$ Å and $c = 17.3$ Å. The polymer chains lie parallel to c. Figure P7.3 shows the line breadths, $\Delta(2\theta)$, of the (100) and (105) peaks, over this temperature range.

(a) Describe *quantitatively* the shape of the crystals for the material crystallized at 150°C.

(b) Interpret Rybnikář's data in terms of the effect of crystallization temperature on the crystallite size and shape.

7.4. The data of Fig. 7.18 were taken using CuK_α radiation ($\lambda = 1.54$ Å). What is the diameter of the particles?

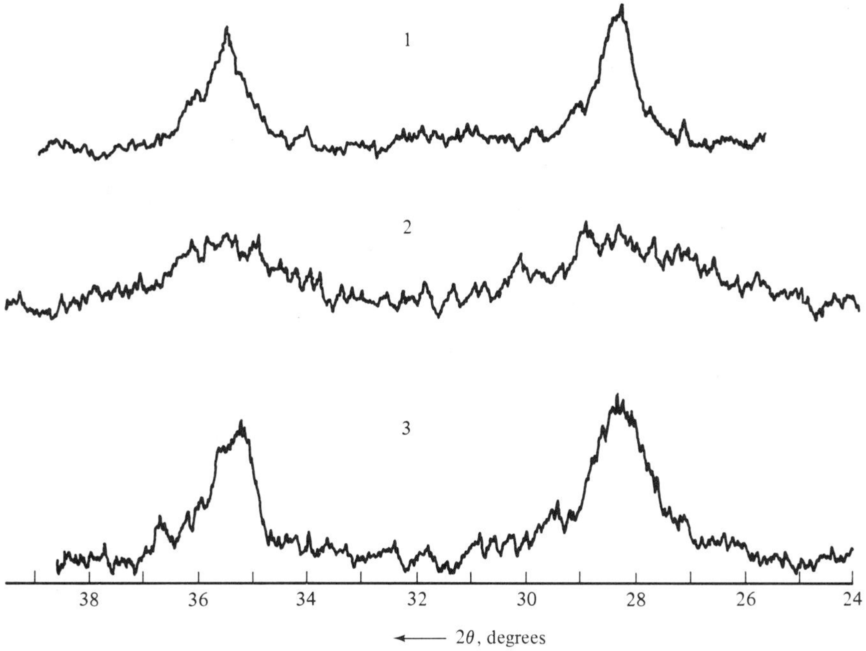

Figure P7.2.

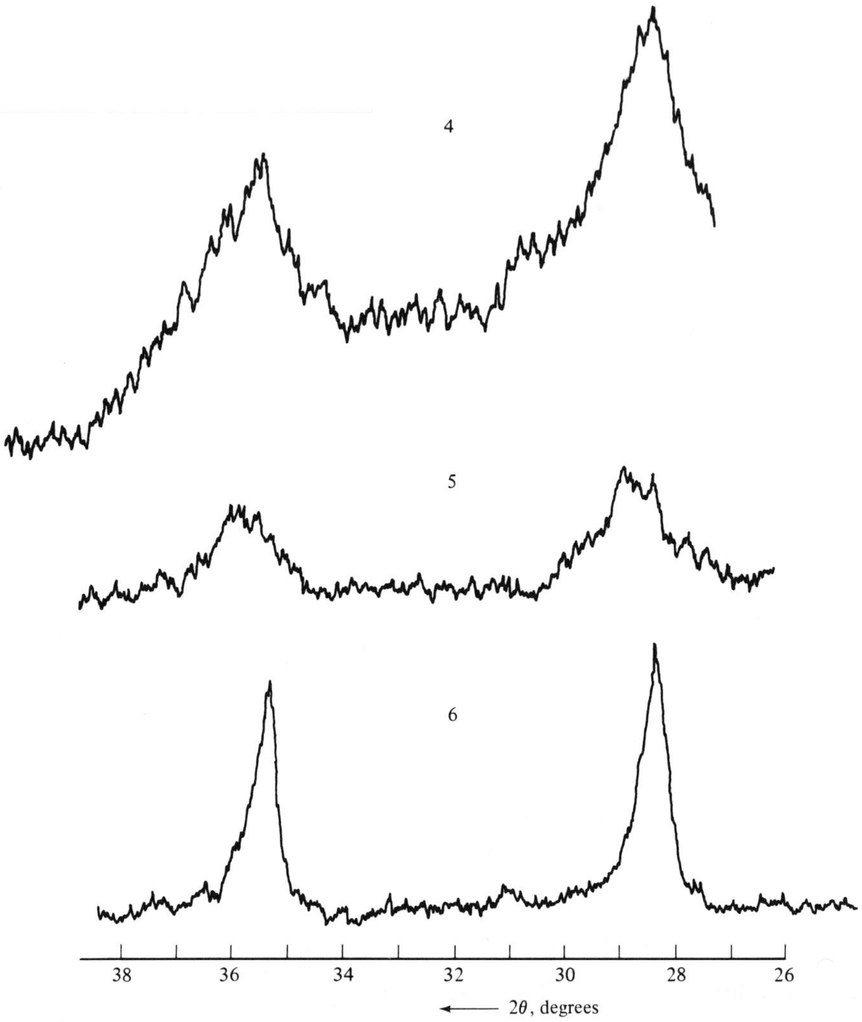

Figure P7.2 (Continued).

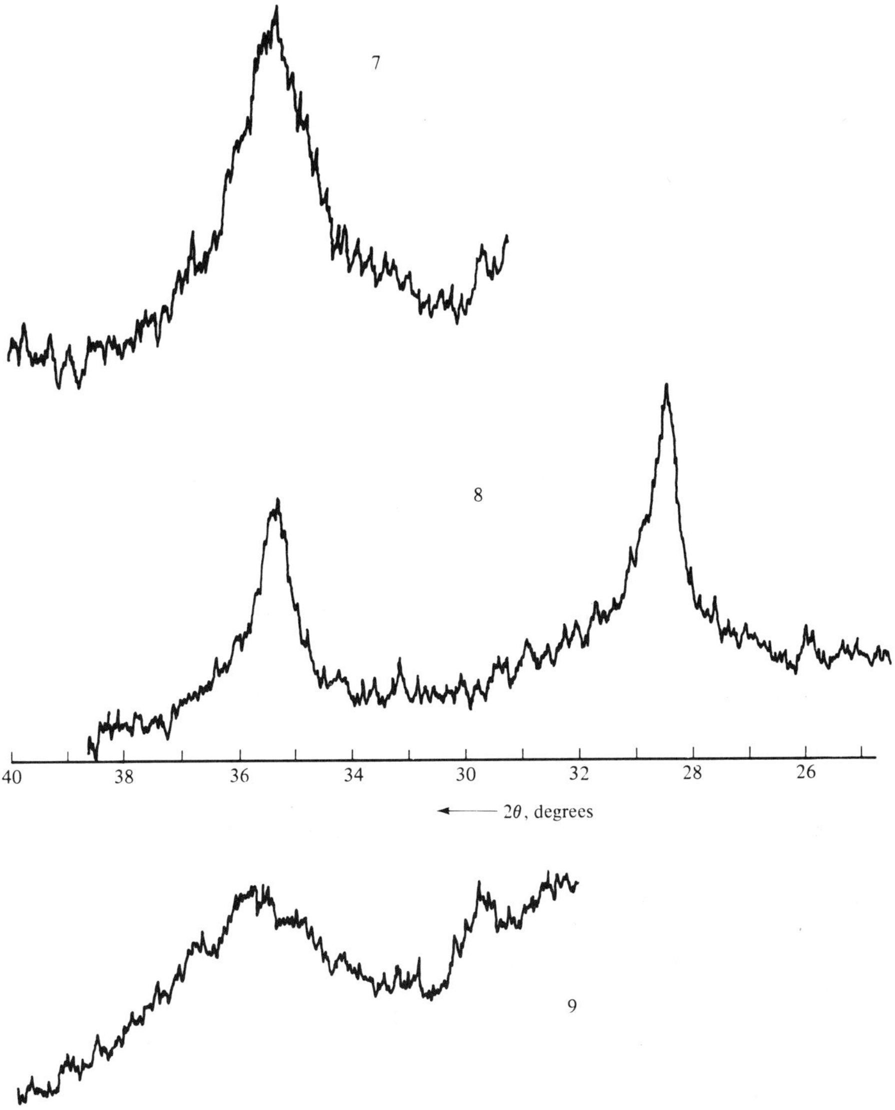

Figure P7.2 (Continued).

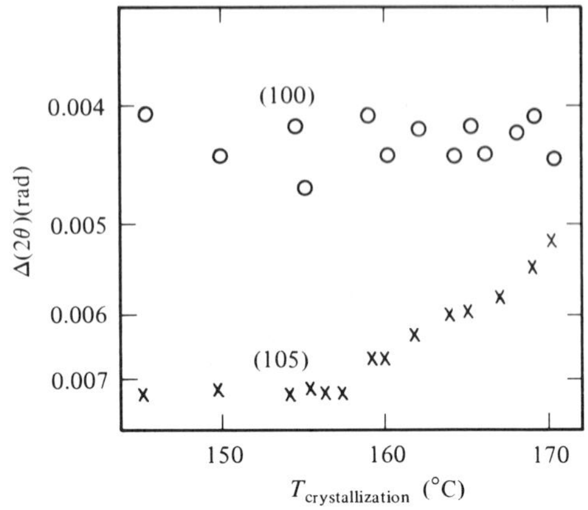

Figure P7.3.

7.5. What is the radius of gyration of the voids which produced the neutron small-angle scattering curves of Fig. 7.16? The wavelength of the incident thermal neutrons is 8.0 Å.

7.6. Figure 7.17 is a neutron small-angle scattering curve for hydrogenated polyethylene molecules, $(CH_2)_n$, in a melt which is otherwise deuterated polyethylene, $(CD_2)_n$. The hydrogenated species comprises 5% of the total. Figure 7.19 is a Guinier plot of the same data. The neutron wavelength is 8.0 Å.
(a) What is the radius of gyration of the hydrogenated polyethylene molecules?
(b) How would the observations have differed had x-ray small-angle scattering been used? Presume FeK_α x-rays ($\lambda = 1.93$ Å).

7.7. Figure 7.21 shows x-ray small-angle scattering data from platelets oriented randomly in the x-ray beam. CuK_α ($\lambda = 1.54$ Å) radiation was used. What is the platelet thickness?

7.8. It has been suggested that the scattering of x-rays at low angles could afford a useful tool for the measurement of the thickness of thin solid films. Develop an intensity expression for the scattering by such a film, given a geometry by which the beam enters and leaves the film at equal angles and on the same surface. How do you get a film thickness from such small angle data?

7.9. A general experimental difficulty in small-angle scattering is the superposition of the main beam. The unscattered forward beam is often some 10^6 more intense than the forward-diffracted ray. Consequently, any "tail" in the main beam will superpose strongly in the scattered waves. Conventionally, this problem is overcome by stringent collimation or by focusing of the main beam (using curved crystal monochrometers or glass mirrors).

A radically different x-ray small angle diffractometer has been designed and built by U. Bonse and M. Hart. The instrument utilizes two highly perfect silicon crystals which have each had a parallel sided groove cut in them. The basic geometry and beam path is described in Fig. P7.9. You are to describe which elements of the system are stationary and which move and in what direction (or about what

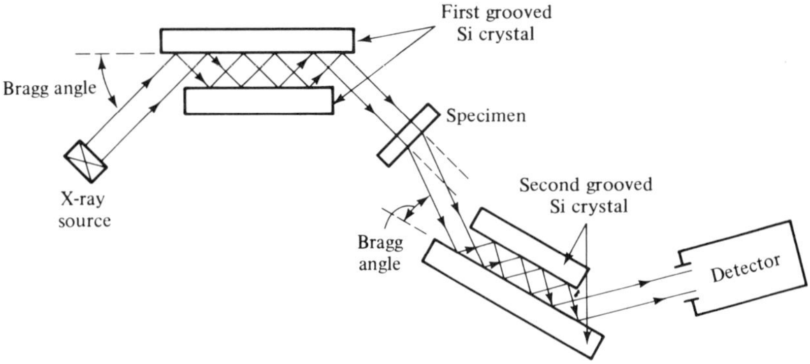

Figure P7.9.

axis)? What should be the special properties of this apparatus and in what ways should it exceed the performance of highly collimated small-angle diffractometers? Upon what specific principle is this instrument based?

BIBLIOGRAPHY

A. Line Broadening

1. B. D. CULLITY, *Elements of X-ray Diffraction 2nd Ed.*, Addison-Wesley Publishing Co., Inc., N.Y. (1976).
2. A. GUINIER, *X-ray Diffraction in Crystals, Imperfect Crystals, and Amorphous Bodies*, Ch. 5, W. H. Freeman & Company Publishers, San Francisco (1963).
3. P. B. HIRSCH, A. HOWIE, R. B. NICHOLSON, D. W. PASHLEY, and M. J. WHELAN, *Electron Microscopy of Thin Crystals*, Ch. 6, Plenum Publishing Corporation, N.Y. (1967).
4. A. GUINIER and G. FOURNET, *Small-Angle Scattering of X-rays*, Wiley, N.Y. (1955).
5. R. W. JAMES, *The Optical Principles of the Diffraction of X-rays*, Ch. 10, G. Bell & Sons, Ltd., London (1954).
6. H. P. KLUG and L. E. ALEXANDER, *X-ray Diffraction Procedures for Polycrystalline and Amorphous Materials, 2nd Ed.*, Ch. 9, Wiley-Interscience, N.Y. (1959).
7. A. J. C. WILSON, *X-ray Optics; the Diffraction of X-rays by Finite and Imperfect Crystals*, Chs. 3 and 4, Methuen, London (1949).

B. Small-Angle Scattering

1. L. E. ALEXANDER, *X-ray diffraction Methods in Polymer Science*, Wiley-Interscience, N.Y. (1969).
2. H. BRUMBERGER, ed., *Small-Angle Scattering*, Gordon & Breach Science Publishers, Inc., N.Y. (1967).

3. P. A. EGELSTAFF, *Thermal Neutron Scattering,* Ch. 17, Academic Press, Inc., N.Y. (1965).

4. A. GUINIER, *X-ray Diffraction in Crystals, Imperfect Crystals, and Amorphous Bodies,* Ch. 10, W. H. Freeman & Company Publishers, San Francisco (1963).

5. A. GUINIER and G. FOURNET, *Small-Angle Scattering of X-rays,* Wiley, N.Y. (1955).

6. M. G. BUERGER, *Introduction to Crystal Geometry,* Chs. 12, 17, 18, McGraw-Hill Book Company, N.Y. (1971).

7. *Journal of Applied Crystallography, Vol. 11,* part 5, Oct., 1978. This is the Proceedings of the Fourth International Conference on Small-Angle Scattering of X-rays and Neutrons and contains excellent reviews of the instrumentation for, and use of, small-angle scattering.

8. M. KAKUDO and N. KASAI, *X-ray Diffraction in Polymers,* Elsevier North-Holland, Inc., N.Y. (1972).

The Scattering of X-rays by One Electron (Thomson Scattering)[1]

A

In this appendix we calculate the intensity of electromagnetic radiation at a point located at r, θ, ϕ relative to the electron which has scattered the radiation. The geometry considered is sketched in Fig. A.1.

The first thing which must be done is to calculate the strength of the electromagnetic field; that is, we must calculate either **B**, the magnetic field, or **E**, the electric field, of the electromagnetic radiation. The analysis is quite formal, beginning with Maxwell's equations

$$\nabla \cdot \mathbf{D} = \epsilon\epsilon_o \nabla \cdot \mathbf{E} = \rho \tag{A1.1}$$

$$\nabla \cdot \mathbf{B} = \mu\mu_o \nabla \cdot \mathbf{H} = 0 \tag{A1.2}$$

$$\nabla \times \mathbf{E} = -\frac{\partial \mathbf{B}}{\partial t} = -\mu\mu_o \left(\frac{\partial \mathbf{H}}{\partial t}\right) \tag{A1.3}$$

$$\nabla \times \mathbf{H} = \mathbf{j} + \frac{\partial \mathbf{D}}{\partial t} = \mathbf{j} + \epsilon\epsilon_o \left(\frac{\partial \mathbf{E}}{\partial t}\right) \tag{A1.4}$$

where **D** is the electric displacement, **E** is the electric field (or electric intensity), **B** is the magnetic induction, **H** is the magnetic field, ϵ_o is a constant of value 8.8×10^{-12} coulomb newton^{-1} meter^{-2}, ϵ is the dielectric constant of medium, μ_o is a constant of value $4\pi \times 10^{-7}$ henry meter^{-1}, μ is the permeability, ρ is the density of free charges in the medium, and **j** is the electrical current density. In these and subsequent equations rationalized mks. units are used.

[1] The treatment given here is based on that used in B. L. Bleaney and B. Bleaney, *Electricity and Magnetism,* Oxford, 1951.

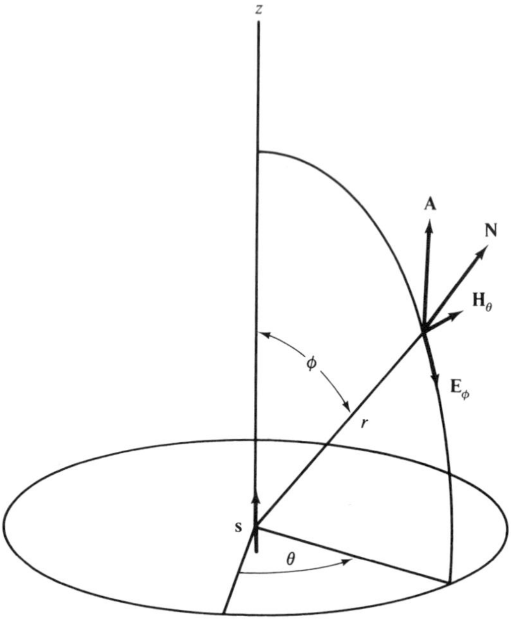

Figure A.1. Geometry used in analysis of the radiation scattered by an electron at the origin.

For calculations to follow it is convenient to put **B** in a new form. In general, a vector **P** must follow this mathematical rule

$$\nabla \cdot \nabla \times \mathbf{P} = 0 \tag{A1.5}$$

since $\nabla \times \mathbf{P}$ is normal to ∇. We may thus express **B** as

$$\mathbf{B} = \nabla \times \mathbf{A} \tag{A1.6}$$

since this is consistent with (A1.2). However, (A1.6) does not uniquely define the new vector **A**, since $\mathbf{B} = \nabla \times (\mathbf{A} + \mathbf{Q})$, where **Q** is an arbitrary vector, also satisfies (A1.2). The vector **A** is termed the *vector magnetic potential.* We require a second condition to define **A.** The second condition is derived from (A1.3), which can now be written

$$\nabla \times \mathbf{E} = \frac{\partial(\nabla \times \mathbf{A})}{\partial t} = -\nabla \times \frac{\partial \mathbf{A}}{\partial t} \tag{A1.7}$$

Rewriting (A1.7) we have

$$\nabla \times \left(\mathbf{E} + \frac{\partial \mathbf{A}}{\partial t}\right) = 0 \tag{A1.8}$$

The solution to (A1.8) is

$$\mathbf{E} = -\frac{\partial \mathbf{A}}{\partial t} + \text{constant} \tag{A1.9}$$

But we know from electrostatics that the scalar potential V is defined by

$$\mathbf{E} = -\nabla V \tag{A1.10}$$

in the absence of time dependence. Thus, the constant in (A1.9), evaluated for the case of no time dependence must be $-\nabla V$, and

$$\mathbf{E} = -\frac{\partial \mathbf{A}}{\partial t} - \nabla V \tag{A1.11}$$

Using (A1.1) and (A1.11), we have

$$\nabla \cdot \mathbf{E} = \frac{\rho}{\epsilon\epsilon_o} = \nabla \cdot \mathbf{A} - \nabla \cdot \nabla V \tag{A1.12}$$

In a vacuum, wherein $\rho = 0$, we recognize Laplace's equation

$$\nabla \cdot \nabla V = \nabla^2 V = 0 \tag{A1.13}$$

Hence, we would expect that $\nabla \cdot \mathbf{A} = 0$. However, it is more convenient to set

$$\nabla \cdot \mathbf{A} = -\mu\mu_o\epsilon\epsilon_o \frac{\partial V}{\partial t} = \frac{-1}{v^2}\frac{\partial V}{\partial t} \tag{A1.14}$$

since this still gives $\nabla \cdot \mathbf{A} = 0$ when V is not time dependent. Equation (A1.14) is called the Lorentz condition. The particular form for (A1.14) was chosen to facilitate the ensuing derivations. Equations (A1.6) and (A1.14) now define the vector **A**, called the vector magnetic potential. (In (A1.14) we have used the relation $v = 1/\sqrt{\mu\mu_o\epsilon\epsilon_o}$, where v is the velocity of propagation of electromagnetic waves in the medium.)

Using the foregoing, we can now find a differential equation which can be solved to obtain **A**. Using (A1.5), we have, from (A1.4)

$$\nabla \times \mathbf{B} = \nabla \times \nabla \times \mathbf{A} = \mu\mu_o\epsilon\epsilon_o\left(\frac{\partial \mathbf{E}}{\partial t}\right) + \mu\mu_o\mathbf{j} \tag{A1.15}$$

But, for an arbitrary vector **P**,

$$\nabla \times \nabla \times \mathbf{P} = \nabla(\nabla \cdot \mathbf{P}) - \nabla^2\mathbf{P} \tag{A1.16}$$

Using (A1.14), (A1.15), and (A1.16) we can now write

$$\nabla \times \mathbf{B} = -\frac{1}{v^2}\nabla\frac{\partial v}{\partial t} - \nabla^2\mathbf{A} = \frac{1}{v^2}\left(\frac{\partial \mathbf{E}}{\partial t}\right) + \mu\mu_o\mathbf{j} \tag{A1.17}$$

Substituting (A1.11) into (A1.17), we have

$$\frac{-1}{v^2}\nabla\frac{\partial v}{\partial t} - \nabla^2\mathbf{A} = -\frac{\partial^2\mathbf{A}}{v^2\partial t^2} - \frac{1}{v^2}\nabla\frac{\partial v}{\partial t} + \mu\mu_o\mathbf{j} \tag{A1.18}$$

Rewritten, (A1.18) becomes

$$-\nabla^2\mathbf{A} + \frac{1}{v^2}\frac{\partial^2\mathbf{A}}{\partial t^2} = \mu\mu_o\mathbf{j} \tag{A1.19}$$

The solution to (A1.19) is

$$\mathbf{A} = \frac{\mu\mu_o}{4\pi}\int\frac{[\mathbf{j}]\,d\tau}{r} \tag{A1.20}$$

Here [**j**] means the current density at a time $(t - r/v)$. If we consider the magnetic potential **A** at the point r, θ, ϕ, at time t, we see that it is the result of a disturbance set up at the origin (at the electron) at time $t - r/v$; i.e., the disturbance has propagated with the velocity v, and has reached the point at a time r/v after its initiation.

We will now determine [**j**] for our system of an oscillating electron. This [**j**] can then be put into (A1.20) and then (A1.20) can be solved to yield **A** and, hence, **B**. First, recall the general definition for a current

$$|\mathbf{i}| = \frac{dq}{dt} \tag{A1.21}$$

This current is found by analogy to the kinetic theory of gases. Consider Fig. A.2. We have an electron oscillating back and forth in one dimension. At any point along its path, it has velocity v_s given by

$$\begin{aligned} v_s &= \frac{ds}{dt} = \frac{d}{dt}[s_o \sin \omega t] \\ &= \omega s_o \cos \omega t \end{aligned} \tag{A1.22}$$

Now dq/dt is just the amount of charge crossing unit area per second. As in the illustration, we consider the effective current crossing the shaded area. If the current were steady, the number of charges crossing that area per unit time would be $q'\,v_s$ coulombs/sec, where q' is the charge per unit length of

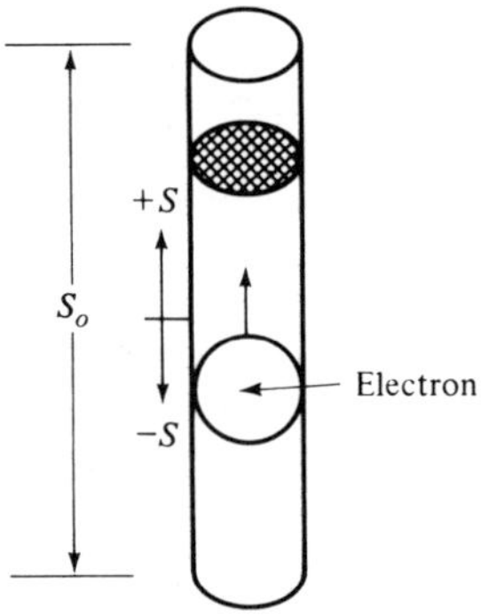

Figure A.2. An oscillating electron.

the cylinder shown. If s_o is the total path traversed by the electron, then $q' = e/s_o$, where e is the charge of one electron. Thus

$$i = \frac{e}{s_o} v_s = e\hat{\mathbf{s}} \cos \omega t \tag{A1.23}$$

where $\hat{\mathbf{s}}$ is a unit vector in the s-direction. And $[i]$ must be

$$[i] = e \cos [\omega t - \omega(r/v)] \tag{A1.24}$$

In our case

$$[\mathbf{j}]\, d\tau = [\mathbf{i}] ds = [i]\, d\mathbf{s} \tag{A1.25}$$

where $d\tau$ is an element of volume. Inserting (A1.24) and (A1.25) into (A1.26), we have

$$\mathbf{A} = \frac{\mu\mu_o}{4\pi} \int_0^{s_o} \frac{[i] d\mathbf{s}}{r} = \frac{\mu\mu_o}{4\pi} \frac{[i]}{r} \int_0^{s_o} d\mathbf{s} = \frac{\mu\mu_o [i] \mathbf{s}_o}{4\pi r} \tag{A1.26}$$

To determine what value of s_o to use, we must solve the equation of motion of the electron. The easiest case to solve is that for which the electron is free—i.e., has no restoring or frictional forces acting on it. For this case, we write

$$m \frac{d^2\mathbf{s}}{dt^2} = e\mathbf{E} = e\mathbf{E}_o\, e^{i\omega t} \tag{A1.27}$$

where m is the mass of the electron and $\mathbf{E}$ is the electric field associated with the electromagnetic radiation incident on the electron. The solution to (A1.27) is

$$\mathbf{s} = -\frac{e}{m\omega^2} \mathbf{E} \tag{A1.28}$$

For s_o, we then have

$$s_o = -\frac{e}{m\omega^2} E_o \tag{A1.29}$$

Inserting (A1.29) in (A1.26), we have

$$\mathbf{A} = \frac{-\mu\mu_o e^2}{4\pi m\omega r} \mathbf{E}_o \cos [\omega(t - r/v)] \tag{A1.30}$$

$\mathbf{A}$ can now, finally, be used to calculate the magnetic field at the point r, θ, ϕ

$$\mathbf{H} = -\frac{1}{\mu\mu_o} \mathbf{B} = -\frac{1}{\mu\mu_o} \nabla \times \mathbf{A} = -\frac{1}{4\pi} \frac{e^2}{m\omega} \mathbf{E}_o \nabla \times \left[\hat{\mathbf{s}} \frac{\cos [\omega(t - r/v)]}{r} \right] \tag{A1.31}$$

where $\hat{\mathbf{s}}$ is a unit vector in the s-direction. Using standard vector calculus, (A1.31) can be transformed as follows

$$\mathbf{H}=\frac{e^2E_o}{4\pi m\omega}\left\{\frac{\cos[\omega(t-r/v)]}{r}\nabla\times\hat{\mathbf{s}}-\hat{\mathbf{s}}\times\nabla\left[\frac{\cos[\omega(t-r/v)]}{r}\right]\right\} \tag{A1.32}$$

But, since $\nabla \times \hat{\mathbf{s}} = 0$ (because $\hat{\mathbf{s}}$ is constrained to lie in a specific direction), we are left with

$$\mathbf{H}=-\frac{e^2E_o}{4\pi m\omega}\hat{\mathbf{s}}\times\nabla\left[\frac{\cos\omega(t-r/v)}{r}\right] \tag{A1.33}$$

and, since

$$\nabla\left[\frac{\cos\omega(t-r/v)}{r}\right]=\hat{\mathbf{r}}\frac{\partial}{dr}\left[\frac{\cos\omega(t-r/v)}{r}\right] \tag{A1.34}$$

we have

$$\mathbf{H}=-\frac{e^2E_o}{4\pi m\omega}(\hat{\mathbf{s}}\times\hat{\mathbf{r}})\left[\frac{\cos\omega(t-r/v)}{r}\right] \tag{A1.35}$$

Now, $\hat{\mathbf{s}} \times \hat{\mathbf{r}} = (\sin\phi)\,\hat{\boldsymbol{\theta}}$, where $\hat{\boldsymbol{\theta}}$ is a unit vector in the θ-direction. We see that the only component of **H** is in the ϕ-direction. Carrying out the differentiation in (A1.35), we have

$$H_\phi=-\frac{e^2E_o}{4\pi m\omega}\left[\frac{\omega/v}{r}\sin\omega(t-r/v)-\frac{1}{r^2}\cos\omega(t-r/v)\right]\sin\phi \tag{A1.36}$$

But we are interested in the intensity at a point quite far removed from the electron. Hence the term in $1/r^2$ can be neglected, and we have

$$H_\phi=-\frac{e^2E_o}{4\pi mvr}\sin\omega(t-r/v)\sin\phi \tag{A1.37}$$

Since, in general, $E = \mu\mu_o vH$, we have

$$E_\theta=\frac{\mu\mu_o e^2E_o}{4\pi mr}\sin\omega(t-r/v)\sin\phi \tag{A1.38}$$

The maximum amplitude of E_θ is

$$(E_\theta)_o=\frac{\mu\mu_o e^2E_o}{4\pi mr}\sin\phi \tag{A1.39}$$

In general, the only restriction on the direction of E_o is that it lie normal to the direction of propagation of the wave. But it may with equal probability lie in *any* direction normal to the direction of propagation. It is convenient to separate into components parallel and normal to the plane; as in Fig. A.3. Hence, as indicated

$$E_\parallel=\frac{e^2\mu\mu_o E_o}{4\pi mr}\sin\phi=\frac{e^2\mu\mu_o E_o}{4\pi mr}\cos 2\theta,\ \text{where} \tag{A1.40}$$

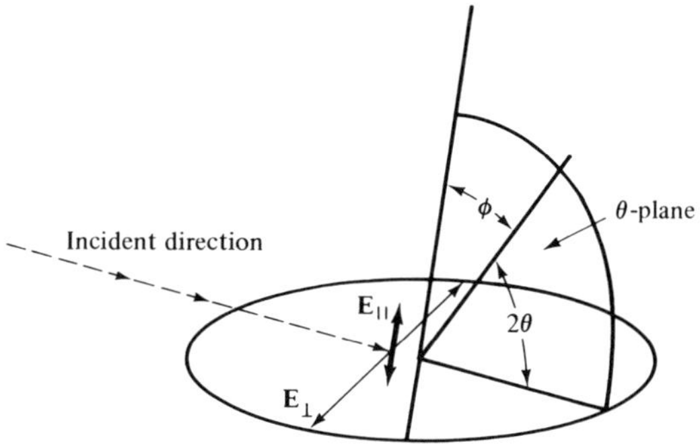

Figure A.3. Separation of the electric vector **E** into components parallel and normal to the θ-plane.

2θ is the scattering angle. And

$$E_{\perp} = \frac{e^2 \mu \mu_o E_o}{4\pi m r} \sin 90^\circ = \frac{e^2 \mu \mu_o E_o}{4\pi m r} \tag{A1.41}$$

In general, $I = E^2$, where I is the intensity of the wave. Thus

$$I_{\parallel} = \left(\frac{\mu \mu_o e^2}{4\pi m r}\right)^2 I_o \cos^2 2\theta$$
$$I_{\perp} = \left(\frac{\mu \mu_o e^2}{4\pi m r}\right)^2 I_o \tag{A1.42}$$

Recombining these components of the incident wave, the intensity of radiation scattered at angle 2θ by one electron—the Thomson intensity I_e—is given by

$$I_e = \frac{1}{2} I_{\parallel} + \frac{1}{2} I_{\perp} = I_o \left(\frac{\mu \mu_o e^2}{4\pi m r}\right)^2 \left(\frac{1 + \cos^2 2\theta}{2}\right) \tag{A1.43}$$

The Thomson classical electron radius r_e is now defined through

$$I_e = I_o \frac{r_e^2}{r^2} \left(\frac{1 + \cos^2 2\theta}{2}\right) \tag{A1.44}$$

The value of the Thomson classical electron radius is 2.81×10^{-15} m.

Tables B

TABLE B.1

X-RAY ATOMIC-SCATTERING FACTORS[a]

$(\sin\theta)/\lambda(A^{-1})$		0.0	0.1	0.2	0.3	0.4	0.5	0.6	0.7	0.8	0.9	1.0	1.1	1.2	1.3	1.4	1.5
H	1	1.0	0.81	0.48	0.25	0.13	0.07	0.04	0.02	0.02	0.01	0.01					
He	2	2.0	1.83	1.45	1.06	0.74	0.52	0.36	0.25	0.18	0.13	0.10	0.07	0.05	0.04	0.03	0.03
Li	3	3.0	2.22	1.74	1.51	1.27	1.03	0.82	0.65	0.51	0.40	0.32	0.26	0.21	0.16		
Be	4	4.0	3.07	2.07	1.71	1.53	1.37	1.20	1.03	0.88	0.74	0.62	0.52	0.43	0.37		
B	5	5.0	4.07	2.71	1.99	1.69	1.53	1.41	1.28	1.15	1.02	0.90	0.78	0.68	0.60		
C	6	6.0	5.13	3.58	2.50	1.95	1.69	1.54	1.43	1.32	1.22	1.11	1.01	0.91	0.82	0.74	0.66
N	7	7.0	6.20	4.60	3.24	2.40	1.94	1.70	1.55	1.44	1.35	1.26	1.18	1.08	1.01		
O	8	8.0	7.25	5.63	4.09	3.01	2.34	1.94	1.71	1.57	1.46	1.37	1.30	1.22	1.14		
F	9	9.0	8.29	6.69	5.04	3.76	2.88	2.31	1.96	1.74	1.59	1.48	1.40	1.32	1.25		
Ne	10	10.0	9.36	7.82	6.09	4.62	3.54	2.79	2.30	1.98	1.76	1.61	1.50	1.42	1.35	1.28	1.22
Na	11	11.0	9.76	8.34	6.89	5.47	4.29	3.40	2.76	2.31	2.00	1.78	1.63	1.52	1.44	1.37	1.31
Na^+	11	10.0	9.55	8.39	6.93	5.51	4.33	3.42	2.77	2.31	2.00	1.79	1.63	1.52	1.44	1.37	1.30
Mg	12	12.0	10.50	8.75	7.46	6.20	5.01	4.06	3.30	2.72	2.30	2.01	1.81	1.65	1.54		
Al	13	13.0	11.23	9.16	7.88	6.77	5.69	4.71	3.88	3.21	2.71	2.32	2.05	1.83	1.69	1.57	1.48
Si	14	14.0	12.16	9.67	8.22	7.20	6.24	5.31	4.47	3.75	3.16	2.69	2.35	2.07	1.87	1.71	1.60
P	15	15.0	13.17	10.34	8.59	7.54	6.67	5.83	5.02	4.28	3.64	3.11	2.69	2.35	2.10	1.89	1.75
S	16	16.0	14.33	11.21	8.99	7.83	7.05	6.31	5.56	4.82	4.15	3.56	3.07	2.66	2.34		
Cl	17	17.0	15.33	12.00	9.44	8.07	7.29	6.64	5.96	5.27	4.60	4.00	3.47	3.02	2.65		
Cl^-	17	18.0	16.02	12.20	9.40	8.03	7.28	6.64	5.97	5.27	4.61	4.00	3.47	3.03	2.65	2.35	2.11
A	18	18.0	16.30	12.93	10.20	8.54	7.56	6.86	6.23	5.61	5.01	4.43	3.90	3.43	3.03		
K	19	19.0	16.73	13.73	10.97	9.05	7.87	7.11	6.51	5.95	5.39	4.84	4.32	3.83	3.40	3.01	2.71
Ca	20	20.0	17.33	14.32	11.71	9.64	8.26	7.38	6.75	6.21	5.70	5.19	4.69	4.21	3.77	3.37	3.03

[a] Reprinted from *International Tables for X-ray Crystallography, Vol. III,* with the permission of the International Union of Crystallography.

TABLE B.1 *(cont.)*

$(\sin\theta)/\lambda(A^{-1})$		0.0	0.1	0.2	0.3	0.4	0.5	0.6	0.7	0.8	0.9	1.0	1.1	1.2	1.3	1.4	1 5
Sc	21	21.0	18.72	15.39	12.39	10.12	8.60	7.64	6.98	6.45	5.96	5.48	5.00	4.53	4.09	3.68	3.31
Ti	22	22.0	19.41	16.07	13.20	10.83	9.12	7.98	7.22	6.65	6.19	5.72	5.29	4.84	4.41	4.01	3.64
V	23	23.0	20.47	17.03	14.03	11.51	9.63	8.34	7.48	6.86	6.39	5.94	5.53	5.10	4.71	4.30	3.93
Cr	24	24.0	21.93	18.37	15.01	12.22	10.14	8.72	7.75	7.09	6.58	6.14	5.74	5.34	4.94	4.55	4.18
Mn	25	25.0	22.61	19.06	15.84	13.02	10.80	9.20	8.09	7.32	6.77	6.32	5.93	5.54	5.18	4.80	4.45
Fe	26	26.0	23.68	20.09	16.77	13.84	11.47	9.71	8.47	7.60	6.99	6.51	6.12	5.74	5.39	5.03	4.69
Co	27	27.0	24.74	21.13	17.74	14.68	12.17	10.26	8.88	7.91	7.22	6.70	6.29	5.91	5.58	5.23	4.90
Ni	28	28.0	25.80	22.19	18.73	15.56	12.91	10.85	9.33	8.25	7.48	6.90	6.47	6.08	5.75	5.41	5.09
Cu	29	29.0	27.19	23.63	19.90	16.48	13.65	11.44	9.80	8.61	7.76	7.13	6.65	6.25	5.90	5.57	5.25
Zn	30	30.0	27.92	24.33	20.77	17.42	14.51	12.16	10.37	9.04	8.08	7.37	6.84	6.42	6.07	5.73	5.43
Ga	31	31.0	28.65	24.92	21.47	18.26	15.38	12.95	11.02	9.54	8.46	7.64	7.05	6.58	6.21	5.88	5.58
Ge	32	32.0	29.52	25.53	22.11	19.02	16.19	13.72	11.68	10.08	8.87	7.96	7.29	6.77	6.37	6.02	5.72
As	33	33.0	30.47	26.20	22.69	19.69	16.95	14.48	12.37	10.67	9.34	8.32	7.57	6.98	6.54	6.17	5.86
Se	34	34.0	31.43	26.91	23.24	20.28	17.63	15.20	13.06	11.27	9.83	8.71	7.86	7.21	6.72	6.31	5.99
Br	35	35.0	32.43	27.70	23.82	20.84	18.27	15.91	13.78	11.93	10.41	9.19	8.24	7.51	6.95	6.51	6.16
Kr	36	36.0	33.44	28.53	24.40	21.34	18.82	16.54	14.44	12.57	10.97	9.66	8.62	7.81	7.19	6.70	6.31
Rb	37	37.0	34.11	28.97	24.75	21.29	18.55	16.30	14.47	12.94	11.66	10.58	9.65	8.84	8.14	7.53	6.99
Sr	38	38.0	35.06	29.83	25.51	21.96	19.15	16.84	14.96	13.39	12.07	10.95	9.99	9.16	8.44	7.80	7.24
Y	39	39.0	36.01	30.68	26.28	22.64	19.76	17.39	15.46	13.84	12.48	11.32	10.34	9.48	8.73	8.08	7.50
Zr	40	40.0	36.96	31.54	27.04	23.32	20.37	17.94	15.95	14.29	12.89	11.70	10.68	9.80	9.03	8.36	7.76
Nb	41	41.0	37.91	32.40	27.81	24.01	20.98	18.49	16.45	14.74	13.31	12.08	11.04	10.13	9.33	8.64	8.02
Mo	42	42.0	38.86	33.25	28.57	24.69	21.60	19.04	16.95	15.20	13.73	12.46	11.39	10.45	9.64	8.92	8.29
Tc	43	43.0	39.81	34.12	29.34	25.38	22.21	19.60	17.46	15.65	14.15	12.85	11.74	10.78	9.94	9.21	8.55
Ru	44	44.0	40.76	34.98	30.12	26.07	22.83	20.16	17.96	16.12	14.57	13.24	12.10	11.11	10.25	9.49	8.82

Rh	45	45.0	41.72	35.84	30.89	26.76	23.46	20.72	18.47	16.58	14.99	13.63	12.46	11.45	10.56	9.78	9.09
Pd	46	46.0	42.67	36.70	31.67	27.46	24.08	21.28	18.98	17.05	15.42	14.02	12.82	11.78	10.87	10.07	9.37
Ag	47	47.0	43.63	37.57	32.44	28.16	24.71	21.85	19.50	17.52	15.85	14.42	13.19	12.12	11.19	10.37	9.64
Cd	48	48.0	44.58	38.44	33.22	28.85	25.34	22.42	20.02	17.99	16.28	14.81	13.56	12.46	11.51	10.66	9.92
In	49	49.0	45.5	39.3	34.0	29.6	26.0	23.0	20.5	18.5	16.7	15.2	13.9	12.8	11.8	11.0	10.2
Sn	50	50.0	46.5	40.2	34.8	30.3	26.6	23.6	21.1	18.9	17.2	15.6	14.3	13.2	12.1	11.3	10.5
Sb	51	51.0	47.5	41.1	35.6	31.0	27.2	24.1	21.6	19.4	17.6	16.0	14.7	13.5	12.5	11.6	10.8
Te	52	52.0	48.4	41.9	36.4	31.7	27.9	24.7	22.1	19.9	18.0	16.4	15.1	13.8	12.8	11.9	11.0
I	53	53.0	49.4	42.8	37.1	32.4	28.5	25.3	22.6	20.4	18.5	16.8	15.4	14.2	13.1	12.2	11.3
Xe	54	54.0	50.3	43.7	37.9	33.1	29.2	25.9	23.2	20.9	18.9	17.2	15.8	14.5	13.4	12.5	11.6
Cs	55	55.0	51.3	44.5	38.7	33.8	29.8	26.5	23.7	21.3	19.4	17.7	16.2	14.9	13.8	12.8	11.9
Ba	56	56.0	52.3	45.4	39.5	34.5	30.4	27.0	24.2	21.8	19.8	18.1	16.6	15.3	14.1	13.1	12.2
Ta	73	73.0	68.6	60.4	53.1	46.9	41.7	37.3	33.6	30.4	27.7	25.4	23.3	21.6	20.0	18.6	17.4
W	74	74.0	69.5	61.3	54.0	47.6	42.3	37.9	34.1	30.9	28.2	25.8	23.7	21.9	20.3	18.9	17.7
Re	75	75.0	70.5	62.2	54.8	48.3	43.0	38.5	34.7	31.4	28.7	26.3	24.2	22.3	20.7	19.3	18.0
Os	76	76.0	71.5	63.1	55.6	49.1	43.7	39.1	35.3	32.0	29.1	26.7	24.6	22.7	21.1	19.6	18.3
Ir	77	77.0	72.4	64.0	56.4	49.8	44.4	39.7	35.8	32.5	29.6	27.1	25.0	23.1	21.4	19.9	18.6
Pt	78	78.0	73.4	64.9	57.2	50.6	45.0	40.3	36.4	33.0	30.1	27.6	25.4	23.5	21.8	20.3	18.9
Au	79	79.0	74.4	65.8	58.0	51.3	45.7	41.0	37.0	33.5	30.6	28.0	25.8	23.9	22.2	20.6	19.3
Hg	80	80.0	75.3	66.7	58.8	52.1	46.4	41.6	37.5	34.1	31.1	28.5	26.3	24.3	22.5	21.0	19.6
Tl	81	81.0	76.3	67.6	59.7	52.8	47.1	42.2	38.1	34.6	31.6	29.0	26.7	24.7	22.9	21.3	19.9
Pb	82	82.0	77.2	68.5	60.5	53.6	47.8	42.9	38.7	35.1	32.1	29.4	27.1	25.1	23.3	21.7	20.3
Bi	83	83.0	78.2	69.3	61.3	54.3	48.5	43.5	39.3	35.7	32.6	29.9	27.5	25.5	23.6	22.0	20.6
Th	90	90.0	85.0	75.6	67.1	59.6	53.3	47.9	43.3	39.4	36.1	33.1	30.5	28.3	26.3	24.5	22.9
U	92	92.0	86.9	77.4	68.7	61.1	54.7	49.2	44.5	40.5	37.1	34.0	31.4	29.1	27.0	25.2	23.6

TABLE B.2

DISPERSION CORRECTIONS FOR X-RAY ATOMIC-SCATTERING FACTORS[a]

CrKα radiation (λ=2·291 Å)

Atom	Δf′ (sin θ)/λ=0	Δf′ 0·4	Δf″ (sin θ)/λ=0	Δf″ 0·3	Δf″ 0·4
5 B	0·0	0·0	0·0	0·0	0·0
6 C	0·0	0·0	0·1	0·1	0·1
7 N	0·0	0·0	0·1	0·1	0·1
8 O	0·1	0·1	0·2	0·1	0·1
9 F	0·1	0·1	0·2	0·2	0·2
10 Ne	0·1	0·1	0·3	0·3	0·3
11 Na	0·2	0·2	0·4	0·4	0·4
12 Mg	0·2	0·2	0·5	0·5	0·5
13 Al	0·2	0·2	0·6	0·6	0·6
14 Si	0·3	0·3	0·8	0·8	0·8
15 P	0·3	0·3	1·0	0·9	0·9
16 S	0·3	0·3	1·2	1·2	1·2
17 Cl	0·3	0·3	1·5	1·5	1·4
18 A	0·2	0·2	1·8	1·7	1·7
19 K	0·0	0·0	2·2	2·1	2·1
20 Ca	−0·2	−0·2	2·7	2·7	2·6
21 Sc	−0·7	−0·7	3·2	3·1	3·1
22 Ti	−1·7	−1·7	3·8	3·7	3·7
23 V	(−4·4)	(−4·4)	0·6	0·6	0·5
24 Cr	−2·2	−2·2	0·7	0·7	0·6
25 Mn	−1·8	−1·8	0·8	0·8	0·7
26 Fe	−1·6	−1·6	0·9	0·9	0·8
27 Co	−1·4	−1·4	1·0	1·0	0·9
28 Ni	−1·2	−1·2	1·2	1·2	1·1
29 Cu	−1·1	−1·1	1·3	1·3	1·2
30 Zn	−1·0	−1·0	1·5	1·5	1·4
31 Ga	−0·9	−0·9	1·7	1·6	1·6
32 Ge	−0·8	−0·8	1·9	1·8	1·8
33 As	−0·7	−0·7	2·2	2·1	2·1
34 Se	−0·7	−0·7	2·4	2·3	2·3
35 Br	−0·6	−0·7	2·7	2·6	2·5
36 Kr	−0·6	−0·7	3·0	2·9	2·8
37 Rb	−0·6	−0·7	3·4	3·3	3·2
38 Sr	−0·6	−0·7	3·8	3·7	3·6
39 Y	−0·6	−0·7	4·2	4·0	3·9
40 Zr	−0·7	−0·8	4·6	4·4	4·3
41 Nb	−0·8	−0·8	5·1	4·9	4·8
42 Mo	−0·9	−0·9	5·6	5·4	5·3
43 Tc	−1·0	−1·1	6·2	6·0	5·9
44 Ru	−1·2	−1·2	6·7	6.5	6·4

[a] Reprinted from *International Tables for X-ray Crystallography, Vol. III,* with the permission of the International Union of Crystallography.

TABLE B.2 *(cont.)*

Atom	$\Delta f'$		$\Delta f''$		
	$(\sin\theta)/\lambda=0$	0·4	$(\sin\theta)/\lambda=0$	0·3	0·4
45 Rh	−1·3	−1·4	7·3	7·1	6·9
46 Pd	−1·6	−1·7	7·9	7·7	7·5
47 Ag	−1·9	−2·0	8·6	8·4	8·2
48 Cd	−2·2	−2·3	9·2	9·0	8·8
49 In	−2·7	−2·8	9·9	9·7	9·5
50 Sn	−3·2	−3·3	10·7	10·5	10·3
51 Sb	−3·9	−4·0	11·6	11·3	11·1
52 Te	−4·9	−5·0	12·4	12·1	11·9
53 I	(−7·1)	(−7·1)	13·6	13·3	13·1
54 Xe	—	—	11	11	10
55 Cs	(−12)	(−12)	12	12	11
56 Ba	(−11)	(−11)	8	8	8
57 La	(−14)	(−14)	3	3	3
58 Ce	−10	−10	3	3	3
59 Pr	−9	−9	4	3	3
60 Nd	−8	−8	4	4	4
61 Pm	−7	−7	5	4	4
62 Sm	−7	−7	5	5	4
63 Eu	−6	−6	5	5	5
64 Gd	−6	−6	6	5	5
65 Tb	−6	−6	6	6	6
66 Dy	−5	−6	7	6	6
67 Ho	−5	−5	7	7	7
68 Er	−5	−5	8	7	7
69 Tm	−5	−5	8	8	8
70 Yb	−5	−5	9	8	8
71 Lu	−5	−5	9	9	9
72 Hf	−5	−5	10	10	9
73 Ta	−5	−5	11	10	10
74 W	−5	−5	11	11	10
75 Re	−5	−5	12	12	11
76 Os	−5	−5	13	12	12
77 Ir	−5	−5	14	13	13
78 Pt	−5	−5	15	14	14
79 Au	−5	−5	15	15	14
80 Hg	−5	−5	16	16	15
81 Tl	−5	−5	17	17	16
82 Pb	−6	−6	18	18	17
83 Bi	−6	−6	19	19	18
84 Po	−7	−7	20	20	19
85 At	−8	−8	21	21	20
86 Rn	−9	−9	23	22	22
87 Fr	−10	−10	24	23	23
88 Ra	−11	−11	25	24	24
89 Ac	−12	−12	27	26	26
90 Th	−13	−13	28	27	27
91 Pa	(−15)	(−15)	29	28	28
92 U	(−17)	(−17)	28	27	27

TABLE B.2 *(cont.)*

CuKα radiation ($\lambda = 1{\cdot}542$ Å)

Atom	$\Delta f'$		$\Delta f''$		
	$(\sin\theta)\,\lambda = 0$	0·6	$(\sin\theta)\,\lambda = 0$	0·4	0·6
7 N	0·0	0·0	0·0	0·0	0·0
8 O	0·0	0·0	0·1	0·1	0·1
9 F	0·0	0·0	0·1	0·1	0·1
10 Ne	0·1	0·1	0·2	0·1	0·1
11 Na	0·1	0·1	0·2	0·2	0·2
12 Mg	0·1	0·1	0·3	0·2	0·2
13 Al	0·2	0·1	0·3	0·3	0·3
14 Si	0·2	0·2	0·4	0·4	0·4
15 P	0·2	0·2	0·5	0·5	0·5
16 S	0·3	0·3	0·6	0·6	0·6
17 Cl	0·3	0·3	0·7	0·7	0·7
18 A	0·3	0·3	0·9	0·9	0·8
19 K	0·3	0·3	1·1	1·1	1·0
20 Ca	0·3	0·3	1·4	1·4	1·3
21 Sc	0·3	0·3	1·6	1·6	1·5
22 Ti	0·2	0·2	1·9	1·9	1·8
23 V	0·1	0·1	2·3	2·3	2·2
24 Cr	−0·1	−0·1	2·6	2·6	2·5
25 Mn	−0·5	−0·5	3·0	2·9	2·9
26 Fe	−1·1	−1·1	3·4	3·3	3·3
27 Co	−2·2	−2·2	3·9	3·8	3·8
28 Ni	(−3·1)	(−3·1)	0·6	0·6	0·5
29 Cu	−2·1	−2·1	0·7	0·7	0·6
30 Zn	−1·7	−1·7	0·8	0·7	0·7
31 Ga	−1·5	−1·5	0·9	0·8	0·8
32 Ge	−1·3	−1·3	1·1	1·0	0·9
33 As	−1·2	−1·2	1·2	1·1	1·0
34 Se	−1·0	−1·1	1·3	1·2	1·1
35 Br	−0·9	−1·0	1·5	1·4	1·3
36 Kr	−0·9	−1·0	1·7	1·6	1·5
37 Rb	−0·8	−0·9	1·9	1·8	1·7
38 Sr	−0·7	−0·8	2·1	2·0	1·8
39 Y	−0·7	−0·8	2·3	2·2	2·0
40 Zr	−0·6	−0·7	2·5	2·4	2·2
41 Nb	−0·6	−0·7	2·8	2·6	2·5
42 Mo	−0·5	−0·6	3·0	2·8	2·7
43 Tc	−0·5	−0·6	3·3	3·1	3·0
44 Ru	−0·5	−0·6	3·6	3·4	3·3
45 Rh	−0·5	−0·6	4·0	3·8	3·6
46 Pd	−0·5	−0·6	4·3	4·1	3·9

TABLE B.2 *(cont.)*

Atom	$\Delta f'$		$\Delta f''$		
	$(\sin\theta)/\lambda=0$	0·6	$(\sin\theta)/\lambda=0$	0·4	0·6
47 Ag	−0·5	−0·6	4·7	4·5	4·3
48 Cd	−0·6	−0·7	5·0	4·8	4·6
49 In	−0·6	−0·8	5·4	5·2	5·0
50 Sn	−0·7	−0·9	5·8	5·6	5·4
51 Sb	−0·8	−1·0	6·3	6·1	5·8
52 Te	−0·9	−1·1	6·7	6·5	6·2
53 I	−1·1	−1·3	7·2	6·9	6·7
54 Xe	−1·4	−1·6	7·8	7·5	7·2
55 Cs	−1·7	−1·9	8·3	8·0	7·7
56 Ba	−2·1	−2·3	8·9	8·6	8·3
57 La	−2·5	−2·7	9·6	9·2	8·9
58 Ce	−2·9	−3·1	10·3	9·9	9·6
59 Pr	−3·4	−3·6	11·0	10·6	10·2
60 Nd	−4·2	−4·4	11·7	11·3	10·9
61 Pm	−5·1	−5·3	12·4	12·0	11·5
62 Sm	−6·6	−6·7	13·3	12·8	12·4
63 Eu	—	—	11·0	10·6	10·2
64 Gd	(−12)	(−12)	12·0	11·6	11·2
65 Tb	(−11)	(−11)	8	7	7
66 Dy	−10	−10	8	8	8
67 Ho	(−13)	(−13)	4	4	3
68 Er	(−9)	(−9)	4	4	3
69 Tm	−8	−8	5	4	4
70 Yb	−7	−8	5	4	4
71 Lu	−7	−7	5	5	4
72 Hf	−6	−7	5	5	4
73 Ta	−6	−6	6	5	5
74 W	−6	−6	6	5	5
75 Re	−5	−6	6	6	5
76 Os	−5	−6	7	6	6
77 Ir	−5	−6	7	7	6
78 Pt	−5	−5	8	7	7
79 Au	−5	−5	8	8	7
80 Hg	−5	−5	9	8	8
81 Tl	−4	−5	9	9	8
82 Pb	−4	−5	10	9	9
83 Bi	−4	−5	10	10	9
84 Po	−4	−5	11	10	10
85 At	−4	−5	11	11	10
86 Rn	−4	−5	12	11	11
87 Fr	−4	−5	12	12	11
88 Ra	−4	−5	13	12	12
89 Ac	−4	−5	14	13	12
90 Th	−4	−5	15	14	13
91 Pa	−4	−5	16	15	14

TABLE B.2 *(cont.)*

Atom	$\Delta f'$		$\Delta f''$		
	$(\sin\theta)/\lambda=0$	0·4	$(\sin\theta)/\lambda=0$	0·3	0·4
92 U	−4	−5	16	16	15
93 Np	−4	−5	17	16	16
94 Pu	−5	−5	18	17	16
95 Am	−5	−5	19	18	17
96 Cm	−6	−6	20	19	18

Mo$K\alpha$ radiation (λ=0·7107 Å)

Atom	$\Delta f'$			$\Delta f''$			
	$(\sin\theta)/\lambda=$ 0·0	0·9	1·3	$(\sin\theta)/\lambda=$ 0·0	0·6	0·9	1·3
10 Ne	0·0	0·0	0·0	0·0	0·0	0·0	0·0
11 Na	0·0	0·0	0·0	0·1	0·1	0·1	0·0
12 Mg	0·0	0·0	0·0	0·1	0·1	0·1	0·1
13 Al	0·1	0·1	0·0	0·1	0·1	0·1	0·1
14 Si	0·1	0·1	0·1	0·1	0·1	0·1	0·1
15 P	0·1	0·1	0·1	0·2	0·2	0·1	0·1
16 S	0·1	0·1	0·1	0·2	0·2	0·2	0·1
17 Cl	0·1	0·1	0·1	0·2	0·2	0·2	0·2
18 A	0·1	0·1	0·1	0·3	0·3	0·2	0·2
19 K	0·2	0·2	0·1	0·3	0·3	0·3	0·3
20 Ca	0·2	0·2	0·2	0·4	0·4	0·3	0·3
21 Sc	0·2	0·2	0·2	0·5	0·5	0·4	0·4
22 Ti	0·3	0·2	0·2	0·6	0·6	0·5	0·5
23 V	0·3	0·3	0·2	0·7	0·7	0·6	0·6
24 Cr	0·3	0·3	0·3	0·8	0·8	0·7	0·7
25 Mn	0·4	0·3	0·3	0·9	0·9	0·8	0·7
26 Fe	0·4	0·3	0·3	1·0	0·9	0·9	0·8
27 Co	0·4	0·3	0·3	1·1	1·0	1·0	0·9
28 Ni	0·4	0·3	0·3	1·2	1·1	1·1	1·0
29 Cu	0·3	0·3	0·3	1·4	1·3	1·3	1·2
30 Zn	0·3	0·3	0·2	1·6	1·5	1·5	1·4
31 Ga	0·2	0·2	0·2	1·7	1·6	1·6	1·5
32 Ge	0·2	0·2	0·2	1·9	1·8	1·8	1·7
33 As	0·1	0·1	0·1	2·2	2·1	2·0	1·9
34 Se	−0·1	−0·1	−0·1	2·4	2·3	2·2	2·1

TABLE B.2 *(cont.)*

Atom	Δf′ (sin θ)/λ= 0·0	0·9	1·3	Δf″ (sin θ)/λ= 0·0	0·6	0·9	1·3
35 Br	−0·3	−0·3	−0·3	2·6	2·5	2·4	2·3
36 Kr	−0·6	−0·6	−0·6	2·9	2·8	2·7	2·6
37 Rb	−0·9	−0·9	−0·9	3·2	3·1	3·0	2·9
38 Sr	−1·4	−1·4	−1·5	3·6	3·5	3·4	3·2
39 Y	−2·3	−2·3	−2·4	3·9	3·8	3·7	3·5
40 Zr	−2·8	−2·8	−2·9	0·8	0·7	0·6	0·5
41 Nb	−2·1	−2·1	−2·2	0·9	0·8	0·6	0·5
42 Mo	−1·7	−1·7	−1·8	0·9	0·8	0·7	0·6
43 Tc	−1·4	−1·5	−1·5	1·0	0·9	0·8	0·6
44 Ru	−1·2	−1·3	−1·3	1·1	1·0	0·8	0·7
45 Rh	−1·1	−1·2	−1·3	1·2	1·1	0·9	0·8
46 Pd	−1·0	−1·1	−1·2	1·3	1·2	1·0	0·8
47 Ag	−0·9	−1·0	−1·1	1·4	1·3	1·1	0·9
48 Cd	−0·8	−0·9	−1·0	1·6	1·4	1·3	1·1
49 In	−0·7	−0·8	−0·9	1·7	1·5	1·4	1·2
50 Sn	−0·6	−0·8	−0·9	1·9	1·7	1·5	1·3
51 Sb	−0·6	−0·8	−0·9	2·0	1·8	1·6	1·4
52 Te	−0·5	−0·7	−0·8	2·2	2·0	1·8	1·5
53 I	−0·5	−0·7	−0·8	2·4	2·2	1·9	1·7
54 Xe	−0·4	−0·6	−0·8	2·5	2·3	2·1	1·8
55 Cs	−0·4	−0·6	−0·8	2·7	2·5	2·3	2·0
56 Ba	−0·4	−0·6	−0·8	3·0	2·8	2·5	2·2
57 La	−0·3	−0·5	−0·7	3·2	2·9	2·7	2·4
58 Ce	−0·3	−0·5	−0·7	3·4	3·1	2·9	2·6
59 Pr	−0·3	−0·5	−0·7	3·7	3·4	3·1	2·8
60 Nd	−0·3	−0·5	−0·7	3·9	3·6	3·3	3.0
61 Pm	−0·3	−0·5	−0·7	4·1	3·8	3·5	3·2
62 Sm	−0·3	−0·5	−0·7	4·3	4·0	3·7	3·4
63 Eu	−0·3	−0·5	−0·7	4·6	4·3	3·9	3·6
64 Gd	−0·3	−0·6	−0·7	4·8	4·5	4·1	3·8
65 Tb	−0·4	−0·6	−0·8	5·1	4·7	4·3	4·0
66 Dy	−0·4	−0·7	−0·8	5·4	5·0	4·7	4·3
67 Ho	−0·4	−0·7	−0·8	5·7	5·3	5·0	4·6
68 Er	−0·4	−0·7	−0·8	6·0	5·6	5·3	4·8
69 Tm	−0·4	−0·8	−0·8	6·3	5·9	5·6	5·1
70 Yb	−0·5	−0·8	−0·9	6·7	6·3	5·9	5·4
71 Lu	−0·6	−0·9	−1·0	7·0	6·6	6·1	5·6
72 Hf	−0·7	−1·0	−1·1	7·3	6·9	6·4	5·9
73 Ta	−0·8	−1·1	−1·2	7·6	7·2	6·7	6·2
74 W	−1·0	−1·3	−1·4	8·0	7·6	7·1	6·5
75 Re	−1·2	−1·5	−1·6	8·3	8·0	7·5	6·9
76 Os	−1·4	−1·7	−1·8	8·8	8·4	7·9	7·3
77 Ir	−1·7	−2·0	−2·1	9·2	8·8	8·3	7·7
78 Pt	−1·9	−2·2	−2·3	9·6	9·1	8·7	8·1
79 Au	−2·2	−2·5	−2·6	10·1	9·6	9·2	8·5

TABLE B.2 *(cont.)*

Atom	Δf′			Δf″			
	(sin θ)/λ=			(sin θ)/λ=			
	0·0	0·9	1·3	0·0	0·6	0·9	1·3
80 Hg	−2·6	−2·9	−3·0	10·6	10·1	9·7	9·0
81 Tl	−3·2	−3·5	−3·6	11·2	10·7	10·2	9·5
82 Pb	−3·8	−4·1	−4·2	11·7	11·2	10·7	9·9
83 Bi	−4·5	−4·8	−4·9	11·2	11·7	11·1	10·3
84 Po	−5·3	−5·5	−5·6	12·8	12·3	11·7	10·8
85 At	—	—	—	10	10	9	9
86 Rn	(−8)	(−8)	(−8)	11	11	10	9
87 Fr	(−8)	(−8)	(−8)	8	7	7	6
88 Ra	−7	−7	−7	8	7	7	6
89 Ac	−6	−7	−7	8	8	8	7
90 Th	−6	−7	−7	9	8	7	7
91 Pa	−7	−7	−7	9	8	8	7
92 U	(−8)	(−8)	(−8)	9	9	8	7
93 Np	—	—	—	6	5	5	4
94 Pu	—	—	—	6	5	5	4
95 Am	—	—	—	6	5	5	4
96 Cm	—	—	—	6	6	5	4

TABLE B.3

ATOMIC SCATTERING AMPLITUDES FOR ELECTRONS f^e IN Å[a]

Self-consistent field calculations.* These values are based on the rest mass of the electron. For electrons of velocity v, multiply by $[1-(v/c)^2]^{-\frac{1}{2}}$

Element	Z	(sin θ)/λ (Å⁻¹)																
		0·00	0·05	0·10	0·15	0·20	0·25	0·30	0·35	0·40	0·50	0·60	0·70	0·80	0·90	1·00	1·10	1·20
H	1	0·529	0·508	0·453	0·382	0·311	0·249	0·199	0·160	0·131	0·089	0·064	0·048	0·037	0·029	0·024	0·020	0·017
He	2	(0·445)	0·431	0·403	0·368	0·328	0·288	0·250	0·216	0·188	0·142	0·109	0·086	0·068	0·055	0·046	0·038	0·032
Li	3	3·31	2·78	1·88	1·17	0·75	0·53	0·40	0·31	0·26	0·19	0·14	0·11	0·09	0·08	0·06	0·05	0·05
Be	4	3·09	2·82	2·23	1·63	1·16	0·83	0·61	0·47	0·37	0·25	0·19	0·15	0·12	0·10	0·08	0·07	0·06
B	5	2·82	2·62	2·24	1·78	1·37	1·04	0·80	0·62	0·50	0·33	0·24	0·18	0·14	0·12	0·10	0·08	0·07
C	6	2·45	2·26	2·09	1·74	1·43	1·15	0·92	0·74	0·60	0·41	0·30	0·22	0·18	0·14	0·12	0·10	0·08
N	7	2·20	2·10	1·91	1·68	1·44	1·20	1·00	0·83	0·69	0·48	0·35	0·27	0·21	0·17	0·14	0·11	0·10
O	8	2·01	1·95	1·80	1·62	1·42	1·22	1·04	0·88	0·75	0·54	0·40	0·31	0·24	0·19	0·16	0·13	0·11
F	9	(1·84)	(1·77)	1·69	(1·53)	1·38	(1·20)	1·05	(0·91)	0·78	0·59	0·44	0·35	0·27	0·22	0·18	0·15	(0·13)
Ne	10	(1·66)	1·59	1·53	1·43	1·30	1·17	1·04	0·92	0·80	0·62	0·48	0·38	0·30	0·24	0·20	0·17	0·14
Na	11	4·89	4·21	2·97	2·11	1·59	1·29	1·09	0·95	0·83	0·64	0·51	0·40	0·33	0·27	0·22	0·18	0·16
Mg	12	5·01	4·60	3·59	2·63	1·95	1·50	1·21	1·01	0·87	0·67	0·53	0·43	0·35	0·29	0·24	0·20	0·17
Al	13	(6·1)	5·36	4·24	3·13	2·30	1·73	1·36	1·11	0·93	0·70	0·55	0·45	0·36	0·30	0·25	0·22	(0·19)
Si	14	(6·0)	5·26	4·40	3·41	2·59	1·97	1·54	1·23	1·02	0·74	0·58	0·47	0·38	0·32	0·27	0·23	(0·20)
P	15	(5·4)	5·07	4·38	3·55	2·79	2·17	1·70	1·36	1·12	0·80	0·61	0·49	0·40	0·33	0·28	0·24	0·21
S	16	(4·7)	4·40	4·00	3·46	2·87	2·32	1·86	1·50	1·22	0·86	0·64	0·51	0·42	0·35	0·30	0·25	0·22
Cl	17	(4·6)	4·31	4·00	3·53	2·99	2·47	2·01	1·63	1·34	0·93	0·69	0·54	0·44	0·37	0·31	0·26	0·23
A	18	4·71	4·40	4·07	3·56	3·03	2·52	2·07	1·71	1·42	1·00	0·74	0·58	0·46	0·38	0·32	0·27	0·24
K	19	(9·0)	(7·0)	5·43	(4·10)	3·15	(2·60)	2·14	(1·90)	1·49	1·07	0·79	0·61	0·49	0·40	0·34	0·29	(0·25)
Ca	20	10·46	8·71	6·40	4·54	3·40	2·69	2·20	1·84	1·55	1·12	0·84	0·65	0·52	0·42	0·35	0·30	0·26

Note. ( ) = interpolation or extrapolation.

[a] Reprinted from *International Tables for X-ray Crystallography, Vol. III,* with the permission of the International Union of Crystallography.

TABLE B.3 *(cont.)*

Thomas-Fermi-Dirac statistical model. For electrons of velocity v, multiply by $[1-(v/c)^2]^{\frac{1}{2}}$

(sin θ)/λ (Å^{-1})

Element	Z	0·00	0·05†	0·10	0·15	0·20	0·25	0·30	0·35	0·40	0·50	0·60	0·70	0·80	0·90	1·00	1·10	1·20	1·30	1·40	1·50
Ca	20	5·4	5·08	4·57	3·85	3·13	2·52	2·06	1·72	1·45	1·07	0·82	0·65	0·53	0·44	0·37	0·31	0·27	0·23	0·20	0·18
Sc	21	5·6	5·27	4·72	3·98	3·24	2·61	2·14	1·78	1·51	1·12	0·86	0·68	0·55	0·45	0·38	0·32	0·28	0·24	0·21	0·19
Ti	22	5·8	5·46	4·88	4·12	3·35	2·70	2·21	1·85	1·57	1·16	0·89	0·71	0·57	0·47	0·40	0·34	0·29	0·25	0·22	0·20
V	23	5·9	5·65	5·03	4·24	3·45	2·79	2·29	1·91	1·62	1·20	0·93	0·74	0·60	0·49	0·41	0·35	0·30	0·26	0·23	0·21
Cr	24	6·1	5·84	5·17	4·37	3·56	2·88	2·36	1·98	1·68	1·25	0·96	0·76	0·62	0·51	0·43	0·37	0·32	0·27	0·24	0·21
Mn	25	6·2	5·93	5·34	4·49	3·66	2·97	2·43	2·04	1·73	1·29	0·99	0·79	0·64	0·53	0·45	0·38	0·33	0·29	0·25	0·22
Fe	26	6·4	6·13	5·48	4·62	3·76	3·05	2·51	2·10	1·79	1·33	1·03	0·82	0·66	0·55	0·46	0·39	0·34	0·30	0·26	0·23
Co	27	6·5	6·32	5·62	4·73	3·87	3·14	2·58	2·16	1·84	1·37	1·06	0·84	0·69	0·57	0·48	0·41	0·35	0·31	0·27	0·24
Ni	28	6·7	6·41	5·74	4·85	3·97	3·22	2·65	2·23	1·89	1·41	1·09	0·87	0·71	0·59	0·49	0·42	0·36	0·32	0·28	0·25
Cu	29	6·8	6·61	5·89	4·97	4·06	3·30	2·72	2·29	1·95	1·45	1·13	0·90	0·73	0·60	0·51	0·43	0·38	0·33	0·29	0·25
Zn	30	7·0	6·70	6·03	5·08	4·16	3·38	2·79	2·35	2·00	1·49	1·16	0·92	0·75	0·62	0·52	0·45	0·39	0·34	0·30	0·26
Ga	31	7·2	6·89	6·15	5·20	4·25	3·46	2·86	2·41	2·05	1·53	1·19	0·95	0·77	0·64	0·54	0·46	0·40	0·35	0·31	0·27
Ge	32	7·3	7·09	6·29	5·32	4·35	3·54	2·93	2·46	2·10	1·57	1·22	0·97	0·79	0·66	0·56	0·47	0·41	0·36	0·31	0·28
As	33	7·5	7·18	6·41	5·43	4·44	3·62	2·99	2·52	2·15	1·61	1·25	1·00	0·82	0·68	0·57	0·49	0·42	0·37	0·32	0·29
Se	34	7·6	7·37	6·56	5·53	4·54	3·70	3·06	2·58	2·20	1·65	1·28	1·02	0·84	0·70	0·59	0·50	0·43	0·38	0·33	0·29
Br	35	7·8	7·47	6·68	5·63	4·63	3·78	3·13	2·64	2·25	1·69	1·32	1·05	0·86	0·71	0·60	0·51	0·44	0·39	0·34	0·30
Kr	36	7·9	7·56	6·80	5·74	4·71	3·85	3·19	2·69	2·31	1·73	1·35	1·08	0·88	0·73	0·62	0·53	0·46	0·40	0·35	0·31
Rb	37	8·0	7·75	6·92	5·85	4·80	3·93	3·26	2·75	2·35	1·77	1·38	1·10	0·90	0·75	0·63	0·54	0·47	0·41	0·36	0·32
Sr	38	8·2	7·85	7·04	5·96	4·89	4·00	3·32	2·80	2·40	1·80	1·41	1·13	0·92	0·77	0·65	0·55	0·48	0·42	0·37	0·33
Y	39	8·3	8·04	7·16	6·06	4·98	4·07	3·38	2·86	2·45	1·84	1·44	1·15	0·94	0·78	0·66	0·57	0·49	0·43	0·38	0·33
Zr	40	8·5	8·14	7·28	6·16	5·06	4·15	3·45	2·91	2·50	1·88	1·47	1·17	0·96	0·80	0·68	0·58	0·50	0·44	0·39	0·34
Nb	41	8·6	8·23	7·40	6·27	5·15	4·22	3·51	2·97	2·54	1·92	1·50	1·20	0·98	0·82	0·69	0·59	0·51	0·45	0·39	0·35
Mo	42	8·7	8·42	7·52	6·36	5·24	4·29	3·57	3·02	2·59	1·95	1·53	1·22	1·00	0·84	0·71	0·60	0·52	0·46	0·40	0·36
Tc	43	8·9	8·52	7·63	6·47	5·31	4·36	3·63	3·08	2·64	1·99	1·56	1·25	1·02	0·85	0·72	0·62	0·53	0·47	0·41	0·37
Ru	44	9·0	8·62	7·75	6·56	5·40	4·43	3·69	3·13	2·68	2·03	1·58	1·27	1·04	0·87	0·74	0·63	0·55	0·48	0·42	0·37
Rh	45	9·1	8·81	7·85	6·66	5·48	4·50	3·75	3·18	2·73	2·06	1·61	1·30	1·06	0·89	0·75	0·64	0·56	0·49	0·43	0·38
Pd	46	9·3	8·90	7·97	6·75	5·56	4·57	3·81	3·23	2·77	2·10	1·64	1·32	1·08	0·90	0·77	0·66	0·57	0·50	0·44	0·39
Ag	47	9·4	9·00	8·07	6·85	5·64	4·64	3·87	3·28	2·82	2·13	1·67	1·34	1·10	0·92	0·78	0·67	0·58	0·51	0·45	0·40
Cd	48	9·5	9·19	8·19	6·95	5·72	4·71	3·93	3·34	2·86	2·17	1·71	1·37	1·12	0·94	0·79	0·68	0·59	0·52	0·46	0·40
In	49	9·6	9·29	8·31	7·03	5·80	4·78	3·99	3·39	2·91	2·20	1·73	1·39	1·14	0·95	0·81	0·69	0·60	0·53	0·46	0·41
Sn	50	9·8	9·38	8·40	7·13	5·88	4·84	4·05	3·44	2·95	2·24	1·76	1·41	1·16	0·97	0·82	0·71	0·61	0·54	0·47	0·42
Sb	51	9·9	9·48	8·50	7·22	5·95	4·91	4·10	3·49	3·00	2·27	1·79	1·44	1·18	0·99	0·84	0·72	0·62	0·55	0·48	0·43
Te	52	10·0	9·57	8·62	7·31	6·03	4·97	4·16	3·54	3·04	2·31	1·81	1·46	1·20	1·00	0·85	0·73	0·63	0·55	0·49	0·44
I	53	10·1	9·77	8·71	7·39	6·11	5·04	4·22	3·59	3·08	2·34	1·84	1·48	1·22	1·02	0·87	0·74	0·64	0·56	0·50	0·44
Xe	54	10·2	9·86	8·81	7·49	6·19	5·10	4·27	3·64	3·13	2·38	1·87	1·51	1·24	1·04	0·88	0·76	0·66	0·57	0·51	0·45
Cs	55	10·4	9·96	8·93	7·57	6·26	5·17	4·33	3·68	3·17	2·41	1·90	1·53	1·26	1·05	0·89	0·77	0·67	0·58	0·52	0·46
Ba	56	10·5	10·05	9·02	7·66	6·34	5·23	4·39	3·73	3·21	2·45	1·93	1·55	1·28	1·07	0·91	0·78	0·68	0·59	0·52	0·47
La	57	10·6	10·15	9·12	7·75	6·40	5·30	4·44	3·78	3·26	2·48	1·95	1·57	1·30	1·09	0·92	0·79	0·69	0·60	0·53	0·47
Ce	58	10·7	10·24	9·21	7·84	6·49	5·36	4·50	3·83	3·30	2·51	1·98	1·60	1·32	1·10	0·94	0·80	0·70	0·61	0·54	0·48

† The second decimal place is not significant in this column.

TABLE B.3 *(cont.)*

Element	Z	0·00	0·05	0·10	0·15	0·20	0·25	0·30	0·35	0·40	0·50	0·60	0·70	0·80	0·90	1·00	1·10	1·20	1·30	1·40	1·50
Pr	59	10·8	10·44	9·31	7·92	6·56	5·42	4·55	3·88	3·34	2·55	2·01	1·62	1·33	1·12	0·95	0·82	0·71	0·62	0·55	0·49
Nd	60	10·9	10·53	9·41	8·01	6·63	5·48	4·60	3·93	3·38	2·58	2·03	1·64	1·35	1·13	0·96	0·83	0·72	0·63	0·56	0·50
Pm	61	11·0	10·63	9·53	8·10	6·70	5·55	4·66	3·97	3·43	2·61	2·06	1·66	1·37	1·15	0·98	0·84	0·73	0·64	0·57	0·50
Sm	62	11·1	10·72	9·62	8·17	6·77	5·61	4·71	4·02	3·47	2·65	2·09	1·69	1·39	1·17	0·99	0·85	0·74	0·65	0·57	0·51
Eu	63	11·2	10·82	9·72	8·25	6·85	5·67	4·77	4·07	3·51	2·68	2·11	1·71	1·41	1·18	1·00	0·86	0·75	0·66	0·58	0·52
Gd	64	11·4	10·92	9·79	8·34	6·91	5·73	4·82	4·11	3·55	2·71	2·14	1·73	1·43	1·20	1·02	0·88	0·76	0·67	0·59	0·53
Tb	65	11·5	11·01	9·88	8·42	6·98	5·79	4·87	4·16	3·59	2·74	2·17	1·75	1·45	1·21	1·03	0·89	0·77	0·68	0·60	0·53
Dy	66	11·6	11·11	9·98	8·50	7·05	5·85	4·92	4·20	3·63	2·78	2·19	1·77	1·47	1·23	1·05	0·90	0·78	0·69	0·61	0·54
Ho	67	11·7	11·20	10·08	8·58	7·12	5·91	4·98	4·25	3·67	2·81	2·22	1·80	1·48	1·25	1·06	0·91	0·79	0·70	0·61	0·55
Er	68	11·8	11·30	10·17	8·66	7·19	5·97	5·03	4·30	3·71	2·84	2·25	1·82	1·50	1·26	1·07	0·92	0·80	0·70	0·62	0·56
Tm	69	11·9	11·49	10·27	8·74	7·26	6·03	5·08	4·34	3·75	2·87	2·27	1·84	1·52	1·28	1·09	0·94	0·81	0·71	0·63	0·56
Yb	70	12·0	11·59	10·36	8·82	7·33	6·09	5·13	4·39	3·79	2·91	2·30	1·86	1·54	1·29	1·10	0·95	0·82	0·72	0·64	0·57
Lu	71	12·1	11·68	10·44	8·90	7·40	6·15	5·18	4·43	3·83	2·94	2·32	1·88	1·56	1·31	1·11	0·96	0·83	0·73	0·65	0·58
Hf	72	12·2	11·78	10·53	8·98	7·46	6·20	5·23	4·48	3·87	2·97	2·35	1·90	1·58	1·32	1·13	0·97	0·84	0·74	0·66	0·58
Ta	73	12·3	11·87	10·63	9·05	7·53	6·26	5·28	4·52	3·91	3·00	2·38	1·93	1·59	1·34	1·14	0·98	0·85	0·75	0·66	0·59
W	74	12·4	11·97	10·72	9·13	7·59	6·32	5·33	4·56	3·95	3·03	2·40	1·95	1·61	1·35	1·15	0·99	0·86	0·76	0·67	0·60
Re	75	12·5	12·06	10·79	9·21	7·66	6·38	5·38	4·61	3·99	3·06	2·43	1·97	1·63	1·37	1·17	1·01	0·87	0·77	0·68	0·61
Os	76	12·6	12·16	10·89	9·29	7·72	6·43	5·43	4·65	4·03	3·09	2·45	1·99	1·65	1·38	1·18	1·02	0·89	0·78	0·69	0·61
Ir	77	12·7	12·26	10·96	9·36	7·79	6·49	5·48	4·70	4·07	3·12	2·48	2·01	1·66	1·40	1·19	1·03	0·90	0·79	0·70	0·62
Pt	78	12·8	12·35	11·06	9·44	7·86	6·55	5·53	4·74	4·11	3·16	2·50	2·03	1·68	1·42	1·21	1·04	0·91	0·80	0·70	0·63
Au	79	12·9	12·45	11·13	9·51	7·92	6·60	5·58	4·78	4·14	3·19	2·53	2·05	1·70	1·43	1·22	1·05	0·92	0·80	0·71	0·64
Hg	80	13·0	12·54	11·23	9·58	7·98	6·66	5·63	4·83	4·18	3·22	2·55	2·07	1·72	1·45	1·23	1·06	0·93	0·81	0·72	0·64
Tl	81	13·1	12·64	11·32	9·66	8·05	6·71	5·68	4·87	4·22	3·25	2·58	2·10	1·74	1·46	1·25	1·07	0·94	0·82	0·73	0·65
Pb	82	13·2	12·69	11·39	9·74	8·11	6·77	5·72	4·91	4·26	3·28	2·60	2·12	1·75	1·48	1·26	1·09	0·95	0·83	0·74	0·66
Bi	83	13·2	12·75	11·49	9·81	8·18	6·82	5·77	4·95	4·30	3·31	2·63	2·14	1·77	1·49	1·27	1·10	0·96	0·84	0·74	0·66
Po	84	13·3	12·83	11·56	9·87	8·24	6·88	5·82	4·99	4·33	3·34	2·65	2·16	1·79	1·51	1·28	1·11	0·97	0·85	0·75	0·67
At	85	13·4	12·93	11·66	9·95	8·30	6·93	5·87	5·04	4·37	3·37	2·68	2·18	1·81	1·52	1·30	1·12	0·98	0·86	0·76	0·68
Rn	86	13·5	13·02	11·73	10·02	8·36	6·98	5·92	5·08	4·41	3·40	2·70	2·20	1·82	1·54	1·31	1·13	0·99	0·87	0·77	0·69
Fr	87	13·6	13·12	11·80	10·10	8·42	7·04	5·96	5·12	4·44	3·43	2·73	2·22	1·84	1·55	1·32	1·14	1·00	0·88	0·78	0·69
Ra	88	13·7	13·22	11·90	10·16	8·49	7·09	6·01	5·16	4·48	3·46	2·75	2·24	1·86	1·56	1·34	1·15	1·01	0·88	0·78	0·70
Ac	89	13·8	13·31	11·97	10·24	8·55	7·14	6·06	5·20	4·52	3·49	2·78	2·27	1·87	1·58	1·35	1·16	1·02	0·89	0·79	0·71
Th	90	13·9	13·41	12·04	10·30	8·61	7·20	6·10	5·24	4·55	3·52	2·80	2·29	1·89	1·59	1·36	1·18	1·03	0·90	0·80	0·71
Pa	91	14·0	13·50	12·14	10·37	8·67	7·25	6·15	5·28	4·59	3·55	2·82	2·31	1·91	1·61	1·37	1·19	1·04	0·91	0·81	0·72
U	92	14·1	13·60	12·21	10·45	8·73	7·31	6·19	5·33	4·63	3·58	2·85	2·33	1·93	1·62	1·39	1·20	1·04	0·92	0·82	0·73
Np	93	14·2	13·69	12·28	10·51	8·79	7·35	6·24	5·37	4·66	3·61	2·87	2·35	1·94	1·64	1·40	1·21	1·05	0·93	0·82	0·73
Pu	94	14·3	13·77	12·38	10·59	8·85	7·41	6·28	5·41	4·70	3·63	2·90	2·37	1·96	1·65	1·41	1·22	1·06	0·94	0·83	0·74
Am	95	14·4	13·83	12·45	10·65	8·91	7·46	6·33	5·45	4·74	3·66	2·92	2·39	1·98	1·67	1·43	1·23	1·07	0·95	0·84	0·75
Cm	96	14·4	13·90	12·52	10·71	8·97	7·51	6·38	5·49	4·77	3·69	2·94	2·41	1·99	1·68	1·44	1·24	1·08	0·95	0·85	0·76
Bk	97	14·5	13·98	12·59	10·79	9·03	7·56	6·42	5·53	4·81	3·72	2·97	2·43	2·01	1·70	1·45	1·25	1·09	0·96	0·85	0·76
Cf	98	14·6	14·08	12·69	10·85	9·09	7·61	6·47	5·57	4·84	3·75	2·99	2·45	2·03	1·71	1·46	1·26	1·10	0·97	0·86	0·77
Es	99	14·7	14·17	12·76	10·92	9·14	7·67	6·51	5·61	4·88	3·78	3·01	2·47	2·04	1·73	1·48	1·28	1·11	0·98	0·87	0·78
Fm	100	14·8	14·27	12·83	10·99	9·20	7·72	6·56	5·65	4·91	3·81	3·04	2·49	2·06	1·74	1·49	1·29	1·12	0·99	0·88	0·78
Md	101	14·9	14·37	12·90	11·05	9·26	7·77	6·60	5·69	4·95	3·84	3·06	2·51	2·08	1·75	1·50	1·30	1·13	1·00	0·88	0·79
No	102	15·0	14·46	12·96	11·12	9·33	7·82	6·64	5·73	4·98	3·87	3·09	2·53	2·10	1·77	1·51	1·31	1·14	1·01	0·89	0·80
	103	15·1	14·56	13·05	11·18	9·37	7·86	6·69	5·76	5·02	3·89	3·11	2·54	2·11	1·78	1·53	1·32	1·15	1·01	0·90	0·80
	104	15·2	14·66	13·12	11·25	9·43	7·91	6·73	5·80	5·05	3·92	3·13	2·56	2·13	1·80	1·54	1·33	1·16	1·02	0·91	0·81

Table B.4
Neutron Scattering Data for Elements and Isotopes[a]

Element	Atomic number	Atomic weight of natural element	Specific nucleus	Nuclear spin	b (10^{-12} cm)	$\mathcal{S}$ (barns)	σ (barns)
H	1		H^1	$\frac{1}{2}$	−0·378	1·79	81·5
			H^2	1	0·65	5·4	7·6
He	2		He^4	0	0·30	1·1	1·1
Li	3	6·94			−0·18	0·4	1·2
			Li^6	1	0·7	6	
			Li^7	$\frac{3}{2}$	−0·25	0·8	1·4
Be	4		Be^9	$\frac{3}{2}$	0·774	7·53	7·54
B	5						4·4
C	6		C^{12}	0	0·661	5·50	5·51
			C^{13}	$\frac{1}{2}$	0·60	4·5	5·5
N	7		N^{14}	1	0·940	11·0	11·4
O	8		O^{16}	0	0·577	4·2	4·24
F	9		F^{19}	$\frac{1}{2}$	0·55	3·8	4·0
Ne	10						2·9
Na	11		Na^{23}	$\frac{3}{2}$	0·351	1·55	3·4
Mg	12	24·3			0·54	3·60	3·70
Al	13		Al^{27}	$\frac{5}{2}$	0·35	1·5	1·5
Si	14	28·06			0·42	2·16	2·2
P	15		P^{31}	$\frac{1}{2}$	0·53	3·5	3·6
S	16		S^{32}	0	0·31	1·2	1·2
Cl	17	35·5			0·99	12·2	15
A	18		A^{40}	0	0·20	0·5	0·9
K	19	39·1			0·35	1·5	2·2
Ca	20	40·1			0·49	3·0	3·2
			Ca^{40}	0	0·49	3·0	3·1
			Ca^{44}	0	0·18	0·4	
Sc	21		Sc^{45}	$\frac{7}{2}$	1·18	17·5	24
Ti	22	47·9			−0·34	1·45	4·4
			Ti^{46}	0	0·48 [6]	2·90	
			Ti^{47}	$\frac{5}{2}$	0·33 [6]	1·37	
			Ti^{48}	0	−0·58 [6]	4·23	
			Ti^{49}	$\frac{7}{2}$	0·08 [6]	0·08	
			Ti^{50}	0	0·55 [6]	3·80	
V	23		V^{51}	$\frac{7}{2}$	−0·051	0·032	5·1
Cr	24	52·0			0·352	1·56	4·1
			Cr^{52}	0	0·490	3·02	
Mn	25		Mn^{55}	$\frac{5}{2}$	−0·36	1·6	2·0
Fe	26	55·8			0·96	11·4	11·8
			Fe^{54}	0	0·42	2·2	2·5
			Fe^{56}	0	1·01	12·8	12·8
			Fe^{57}		0·23	0·64	2
Co	27		Co^{59}	$\frac{7}{2}$	0·28	1·0	6
Ni	28	58·7			1·03	13·4	18·0
			Ni^{58}	0	1·44	25·9	
			Ni^{60}	0	0·30	1·1	
			Ni^{62}	0	−0·87	9·5	

[a] Reprinted from *International Tables for X-ray Crystallography, Vol. III*, with the permission of the International Union of Crystallography.

TABLE B.4 *(cont.)*

Element	Atomic number	Atomic weight of natural element	Specific nucleus	Nuclear spin	b (10^{-12} cm)	$\mathscr{S}$ (barns)	σ (barns)
Cu	29	63·6			0·79 [1]	7·8	8·5
			Cu^{63}	$\frac{3}{2}$	0·67 [1]	5·7	
			Cu^{65}	$\frac{3}{2}$	1·11 [1]	15·3	
Zn	30	65·4			0·59	4·3	4·2
Ga	31	69·7					7·5
Ge	32	72·6			0·84	8·8	9·0
As	33		As^{75}	$\frac{3}{2}$	0·63	5·0	8
Se	34	79·0			0·89	10·0	
Br	35	79·9			0·67	5·7	6·1
Kr	36	82·9					
Rb	37	85·5			0·55	3·8	5·5
Sr	38	87·6			0·57	4·1	10
Y	39		Y^{89}	$\frac{1}{2}$			
Zr	40	91·2			0·62	4·9	6·3
Nb	41		Nb^{93}	$\frac{9}{2}$	0·691	6·0	6·6
Mo	42	95·9			0·661	5·5	6·1
Tc	43						
Ru	44	101·7			0·73 [4]	6·68	6·81
Rh	45		Rh^{103}	$\frac{1}{2}$	0·60	4·5	5·6
Pd	46	106·7			0·63	5·0	4·8
Ag	47	107·9			0·61	4·6	6·5
			Ag^{107}	$\frac{1}{2}$	0·83	8·7	10
			Ag^{109}	$\frac{1}{2}$	0·43	2·3	6
Cd	48	112·4			$0{\cdot}38+i.0{\cdot}12$ [7]		
In	49	114·8			0·36 [4]	1·63	
Sn	50	118·7			0·61	4·6	4·9
Sb	51	121·8			0·54	3·7	4·2
Te	52	127·5			0·56	4·0	4·5
			Te^{120}		0·52	3·4	
			Te^{123}	$\frac{1}{2}$	0·57	4·2	
			Te^{124}		0·55	3·9	
			Te^{125}	$\frac{1}{2}$	0·56	4·0	
I	53		I^{127}	$\frac{5}{2}$	0·52	3·4	3·8
Xe	54	130·2					
Cs	55		Cs^{133}	$\frac{7}{2}$	0·49	3·0	7
Ba	56	137·4			0·52	3·4	6
La	57		La^{139}	$\frac{7}{2}$	0·83	8·7	9·3
Ce	58	140·25			0·46	2·7	2·7
			Ce^{140}		0·47	2·8	2·6
			Ce^{142}		0·45	2·6	2·6
Pr	59		Pr^{141}	$\frac{5}{2}$	0·44	2·4	4·0
Nd	60	144·3			0·72	6·5	16
			Nd^{142}		0·77	7·5	7·5
			Nd^{144}		0·28	1·0	1·0
			Nd^{146}		0·87	9·5	9·5

TABLE B.4 (cont.)

Element	Atomic number	Atomic weight of natural element	Specific nucleus	Nuclear spin	b (10^{-12} cm)	$\mathcal{S}$ (barns)	σ (barns)
Il	61						
Sm	62	150·4					
			Sm^{152}		−0·5	3	
			Sm^{154}		0·8	8	
Eu	63	152·0					
Gd	64	157·3					
Tb	65		Tb^{159}	$\frac{3}{2}$			
Dy	66	162·5					
Ho	67		Ho^{165}	$\frac{7}{2}$	0·85 [2]	9·1	~13
Er	68	167·6			0·79	7·8	15
Tm	69		Tm^{169}				
Yb	70	173·0					
Lu	71	175·0					
Hf	72	178·6			0·88	9·7	
Ta	73		Ta^{181}	$\frac{7}{2}$	0·70	6·1	6
W	74	183·9			0·466	2·74	5·7
Re	75	186·2			0·92 [5]	10·6	
Os	76	190·2			1·08 [3]	14·7	14·9
Ir	77	192·2			0·36 [4]	1·63	1·66
Pt	78	195·2			0·95	11·2	12
Au	79		Au^{197}	$\frac{3}{2}$	0·76	7·3	9
Hg	80	200·6			1·3	22	26·5
Tl	81	204·4			0·89 [3]	10·0	10·1
Pb	82	207·2			0·96	11·5	11·4
Bi	83		Bi^{209}	$\frac{9}{2}$	0·864	9·35	9·37
Po	84	210					
At	85						
Rn	86	222					
Fr	87						
Ra	88	226					
Ac	89	227					
Th	90		Th^{232}	0	1·01	12·8	12·6
Pa	91	231					
U	92		U^{238}		0·85	9·0	
Np	93						
Pu	94						

Values are taken mainly from earlier compilations:
SHULL, C. G. and WOLLAN, E. O. *Phys. Rev.*, **81**, 527, 1951;
BACON, G. E. *Neutron Diffraction.* (Clarendon Press, Oxford, 1955);
SHULL, C. G., unpublished, 1955;
together with recent additions indicated by numbered references as follows:

[1] KEATING, D. T. *et al. Phys. Rev.*, **111**, 261, 1958.
[2] KOEHLER, W. C., WILKINSON, M. K., and WOLLAN, E. O. Unpublished.
[3] LE ROY HEATON and SIDHU, S. S. *Phys. Rev.*, **105**, 216, 1957.
[4] SIDHU, S. S., LE ROY HEATON, and MUELLER, M. H. ANL paper No. 1196, 1958.
[5] WILKINSON, M. K., CABLE, J. W. *et al.* Unpublished.
[6] WILKINSON, M. K. and SHULL, C. G. Unpublished.
[7] for a wavelength $\lambda = 1{\cdot}075$ Å: PETERSON, S. W. and SMITH, H. G. Unpublished.

b is the coherent scattering amplitude in units of 10^{-12} cm.

Table B.5

Mass Absorption Coefficients for X- and γ-Rays[a]

Radiation traversing a layer of substance is reduced in intensity by a constant fraction μ per centimeter. After penetrating to a depth x the intensity is $I = I_0 e^{-\mu x}$ where I_0 is the intensity at the surface. μ/ρ is the mass absorption coefficient where ρ is the density of the material.

Values of μ/ρ for λ = .005 Å to λ = 44.6 Å. Where two values of μ/ρ for one value of λ occur they represent the maximum and minimum values at an absorption discontinuity.

Compiled by **S. J. M. Allen**

λ = 44.6 – 2.74 Å

λ, Å	He	C	N	O	Ne	Al	S	Cl	A
44.6	3600				13100	....	....	....	45700
11.88	....	2170	3850	5765	6850	850	....	....	....
9.87	....	1063	1796	2540	4310	500	1320	1570	1860
8.32	....	656	1109	1585	2750	330	794	962	1160
7.94	....	......	.....	......	.....	280	....	....	.....
						3700			
6.97	....	390	645	976	1727	2800	500	610	748
5.39	....	185	312	476	865	1450	249	310	360
5.17	....	160	273	413	763	1350	221	277	324
5.01	....	......	....	.....	.....	....	210	....	.....
							2260		
4.38	....	......	.....	.....	.....	....	....	178	.....
								1830	
4.36	....	97.8	166	258	478	815	1570	1800	202
4.15	....	84.6	144	222	416	720	1350	1476	174
3.93	....	71.0	121	189	356	635	1175	1256	153
3.87	....	......	.....	...	.....	....	....	....	148
									1460
3.69	....	......	......	.....	.....	....	....	....	.....
3.59	....	55.2	96	150	279	500	928	966	1215
3.51	....	......	.....	.....		....	....	....	.....
3.38	....	46.0	79.5	117	231	425	795	880	1025
3.35	....	43.0	.....	.....	.....	417	780	870	1015
3.24	....	......	.....	.....	.....	....	....	....	.....
3.03	....	35.0	......	84.0	175	323	595	670	760
2.74	....	25.0	......	60.0	135	250	454	520	600

λ = 44.6 – 2.74 Å

λ, Å	Fe	Ni	Cu	Zn	Kr	Ag	Sn	Xe	Pt	Au
44.6	...	.....	....	...	31800	....	...	6740	....	12500
11.88	...	6900	7550	...	......	....	...	....	....	....
9.87	...	4540	5030	...	......	2700	...	....	2440	....
8.32	...	3140	3450	...	......	1800	...	....	1560	....
7.94	..	....	....	...	......	....	...	....	....	....
6.97	...	2000	2130	...	......	1300	...	....	1190	....
5.39	..	1250	1290	...	...	845	...	....	1645	....
5.17	...	1150	1190	...	......	790	...	....	....	....
5.01	...	....	....	...	......	....	...	....	....	
4.38	...	...	....	...	......	....	...	....	....	
4.36	610	715	760	910	......	535	640	....	....	
4.15	540	630	690	820	......	461	550	....	1290	
3.93	470	555	610	715	......	408	490	....	..	
3.87	...	....	....	...	......	....	...	....	....	
3.69	...	....	....	...	......	354	...	....	....	
						1410				
3.59	375	450	495	575	......	1360	...	....	1370	
3.51	...	....	....	..	......	1300	...	....	....	
						1510				
3.38	320	380	....	495	......	...	...	....	....	
3.35	312	375	404	480	......	1310	...	....	1120	..
3.24	..	....	....	...	......	1230	...	....	....	....
						1440				
3.03	245	290	315	375	......	1290	...	....	939	....
2.74	185	239	262	283	......	925	...	....	756	....

[a] Reprinted with permission from *CRC Handbook of Chemistry and Physics, 51st Edition.* Copyright The Chemical Rubber Company, CRC Press, Inc.

λ = 2.50 – .900 Å

λ, Å	H	Li	Be	B	C	N	O	Ne	Na
2.50	.52	4.0	6.1	9.1	17.8	.....	44.5	100	128
2.29	..	....	..	....	15.0	.....	36.4	75.5	.....
1.93	.50	2.10	3.05	4.7	8.75	14.0	21.7	49.0	61.3
1.74	..	....	....	....	.....	.....	.....	.....	.....
1.656	...	....	....	....	.....	.....	.....	.....	.....
1.539	.48	1.10	1.60	2.45	4.52	7.45	11.1	24.0	32.1
1.484	...	..	...	.....	.....	.....	.....	.....	.....
1.432	...	...	...	.....	.....	.....	.....	.....	.....
1.389	.47	86	1.25	1.87	3.35	5.50	8.1	17.0	23.4
1.377	.	.		..	.....	.....	.....	.....	.....
1.293	..	...	..	..	.....	.....	.....	.....	....
1.280	..	...	..	...	.....	.....	.....	.....	.....
1.235	.46	67	.95	1.35	2.42	3.95	5.7	12.4	17.1
1.104		.	..	...	.....	.....	.....	.....	.....
1.071	..	...	...	.....	.....	.....	.....	.....	.....
1.038	.	...	...	.....	.....	.....	.....	.....	.....
1.000	45	.43	.55	.76	1.36	2.10	3.13	6.5	8.8
.980	.	...	..	.....	...	.....	.....	.....	.....
.949	...	....	....	....	1.20	.....	.....	.....	.....
.932	...	....	....	....	..	.....	.....	.....	.....
.900	...	....	....	....	1.05	.....	.....	.....	.....

λ = 2.50 – .900 Å

λ, Å	Mg	Al	S	Cl	A	Ca	Fe	Ni	Cu
2.50	161	193	355	400	475	620	147	180	197
2.29	.....	150	285	315	355	480	115	137	153
1.93	77.2	93.5	173	198	235	306	71.2	89.5	96.2
1.74	.....	83.0	.....	....	.....	...	54 465	.....	.....
1.656	.....	60.7	110	126	143	195	410	59.2	63.5
1.539	40.8	49.0	91	103	114	163	325	48.0	50.9
1.484	.....	......	.....	.....	.....	...	.....	40.5 338	.....
1.432	.....	40.0	75	85	93	130	285	325	42
1.389	31.5	36.8	68.5	76.7	85.7	125	252	275	38.5
1.377	.....	......	.....	.....	.....	...	.....	.....	37.0 307
1.293	.....	29.8	55.3	60	72	102	212	233	260
1.280	.....	28.8	.....	.....	.....	...	.....	225	252
1.235	21.4	26.3	49.5	55.5	62.5	90	181	208	230
1.104	.....	18.6	38.0	44	50	67	135	155	175
1.071	.....	......	.....	.....	.....	...	.....	.....	.....
1.038	.....	......	.....	.....	.....	...	.....	.....	.....
1.000	11.8	14.12	26.7	29.7	34.5	49	100	121	130
.980	.....	......	.....	.....	.....	...	.....	.....	.....
.949	.....	12.0	22.0	24.5	.....	42	86	99	114
.932	.....	......	.....	.....	.....	...	.....	.....	.....
.900	.....	10.4	.....	.....	.....	...	74.5	86.5	98.5

λ = 2.50 – .900 Å

λ, Å	Zn	Br	Mo	Ag	Sn	I	W	Pt	Au	Pb
2.50	228	...	..	710	850	...	...	596	.....	...
2.29	180	...	..	550	670	...	...	480	.....	...
1.93	110	...	..	405	470	...	300	358	385	428
1.74	.....	...	..	.....	.....	...	...	.....	.....	...
1.656	72.5	...	..	285	.....	...	...	228	.....	...
1.539	58.6	89	..	217	247	290	176	202	213	230
1.484	.....	...	..	.....	.....	...	...	.....	.....	...
1.432	49.3	...	..	192	220	...	130	172	179	202
1.389	45.2	...	..	174	209	...	...	155	166	185
1.377	.....	...	..	.....	.....	...	...	.....	.....	...
1.293	39	...	..	146	176	...	...	132	138	154
1.280	36	...	..	127	146	...	...	.....	.....	...
1.235	287 250	...	..	125	140	...	95	115	122	137
1.104	208	...	..	96.5	115	...	...	99	107	120
1.071	.....	...	..	.....	.....	...	...	77.5 198	.....	...
1.038	.....	..	..	.....	.....	...	...	.....	76.5 194	...
1.000	145	...	52	73.0	86.0	...	...	165	174	75
.980	.....	...	..	.....	.....	...	...	155	168	73
.949	120	...	..	63.0	75.5	...	...	146	156	68 168
.932	.....	...	..	.....	.....	...	...	136 184	148	159
.900	112	150	..	54.2	65.0	...	...	168	134 182	145

TABLE B.5 *(cont.)*

λ = .892 − .184 Å

λ, Å	H	Li	Be	B	C	N	O	Ne	Na	Mg	Al
.892	....	....	....	....	....	....	....	....	....	....	....
.880	.440	.350	.425	.580	.990	1.50	2.20	4.55	6.10	8.34	9.75
.862	....	....	....	....	....	....	....	....	....	....	....
.850	....	....	....	....	.907	....	....	....	....	....	8.85
.814	....	....	....	....	.814	....	....	....	....	....	7.85
.780	....	....	....	....	.750	....	....	....	....	....	6.86
.710	.435	.260	.315	.365	.598	.870	1.22	2.50	3.30	4.30	5.22
.680	....	....	....	....	.550	....	....	....	....	....	4.52
.631	.435	.225	.255	.305	.467	.610	.900	1 80	2.30	3.0	3.73
.618	....	....	....	....	....	....	....	....	....	....	....
.560	....	....	....	....	.370	....	....	....	....	....	2.60
.497	.435	.198	.210	.220	.315	.400	.520	.930	1.18	1.52	1.90
.485	....	....	....	....	.308	....	....	....	....	....	1.77
.476	.430	....	....	.215	.304	....	.485	....	....	....	1.74
.424	....	....	....	....	....	....	....	....	....	....	1.23
.417	.390	.180	.185	.198	.256	.310	.372	.580	.750	.940	1.170
.380	....	....	....	....	.230	....	....	....	....	....	.950
.331	....	....	....	....	....	....	....	....	....	....	....
.260	.385	.156	.166	.175	.185	.200	.210	.270	.305	.343	.402
.220	....	....	....	....	.178	....	....	....	....	....	.300
.200	.375	.151	.160	.165	.175	.180	.183	.210	.225	.250	.270
.184	....	....	....	....	.166	....	....	....	....	....	.246

λ = .892 − .184 A

λ, Å	S	Cl	A	Ca	Fe	Ni	Cu	Zn	Br	Sr	Mo
.892	....	....	....	....	....	....	....	....	....	....	....
.880	18.2	20.7	24.0	34.8	69.5	82	91.2	103	....	....	36.0
.862	....	....	....	....	63.5	74	84.5	96.5	....	....	....
.850	....	....	....	....	57	66	75.7	86	....	....	28
.814	....	....	....	....	50.5	59.5	67.5	77	....	....	....
.780	....	....	....	....	....	....	....	....	....	....	....
.710	9.90	11.6	13.0	18.6	38.5	48.1	51.0	59.0	80	106	19.9
.680	....	....	....	....	32.7	41	45.3	52.7	....	....	....
.631	6.90	8.40	9.80	13.3	27.0	34	36.2	41.0	56.8	72.5	15.0
.618	....	....	....	....	....	....	....	....	....	....	12.5 88.0
.560	....	....	....	....	18.2	24	25.5	30.7	....	....	....
.497	3.50	4.20	5.0	6.60	13.9	17.9	18.4	21.0	32.0	40.5	50.2
.485	....	....	....	....	12.4	15.4	16.9	19.5	....	....	....
.476	....	....	....	....	....	....	16.6	....	....	....	42
.424	....	....	....	....	....	....	....	....	....	....	....
.417	2.10	2.47	2.95	3.97	8.45	10.5	11.45	12.3	19.0	24.0	30.0
.380	....	....	....	....	6.32	7.70	8.42	9.95	....	....	22
.331	....	....	....	....	....	....	....	....	....	....	....
.260	.650	.750	.850	1.10	2.28	2.89	3.16	3.58	5.30	6.50	8.20
.220	....	....	....	....	1.42	1.80	2.00	2.32	....	....	....
.200	.400	.445	.500	.630	1.10	1.45	1.55	1.78	2.4	3.32	4.30
.184	....	....	....	....	....	1.24	....	....	....	....	....

λ, Å	Ag	Sn	I	Ba	Ta	W	Pt	Au	Pb	Bi	U
.892	....	....	....	....	....	....	165 201	178	142	....	....
.880	50	60	....	....	....	....	195	170	135	....	....
.862	....	....	....	....	....	....	185	163 193	130	....	....
.850	46	56	....	....	....	....	179	186	124	....	....
.814	41	49.5	....	....	....	....	160	167	111 150	....	....
.780	36	44.5	....	....	....	....	144	150	136 166	....	....
.710	27.5	34.0	38.5	42.0	100	104	115	120	136	....	..
.680	23.5	28.4	....	....	....	....	102	108	120	....	....
.631	19.6	23.0	26.4	31.1	72	75	84.5	87	98	....	....
.618	....	....	....	....	....	....	....	....	....	....	....
.560	13.3	16.2	....	....	....	....	62	66	75	....	....
.497	10.5	11.8	15.6	17.8	36	38	47	48.5	52.8	....	....
.485	9.8 62.5	11.1	....	....	....	....	....	....	....	....	....
.476	60	....	....	....	....	....	42	....	47.5	....	....
.424	43.5	8.0 46.6	....	....	....	....	....	....	....	....	....
.417	41	45	9.2	10.5	21.5	22.5	27.4	28.4	32.0	....	..
.380	31.2	34	....	....	....	17.3	21.1	22	26.4	27.8	....
.331	21.7	24.5	....	5.4 28.0	....	....	....	....	18.1	19.5	....
.260	11.4	12.8	14.2	16.1	6.7	6.85	8.0	8.3	10.0	11.0	....
.220	7.05	7.80	....	....	....	4.25	5.25	5.50	5.92	6.4	....
.200	5.48	6.20	7.0	8.0	3.4	3.50	4.25	4.40	4.90	5.1	5.40
.184	4.45	....	....	....	2.8 11.8	....	3.45	3.60	4.05	4.2	....

TABLE B.5 (cont.)

λ = .178 – .005 A

λ, Å	H	Li	Be	B	C	N	O	Ne	Na	Mg	Al
.178	....	....	....	....	.164	....	....	....	....	....	.235
.175	.360	.144	.150	.155	.163	.166	.169	.185	.195	.205	.228
.158	....	....	....	....	.160	....	....	....	....	....	.208
.155	....	....	....	....	....	....	....	....	....	....	....
.146	.340	....	....	....	.155	....	.162	....	.170	.176	.195
.142	.330	....	....	....	.153	....	....	....	....	....	.191
.130	.320	.132	....	.149	.152	....	.157	....	.160	.168	.186
.120	....	....	....	....	.150	....	.154	....	....	.163	.172
.113	.310	....	....	....	.147	....	.153	....	.155	.160	.166
.107	....	....	....	....	....	....	....	....	....	....	....
.098	.280	.125	....	.138	.142	....	.144	....	.150	.152	.156
.080	.255	....	....	....	.137	....	....	....	....	....	.146
.072	.250	.118	....	.132	.136	....	.137	....	.139	.140	.143
.064	.245	.110	....	.126	.130	....	.130	....	.130	.130	.130
.050	....	....	....	....	....	120	....	....	....	....	.115
.040	.205	....	....	....	.110	....	....	....	....	....	.106
.030	.180	....	....	....	.095	....	....	....	....	....	.093
.024	.165	....	....	....	.080	....	....	....	....	....	.079
.010	.117	....	....	....	.059	....	....	....	....	....	.058
.005	.078	....	....	....	.0385	....	....	....	....	....	.0380

λ, Å	S	Cl	A	Ca	Fe	Ni	Cu	Zn	Br	Sr	Mo
.178	....	....	....	....	....	....	1.15	....	....	....	....
.175	.335	.341	.400	.460	.800	1.05	1.12	1.26	1.90	2.24	2.95
.158	....	....	....	....	.640	.815	.862	.990	....	....	....
.155	....	....	....	....	....	....	....	....	....	....	....
.146	.249	.280	....	.345	.520	....	.680	....	....	....	....
.142	....	....	....	....	.515	.630	.670	.780	....	....	1.55
.130	.220	.230	....	.290	.424	....	.551	....	....	....	....
.120	.200	....	....	....	.368	.430	.455	.537	....	....	....
.113	.189	.195	....	.230	.337	....	.422	....	....	....	....
.107	....	....	....	....	....	....	....	....	....	....	....
.098	.166	.176	....	.200	.265	....	.325	....	....	....	.790
.080	....	.164	....	....	.235	.264	.268	.308	....	....	....
.072	.150	.158	....	.180	.202	....	.232	....	....	....	....
.064	.139	.142	....	.155	.178	....	.198	....	....	....	.413
.050	....	....	....	....	.140	....	.155	....	....	....	....
.040	....	....	....	....	.118	....	.126	....	....	....	....
.030	....	....	....	....	.095	....	.100	....	....	....	....
.024	....	....	....	....	.080	....	.081	....	....	....	....
.010	....	....	....	....	.058	....	.057	....	....	....	....
.005	....	....	....	....	....	....	.0380	....	....	....	....

λ, Å	Ag	Sn	I	Ba	Ta	W	Pt	Au	Pb	U
.178	....	....	....	....	....	2.7 11.3	3.16	3.30	3.55	....
.175	3.96	4.50	5.10	5.70	10.0	10.5	2.97	3.13	3.48	3.95
.158	3.00	3.40	....	....	....	8.6	2.45 9.40	2.43	2.60	....
.155	....	....	....	....	....	....	....	2.30 8.80	....	....
.146	2.48	2.66	....	....	6.75	....	7.60	7.85	2.35	2.70
.142	2.31	2.64	....	....	....	6.75	7.20	7.33	2.10 7.75	....
.130	1.97	2.12	....	....	5.10	....	6.30	6.40	6.55	2.20
.120	1.61	1.77	....	2.20	....	4.60	4.92	4.98	5.20	1.90
.113	1.47	1.60	....	....	3.80	....	4.40	4.50	4.75	....
.107	....	....	....	....	....	....	....	....	....	1.62 4.65
.098	1.05	1.17	....	....	2.80	....	3.15	3.21	3.50	3.90
.080	.73	.79	....	....	....	2.30	2.40	2.42	2.50	2.70
.072	.584	.614	....	....	1.75	....	2.00	2.05	2.10	2.25
.064	.465	.490	....	....	1.35	....	1.52	1.55	1.64	1.80
.050	....	.320	....	....	....	....	.86	.88	1.00	....
.040	....	.21	....	....	....	....	....	....	.62	....
.030	....	.13	....	....	....	....	....	....	.38	....
.024	....	.10	....	....	....	....	....	....	.21	....
.010	....	.060	....	....	....	....	....	....	.071	.082
.005	....	.0385	....	....	....	....	....	....	.0425	.044

Derivation of the Fundamental Equation of the Dynamical Theory

C

Substituting (4.31) into $(1 - \psi)\mathbf{D}$, we obtain for the total electrical displacement $\mathbf{E}$ within the crystal

$$\begin{aligned}\mathbf{E} = (1-\psi)\mathbf{D} &= \left(1 - \sum_H \psi_H e^{2\pi i \mathbf{r}_H^* \cdot \mathbf{r}}\right) \sum \mathbf{D}_L e^{i\omega_0 t + 2\pi i \mathbf{k}_L \cdot \mathbf{r}} \\ &= e^{i\omega_0 t}\left[\sum_L \mathbf{D}_L e^{+2\pi i \mathbf{k}_L \cdot \mathbf{r}} - \sum_H \sum_L \psi_H \mathbf{D}_L e^{2\pi i (\mathbf{r}_H + \mathbf{k}_L)\cdot \mathbf{r}}\right.\end{aligned} \tag{C1.1}$$

Setting

$$\mathbf{r}_H^* + \mathbf{k}_L = \mathbf{k}_{H+L} \tag{C1.2}$$

and

$$H + L = M \tag{C1.3}$$

we have

$$(1+\psi)\mathbf{D} = e^{i\omega_0 t} \sum_L \mathbf{D}_L e^{2\pi i \mathbf{k}_L \cdot \mathbf{r}} - \sum_M \left(\sum_L \psi_{M-L} \mathbf{D}_L\right) e^{2\pi i \mathbf{k}_M \cdot \mathbf{r}} \tag{C1.4}$$

Defining the quantity C_M by

$$C_M = \sum_L \psi_{M-L} \mathbf{D}_L \tag{C1.5}$$

we have

$$(1-\psi)\mathbf{D} = e^{i\omega_o t}\sum_H (\mathbf{D}_H - \mathbf{C}_H)e^{2\pi i\mathbf{k}_H\cdot\mathbf{r}} \tag{C1.6}$$

Substituting (C1.6) into our basic equation (4.33), we have

$$\nabla\mathbf{x}[\nabla\mathbf{x}(1-\psi)\mathbf{D}] = e^{i\omega_o t}\nabla\mathbf{x}\left[\nabla\mathbf{x}\sum_H (\mathbf{D}_H - \mathbf{C}_H)e^{2\pi i\mathbf{k}_H\cdot\mathbf{r}}\right] = -\frac{1}{c^2}\frac{\partial^2\mathbf{D}}{\partial t^2} \tag{C1.7}$$

The first cross product is

$$\begin{aligned}\sum_H [\hat{\mathbf{i}}\{&(\mathbf{D}_H - \mathbf{C}_H)_z e^{2\pi i\mathbf{k}_H\cdot\mathbf{r}}[2\pi i(\mathbf{k}_H)_y] - (\mathbf{D}_H - \mathbf{C}_H)_y e^{2\pi i\mathbf{k}_H\cdot\mathbf{r}}[2\pi i(\mathbf{k}_H)_z]\}\\ &+\hat{\mathbf{j}}\{(\mathbf{D}_H - \mathbf{C}_H)_x e^{2\pi i\mathbf{k}_H\cdot\mathbf{r}}[2\pi i(\mathbf{k}_H)_z] - (\mathbf{D}_H - \mathbf{C}_H)_z e^{2\pi i\mathbf{k}_H\cdot\mathbf{r}}[2\pi i(\mathbf{k}_H)_x]\}\\ &+\hat{\mathbf{k}}\{(\mathbf{D}_H - \mathbf{C}_H)_y e^{2\pi i\mathbf{k}_H\cdot\mathbf{r}}[2\pi i(\mathbf{k}_H)_x] - (\mathbf{D}_H - \mathbf{C}_H)_x e^{2\pi i\mathbf{k}_H\cdot\mathbf{r}}[2\pi i(\mathbf{k}_H)_y]\}]\end{aligned}$$

$$= \sum_H e^{2\pi i\mathbf{k}_H\cdot\mathbf{r}}\,\mathbf{k}_H \times (\mathbf{D}_H - \mathbf{C}_H) \tag{C1.8}$$

Substituting (C1.8) into (C1.7), and working out the second cross product in the same way, we have

$$e^{i\omega_o t}\sum_H e^{2\pi i\mathbf{k}_H\cdot\mathbf{r}}\,\mathbf{k}_H \times [\mathbf{k}_H \times (\mathbf{D}_H - \mathbf{C}_H)] = e^{i\omega_o t}\kappa^2\sum_H \mathbf{D}_H e^{2\pi i\mathbf{k}_H\cdot\mathbf{r}} \tag{C1.9}$$

where we have used

$$\kappa = \frac{\omega_o}{c} \tag{C1.10}$$

for the wave vector of the incident (vacuum) wave. Equating coefficients, we have the set of equations

$$-\mathbf{k}_H \times [\mathbf{k}_H \times (\mathbf{D}_H - \mathbf{C}_H)] = \kappa^2\mathbf{D}_H \tag{C1.11}$$

Recall that for any three noncolinear vectors $\mathbf{V}_1$, $\mathbf{V}_2$, and $\mathbf{V}_3$

$$\mathbf{V}_1 \times (\mathbf{V}_2 \times \mathbf{V}_3) = \mathbf{V}_2(\mathbf{V}_1\cdot\mathbf{V}_3) - \mathbf{V}_3(\mathbf{V}_1\cdot\mathbf{V}_2) \tag{C1.12}$$

Thus

$$\kappa^2\mathbf{D}_H = \mathbf{k}_H[\mathbf{k}_H \cdot (\mathbf{D}_H - \mathbf{C}_H)] - (\mathbf{D}_H - \mathbf{C}_H)(\mathbf{k}_H \cdot \mathbf{k}_H) \tag{C1.13}$$

Substituting in our expression for $\mathbf{C}_H$, we have

$$\kappa^2\mathbf{D}_H = \mathbf{k}_H\left[\mathbf{k}_H \cdot \left(\mathbf{D}_H - \sum_L \psi_{H-L}\mathbf{D}_L\right)\right] - (\mathbf{D}_H - \Sigma\psi_{H-L}\mathbf{D}_L)k_H^2 \tag{C1.14}$$

Rearranging terms, our final relationship is

$$\sum_L (\psi_{H-L}(\mathbf{k}_H \cdot \mathbf{D}_L)\mathbf{k}_H - \psi_{H-L}k_H^2\mathbf{D}_L) = (\kappa^2 - k_H^2)\mathbf{D}_H \tag{C1.15}$$

Index